The Art of Recording

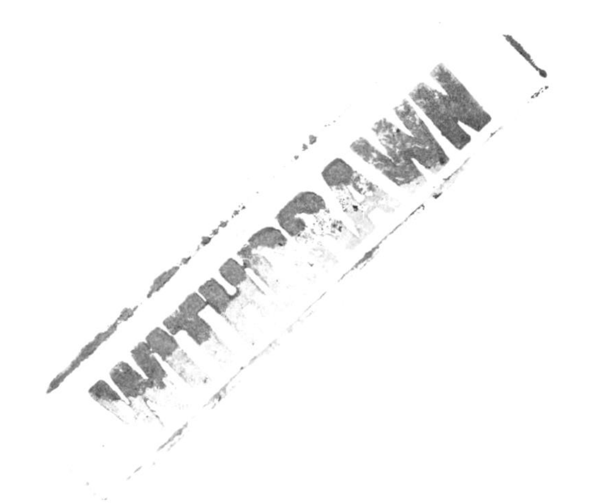

The Art of Recording

Understanding and Crafting the Mix

William Moylan

Focal Press
An Imprint of Elsevier

Amsterdam Boston London New York Oxford Paris
San Diego San Francisco Singapore Sydney Tokyo

Focal Press is an imprint of Elsevier

Recognizing the importance of preserving what has been written, Elsevier prints its books on acid-free paper whenever possible.

Library of Congress Cataloging-in-Publication Data
Moylan, William.
The art of recording : understanding and crafting the mix / William Moylan
p.cm.
Includes bibliographical references and index.
ISBN-13: 978-0-240-80483-5 ISBN-10: 0-240-80483-X (alk. paper)
1. Sound—Recording and reproducing. 2. Acoustical engineering. 3. Music Theory. 4. Music—Editing I. Title.

TK17881.4 .M69 2002
621.389'3—dc21
ISBN-13: 978-0-240-80483-5 2001055611
ISBN-10: 0-240-80483-X
British Library Cataloguing-in-Publication Data
A catalogue record for this book is available from the British Library.

The publisher offers special discounts on bulk orders of this book.
For information, please contact:

Manager of Special Sales
Elsevier
225 Wildwood Avenue
Woburn, MA 01801-2041
Tel: 781-904-2500
Fax: 781-904-2620

For information on all Focal Press publications available, contact our World Wide Web home page at www.focalpress.com

10 9 8 7 6 5 4

Printed in the United States of America

Table of Contents

List of Exercises

These exercises are designed to develop listening and evaluation skills

Foreword

Nowhere within the mystery of creation is the concept of infinity more closely demonstrated than in the human response to sound.

Sounds barely audible in the quiet solitude of a forest glade contain information about direction, height, distance, and character, which unconsciously provide awareness of our surroundings.

Sounds as you step into a great cathedral, hearing nearby soft footfalls on ancient flag stones, the singing of a distant choir provides clues of distance and perspective, and even invite sharing of mood.

Sounds in a concert hall, an office, a bathroom or a recording studio; each sound has its own message prompting our response.

Blindfold, the smallest sample provides amazing awareness of our environment.

The way sound behaves within a space paints a picture. We don't have to analyze, measure, or evaluate. Created in the image of a Communicating God, we communicate with speech to express what we think and with music to express what we feel.

Music is as old as man. References to music, song, poetry, and musical instruments go back thousands of years. We see Egyptian bas-reliefs and know that music played an enormous part in their culture and religion. Near the end of King David's reign, the Hebrews had professional Temple choirs and a 4,000 strong orchestra.[1] We read of strolling minstrels and court musicians of later ages. But what did they sound like?

Every age of man is recorded in writing, painting, and sculpture but there are no sound recordings.

Today we can choose from vast libraries of music from any part of the world that have become the benchmarks of artistic and technical quality for many thousands of people who may never have even been to a concert.

Technical quality has improved enormously over the years but surprisingly, we can still enjoy the very earliest recordings with all their imperfections of noise, distortion, limited bandwidth, and poor dynamic range. Why, then, is *The Art of Recording* important?

As technology advances, realism has become more accurate but it seems that the artistic qualities of a performance have become more elusive. Neither accuracy nor artistry can be tabulated in some book of rules. Unlike hardware design, for example, computers, where precision and speed predominate and the figures accurately define the performance; in the field of Sound Recording, specifications and measurements cannot describe the sounds that we hear and neither can they predict the effect of those sounds on us. The recordist must provide the vital link as our interpreter.

In the 1953, 4th edition of his *Radio Designer's Handbook*, Langford-Smith states: "It is common practice to regard the ear as the final judge of fidelity, but this can only give a true judgment when the listener has acute hearing, a keen ear for distortion and is not in the habit of listening to distorted music. A listener with a keen ear for distortion can only cultivate this faculty by making frequent direct comparisons with the original music in the concert hall."[2]

We must cultivate "A Point of Reference."

Truly successful recordists and producers have developed very high degrees of refinement and can perceive qualities (or the lack of them!) in the sound recording and reproducing chain, which seem to defy reason and sometimes contradict our current state of knowledge.

In 1977, Geoff Emmerick, who, with George Martin, recorded The Beatles at Abbey Road and later at Air Studios, London, showed me that he could hear a difference between two identical channels on a recently delivered new console. After some hours of listening with him, I agreed that I could hear a subtle difference. When we measured I found that out of 48 channels, three had been incorrectly terminated and displayed a rise of 3 dB at 54 kHz. The limit of hearing for most humans does not extend beyond 20 kHz and this small resonance, whilst obviously an oversight in the factory, would not normally have been regarded as important.

One of the significant features of this episode was that Geoff was deeply "unhappy," even "distressed" at what he was hearing or perceiving.

Since then I have seen much more evidence that the range beyond 20 kHz is part of human awareness. Newly introduced designs which transmit frequencies to beyond 100 kHz (with low distortion and noise), surprisingly, sound warmer, sweeter, and fuller.

In 1987 when addressing the Institute of Broadcast Sound in London, I carried out a simple experiment for the first time, with the object of discovering what effect frequencies above 20 kHz might have on a professionally aware audience.

A generator capable of switching between a sine and a square wave was fed through an "ordinary" amplifier to an "ordinary" monitor loudspeaker. The frequency was set at 3 kHz. The audience confirmed that they could hear the third harmonic as a superimposed 9 kHz tone or "whistle," when the generator was switched to square wave. (A square wave contains predominantly odd harmonics. When the first of these exceeds the limit of hearing, the sine and square wave should sound the same.)

The frequency was then progressively raised and older members soon admitted that they could no longer hear the third harmonic "whistle." But all could still hear a *difference* in quality as the generator was switched between sine and square. As the frequency continued to be slowly raised, some could still identify a difference when the fundamental had reached 15 kHz.

This experiment has been repeated many times in different parts of the world without any real attempt at scientific control. "Ordinary" equipment provided by my host was used on every occasion. Results have been surprisingly consistent: some 35 to 45 percent of those present being able to identify a quality difference when the fundamental was as high as 15 kHz.

There was one exception. At the University of Massachusetts Lowell, some 60 percent of the audience was still identifying a difference when the fundamental frequency had exceeded 17 kHz.

At the 83rd Convention of the A.E.S., Dr. William Moylan proposed a "Systematic Method for the Aural Analysis of Sound Sources."[3] Dr. Moylan uses his method at the University of Massachusetts Lowell Department of Music. I think this points to the success of Dr. Moylan's training in Aural Analysis! This same method has led to this book.

No measurements or formulae can ever replace the recordist, but he must develop a reliable Point of Reference and learn *"The Art of Recording."*

The learning process is endless. We never arrive at an ultimate state of knowledge. We are, even now, only scratching the surface and this is especially so as we explore new formats to convey not only technical excellence but a whole listening experience of music and its environment.

Can there be a perfect recording?

Only if we could arrive at perfect knowledge. How then, should such knowledge be used? Could it change for example, the world in which we live?

Stephen Hawking, in *A Brief History of Time*,[4] seeks a unified theory—drawing together the general theory of relativity and of quantum mechanics—which would lead to a complete understanding of the events around us and of our own existence. He says: "If we find the answer to that, it would be the ultimate triumph of human reason—for then we would know the mind of God."

We need open minds to *envision* direction, responsible minds to *choose* direction and a Point of Reference beyond ourselves to Whom we are ultimately answerable.

Well over 100 years ago, Lord Rayleigh told us that the ears are the final arbiter of sound:

"Directly or indirectly, all questions connected with this subject must come for decision to the ear, as the organ of hearing; and from it there can be no appeal. But we are not, therefore, to infer that all acoustical investigations are conducted with the unassisted ear. When once we have discovered the physical phenomena, which constitute the foundation of sound, our explorations are, in great measure transferred to another field lying within the dominion of the principles of Mechanics. Important laws are in this way arrived at, to which the sensations of the ear cannot but conform."[5]

To follow William Moylan's *"important laws"* will prepare your ears and your mind for the true *"Art of Recording."* His approach is proven and will lead the reader's ears to the refined levels of "keen-ness" and "acute-ness" required today. Dr. Moylan's book also gives us insight into what makes *"Recording," "Art,"* and provides ways to bring artistic sensibility into our work.

Rupert Neve
Wimberley, TX
September 2001

Footnotes

[1]*Bible*. 1 Chronicles 23:5. " . . . and four thousand are to praise the LORD with the musical instruments I have provided for that purpose."

[2]F. Langford-Smith, ed. *Radio Designer's Handbook*, 4th edition. Sydney, New South Wales: Wireless Press for Amalgamated Wireless Valve Company Pty. Ltd, 1953, Ch.14, Section 12, (iii).

[3]William Moylan, "A Systematic Method for the Aural Analysis of Sound Sources in Audio Reproduction/Reinforcement, Communications, and Musical Contexts," Paper read at the 83rd Convention of the Audio Engineering Society, October, 1987.

[4]Stephen W. Hawking. *A Brief History of Time*. Toronto: Bantam Press, 1992, pp.173–175.

[5]Lord Rayleigh. *The Theory of Sound,* 1st edition, 1877. New York: Dover Publications, 1945.

Preface

The Art of Recording: Understanding and Crafting the Mix is the product of my experiences as an educator in sound recording technology (music and technology), and of my thought processes and observations as a composer of acoustic music and of music for recordings (recording productions and electronic music). It includes what I have learned through my creative work as a recording producer, and through my attempts to be a facile and transparent recording engineer. It incorporates in-depth research into how we hear music and sound as reproduced through loudspeakers (aural perception, music cognition), and into the use of the recording medium to enhance artistic expression, especially in music.

This book has evolved substantially since my initial research in the early 1980s. It has been greatly shaped by nearly 20 years of instructional methods and materials in my courses at the University of Massachusetts Lowell, and by my interaction with many other audio educators and observations of other recording programs worldwide. This evolution has been enhanced by many other people and experiences as well—closely working with numerous, talented production professionals in the audio industry representing many different types and sizes of facilities in many locations, and many conversations with individuals and companies engaged in audio product development, design, and manufacturing.

The concepts and methods of this book have gone through many stages of development. They will continue to change with new technologies and production techniques, and will continue to be refined as we learn more about what we hear, and how we perceive sound and understand art.

I continue to be fascinated by how audio recording can be used to add unique artistic qualities to music, and look forward to the next new music recording and the next new development in audio technology.

Purpose

The Art of Recording seeks to bring the reader to understand how recorded sound is different from live sound, and how those differences can enhance music. It will bring the reader to explore how those sound characteristics appear in significant readings by The Beatles and others. The book also presents a system for the development of critical and analytical listening skills necessary to recognize and understand these sound characteristics.

This leads to the production process itself. The book seeks to move the reader to consider audio recording as a creative process. Techniques and technologies are purposefully not covered. Instead, the book explores the recording process as an act of creating art, and helps the reader envision recording devices as musical instruments. It seeks to develop an artistic sensitivity that will lead readers to find and create their own unique artistic voice in shaping and creating music recordings.

Intended Uses and Audience

The Art of Recording is intended to be used (1) as a resource book for all people involved in audio production; (2) as a textbook for audio production-related courses and/or listening skills-related courses in sound recording technology (music engineering), media/communications, or related programs; or (3) as a self-learning text for the motivated student or interested amateur. It has been written to be accessible by people with limited backgrounds in acoustics, engineering, physics, math, and music.

The intended reader might be an active professional in any one of the many areas of the audio recording industry, or a student studying for a career in the industry. The reader might also be learning recording through self-directed study, or be anyone interested in learning about recording and recorded sound, perhaps an audiophile.

The portions of the book addressing sound quality evaluation will be directly applicable to all individuals who work with sound. Audio engineers in technical areas will benefit from this knowledge, and the related listening skills, as much as individuals in creative, production positions. All those people who talk about sound can make use of the approach to evaluating sound that is presented.

Finally, people interested in the music of The Beatles might gain new insights. Discussions and examinations of their recordings appear throughout the book. The timelessness of their music, and their use of then experimental techniques and technologies make these recordings especially useful and appropriate. The reader will find these examples interesting points of departure for further study of their music and their creative use of recording.

As a resource book, *The Art of Recording* is designed to contribute to the professional development of recordists. The book seeks to clearly define the dimensions of audio recording and brings the reader to approach recording production creatively. It further seeks to expand the current professionals' creative thinking, their skills in critical and analytical listening, and their skill in and sensitivity to accurate and meaningful communication about sound quality.

Many people actively engaged in the creative and artistic roles of the industry are hard-pressed to describe their creative thought processes. The actual materials they are crafting have not previously been well defined. Current professionals have likely had little guidance in identifying the skills required in audio-recording production. It is likely their current skills were developed intuitively, and with little outside assistance. The recordist may already have highly developed creative abilities and listening skills, yet be unaware of the dimensions of those skills. This book will address these areas, and assist the current and future professional in discovering new dimensions within their unique and personal creative voice.

Many excellent books exist on recording techniques and audio technologies. Articles in many excellent magazines, journals, and serials exist that cover recording devices and techniques, audio technologies, and acoustic concerns. These areas are not addressed here.

No sourcebooks or textbooks currently exist that discuss the creative aspects of recording music or audio production. Few books exist that discuss and develop the listening skills required to evaluate recordings for technical quality, and none to evaluate the artistry of recordings. *The Art of Recording* seeks to fill this void. It may be used as a sourcebook or a textbook in all of these areas, from beginning through the most advanced levels.

The book may be used in a wide variety of courses, in many college and university degree programs, and in vocational-type programs in audio and music recording. Music engineering or music production (sound recording technology) programs; communications, multimedia, media, film, radio/television, or telecommunications programs; and music composition programs emphasizing electronic/computer music composition will all have courses that speak to the artistic aspects of recording music (and sound).

The book is well suited to developing the student's music production skills and artistry. It is designed to stimulate thought about the recording process as being a collection of creative resources. These skills and creative ideas can then be applied to the act of crafting the music recording artistically.

As a textbook for sound evaluation and listening skill development, *The Art of Recording* can be used for instruction at various skill levels.

The instructor may determine the level of proficiency required of the students in performing listening evaluations. The development of listening skills is a lengthy and involved process that will go through many stages of accomplishment. The book is useable by students at the beginning of their studies or at the most advanced levels, graduate students as well as undergraduates.

Graphing the activity of the various artistic elements is important for developing listening and evaluation skills, especially during beginning studies. It is also valuable for performing in-depth evaluations of recordings that allow us to study how the artistic aspects of recording have been used by accomplished recording producers. This process of graphing the activity of the various artistic elements is also a useful documentation tool. Working professionals through beginning students will find the process useful.

Finally, it is hoped this book will clarify communication in some small way and in some small segment of our industry.

All audio professionals are required to communicate about sound. The many artistic and technical people of the industry need to communicate clearly. We in the industry presently function without this meaningful exchange of ideas. We often do not communicate well or accurately. In order for communication to occur, a vocabulary must be present. Terms or descriptions must mean the same thing to the people involved, and the terms must apply to something specific within the sound. This book might serve as a meaningful point of departure for an audio recording vocabulary to be devised.

Acknowledgments

Terri Jadick and Marie Lee brought me into the Focal Press family before moving on. I am very grateful for their support and efforts, and for their faith in my book. My deep appreciation to Tricia Tyler for taking over the helm and my project; her great support made this project much easier and very enjoyable. Thanks also to Jennifer Plumley for her patience and work in production, to Jenny Nam, and especially to Diane Wurzel for her assistance as editor and her guidance during the important final stages of this book.

My students in the Sound Recording Technology program at the University of Massachusetts Lowell have worked through the materials and concepts of this book in a number of forms, and have taught me a great deal. My special thanks to these serious and gifted young people for their (largely unknowing) contributions to this project.

Many other people have provided support for this book, or assisted in shaping my thoughts about these materials. While they are far too many to mention, they are all very important to me. Only a very few follow.

A big thank you to Alison Tolman-Rogers for her superb work on the graphs and artwork. Thanks for coming to my rescue!

Thanks to Jenny Kilpatrick and especially Mark Whittaker for their work in reviewing my writing and assisting with sound evaluations and graphs, and to Kerry Kilpatrick for her work.

Thanks to Cleve L. Scott and Pierre Schaeffer for their guidance in the initial directions of my research, and to Vicki and Zachary for their patience and understanding while I was writing this book.

I am deeply indebted to Mr. Rupert Neve for his generosity in providing time, effort, and great insight to this book by writing its Foreword.

Introduction

What makes recording music an art?

What makes the music recording a unique medium for artistic expression?

What is different between a music recording and a live music performance?

Why has the recordist (recording producer or engineer) become recognized as an artist?

How does the recording process (recording techniques and technologies) shape a piece of music?

It is widely recognized that the recording process shapes music. Recording techniques and technologies change the qualities of acoustic sound and impart new sound characteristics. These sound qualities are under the control of an individual that shapes the music recording—the recordist.

The changes in sound quality that are created by recording do not occur in nature. They are unique to audio recordings and give recorded music (or music reproduced over loudspeakers) a set of unique sound characteristics. These characteristics may be very different from a live, unamplified performance. These sound qualities have become accepted as part of the experience of listening to recorded and reproduced sound, and of music.

The unique sound qualities of recording contribute to the character of the piece of music and become part of that piece. How a piece of music can be shaped has thus been extended to include the unique sound qualities of music recordings. The person controlling, or creating, these sound qualities (the recordist) is functioning as a creative artist. This person is a musician of sorts—"conducting" by encouraging and ensuring quality performances, "performing" recording, mixing and processing devices, and "composing" the mix.

Overview of Organization and Materials

The above questions on the previous page will be answered during the course of this book. The elements of sound that shape the artistic qualities of recordings have not previously been defined. The definitions offered herein will allow for the understanding necessary to define the above questions. This book will demonstrate how these aspects of sound are shaped in the recording process and will examine their appearance in well-known recordings by The Beatles. A system for developing listening skills and understanding the artistry of music recordings will also be presented. Through this process and its many related exercises, the reader will be brought to gain understanding of and control over shaping the artistic aspects of audio recordings.

Accordingly, the book is divided into three parts:

- Part 1 defines the artistic elements
- Part 2 evaluates how the artistic elements have been commonly used in production practice and presents a system for listening skill development
- Part 3 explores the artistry of crafting a music recording

Part 1

Part 1 is divided into three chapters. To begin defining the artistic dimensions of music recordings, sound must be understood. The states of sound in air, in human perception, and as applied to music are followed in the sequence of understanding the meaning of sounds. The processes that occur in moving from one state of sound to another are explored. The anomalies that occur in the transfer processes are recognized and evaluated.

Sound as a resource for artistic expression is the basis for Chapter 2. This encompasses the unique sound qualities of recordings and the potential of those qualities to be used in artistic expression. This is followed by an examination of the musical message itself, leading to how musical materials are perceived by the listener, and how that perception leads to the understanding of musical messages and to communication.

Part 1 centers around understanding sound and the listening process itself. It brings to light the importance of the listener in the communication of musical materials.

People listen to sound and music at various levels of intellectual involvement. They will at times listen passively and take an undirected journey through the sensual and emotive states of the music. They might seek an understanding of any literary or extra-musical ideas in the music (with the presence and influences of any text or other image-inducing associations). Listening can also take the form of an aesthetic listening experience (appreciating the interrelationships of the abstract musical materials). These and others are basically recreational, entertainment, therapeutic, or self-enrichment activities. The listening process is approached much differently when it is part of one's work.

When the listening process is approached as part of the professional recordist's responsibilities, it is always an active process that has the listener consciously engaged. How one listens, and what one listens for, is a central concern for the recordist. Listening is one of the primary responsibilities of the recordist. Any person wishing to enter the recording industry must develop their listening skills in one way or another.

Different positions might require very different skills and responsibilities, but nearly all careers in audio involve listening to, evaluating and/or describing sound.

Part 2

Part 2 presents a complete system for evaluating the dimensions of sound in music and audio recordings. The need for sound evaluation and the contexts for sound evaluation in music recordings are discussed at the beginning of this section.

Each element of sound is evaluated separately. How each element is used to shape music recordings is presented. A method of evaluation has been specifically devised for each individual element and reflects actual use of the elements in music recordings. Recordings by The Beatles are used as examples of the qualities of sound being discussed. They provide excellent examples of how the unique sound qualities of recordings have enhanced the music and have at times contributed in fundamental ways to shaping musical ideas.

The methods of evaluation for each individual sound quality are accomplished in relation to a complete, interrelated system of evaluating sound in music recordings. A series of listening exercises is presented throughout the course of the book to guide the reader in developing sound evaluation and listening skills.

The system progresses from simple concepts and listening processes to the most complex. It builds on experiences that are most easily learned and evolves systematically to the most difficult. Listening experiences that many audio professionals or intermediate-level musicians may have already acquired (at least intuitively) are incorporated.

The system will develop the reader's listening skills and will provide the basis for meaningful and accurate communication on sound content and quality.

An objective vocabulary and a way to evaluate sound have been devised to allow precise information about sound to be recognized and communicated. This information will be an actual account of the states or values of the sound material. Subjective impressions of the sound's quality (a very common way musicians and recordists attempt to communicate about sound) are always avoided. They do not allow communication to be accurate and limit its value. The method avoids any personal impressions about the sound quality, and addresses only the physical dimensions of sound as they are perceived and as they appear in the music. The method for evaluating sound will allow individuals to talk about sound in meaningful ways. It will allow people to exchange precise and conclusive information about the sound qualities, once they acquire the skills required to recognize those qualities.

The reader will eventually gain the experience and the knowledge that will allow them to perform quite complex listening and evaluation tasks, and to describe sounds to others.

Time, practice, understanding, repetition, and concentrated effort are all required to develop listening skills. Auditory memory will increase as the listener becomes more accustomed to the sound material, more aware of patterns of levels and changes within all of the artistic elements, and more aware of how to focus attention on the aspects of sound we are conditioned from birth to ignore. Developing refined listening skills is a long-term project, and an individual's listening skills will likely continue to develop throughout their career.

Part 3

Part 3 applies the artistic elements to the recording production process. It presents the concepts and thought processes of recording production, and relates them to the artistic aspects of recording and the artistic elements of sound. It will explore the concepts the recordist will work through during the creative processes of making a music recording.

Part 3 defines general principles. It will not present specific ways to record, nor will it address specific pieces of equipment or specific recording techniques. The principles covered will allow the reader to conceptualize music recording as a creative process. It will place the reader in the position of guiding the artistic product from its beginning as an idea, through its development during the many stages of the recording sequence, to its final form. It is intended to separate tools from technique, to bring the reader to conceptualize the recording project without concern for the ever-changing flow of devices.

This approach will explore the creative potentials of the audio recording medium independently from equipment and technology concerns. How the recording process shapes the characteristics of sounds will be explored, and ways the reader can establish and exercise control of the artistry of recording will be presented. The resources of the recording process are considered for their potential to capture sound qualities, to shape the music recording, and to generate the relationships of a piece of music. The reader will be encouraged to use their new skills at evaluating sound to explore the productions of others, and try to emulate similar ideas in their own productions—to develop their craft and their art.

Two possible recording production sequences are presented and contrasted. Through these two scenarios, the artistic concerns of the recording production process are evaluated. The artistic roles (or functions) of the recordist are contrasted in relation to the sequences.

It is a goal of *The Art of Recording* to bring the reader to explore the creative potentials of the medium's tools (equipment) as musical instruments and to develop an artistic sensitivity through the study of the creative works of others (Part 2).

Applying the artistic elements of recording to the recording process (Part 3) and evaluating the artistic elements in the recording process (Part 2) will often occur simultaneously in production practice. They are, however, two distinct processes. They are presented separately here for clarity and for a thorough presentation. The two are interdependent when considering the evaluation of sound that occurs during the recording production process.

Summary

Part 1 provides the necessary background to recognize the qualities of sound. Part 2 will lead the listener to develop listening and sound evaluation skills and to learn how to communicate objective and meaningful information about sound. Part 3 will bring the reader to gain control of the craft of recording music and shaping sound, and to use the recording process creatively.

The goal of *The Art of Recording: Understanding and Crafting the Mix* is to bring the reader to an awareness of the dimensions of recorded sound, and to an understanding of how the dimensions have been used in music recordings, through the development of listening and evaluation skills. It will seek to lead the reader to find and create their own unique artistic voice in crafting music recordings.

Part One

Understanding Sound and the Aesthetic Qualities of Audio Recordings

1 The Elements of Sound and Audio Recording

Audio recording is the recording of sound. It is the act of capturing the physical dimensions of sound, then reproducing those dimensions either immediately or from a storage medium (magnetic, solid, electronic, digital), and thereby returning those dimensions to their physical, acoustic state. The process moves from physical sound, through the recording/reproduction chain, and back to physical sound.

The "art" in recording centers around the artistically sensitive application of the recording process. The recording process is being used to shape or create sound as an artistic statement (piece of music), or supporting artistic material. To be in control of crafting the artistic product, one must be in control of the recording process, be in control of the ways in which the recording process modifies sound, and be in control of communicating well-defined creative ideas.

These areas of control of the artistic process all closely involve a human interaction with sound. Inconsistencies between the various states of sound are present throughout the audio recording process. Many of these inconsistencies are the result of the human factor: the ways in which humans perceive sound and interpret or formulate its meanings. In order for material to be under their control, the artist (audio professional) must understand the substance of their material: sound, in all its inconsistencies.

The States of Sound

In audio recording, sound is encountered in three different states. Each of these three states directly influences the recording process, and the creation (or capturing) of a piece of art. These three states are:

1. Sound as it exists physically (having physical dimensions);

2. Sound as it exists in human perception (psychoacoustic conception); sound being perceived by humans after being transformed by the ear and interpreted by the mind (the perceived parameters of sound being human perceptions of the physical dimensions); and
3. Sound as idea; sound as it exists as an aural representation of an abstract or a tangible concept, as an emotion or feeling, or representing a physical object or activity (this is how the mind finds meaning from its attention to the perceived parameters of sound); sounds as meaningful events, capable of communication, provide a medium for artistic expression; sounds hereby communicate, have meaning.

The audio recording process ends with sound reproduced over loudspeakers, as sound existing in its physical state, in air. Often the audio recording process will begin with sound in this physical state, to be captured by a microphone.

Humans are directly involved in all facets of the audio recording process through listening to sound. They evaluate the audio signal at all stages while the recording is being made (including the *recordist*—the person making the recording—and all others involved in the industry), and what is heard by the end listener is the reason for making the recording. Humans translate the physical dimensions of sound into the perceived parameters of sound through the listening process (aural perception).

This translation process involves the hearing mechanism functioning on the physical dimensions of sound, and the transmission of neural signals to the brain. The process is nonlinear and alters the information; the hearing mechanism does not produce nerve impulses that are exact replicas of the applied acoustic energy.

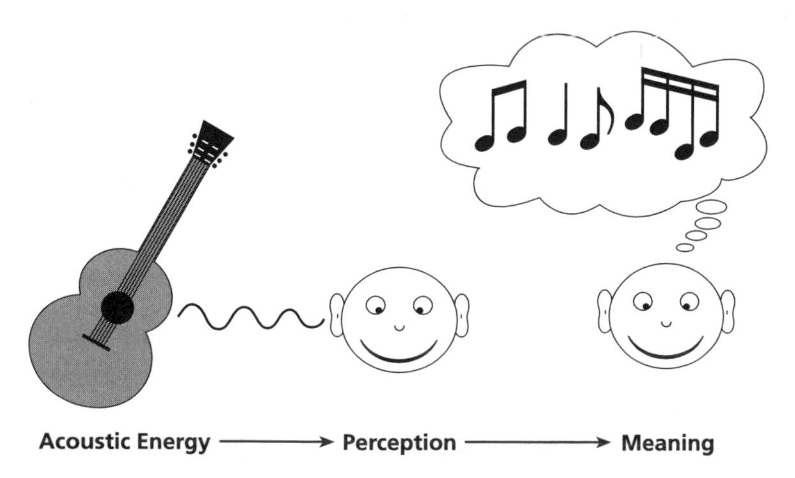

Figure 1-1 Three states of sound: in air, in perception, as message.

Certain aspects of the distortion caused by the translation process are, in general, consistent between listeners and between hearings; they are related to the physical workings of the inner ear or the transfer of the perceived sounds to the mind/brain. Other aspects are not consistent between listeners and between hearings; they relate to the listener's unique hearing characteristics and their experience and intelligence.

The final function occurs at the brain. At a certain area of the cortex, the neural information is processed, identified, consciously perceived, and stored in short-term memory; the neural signals are transferred to other centers of the brain for long-term memory. At this point, the knowledge, experience, attentiveness, and intelligence of the listener become factors in the understanding and perception of sound's artistic elements (or the meanings or message of the sound). The individual is not always sensitive or attentive to the material or to the listening activity, and the individual is not always able to match the sound to their previous experiences or known circumstances.

The physical dimensions (1) are interpreted as perceived parameters of the sound (2). The perceived parameters of sound (2) provide a resource of elements that allow for the communication and understanding of the meaning of sound (and artistic expression) (3).

The audio recording process communicates ideas, and can express feelings and emotions. Audio might take the forms of music, dialog, motion picture action sounds, whale songs, or others. Whatever its form, audio is sound that has some type of meaning to the listener. The perceived sound provides a medium of variables that are recognizable and have meaning, when presented in certain orders or patterns. Sound, as perceived and understood by the human mind, becomes the resource for creative and artistic expression. The artist uses the perceived parameters of sound as the artistic elements of sound, to create and ensure the communication of meaningful (musical) messages.

The individual states of sound as physical dimensions and as perceived parameters will be discussed individually, in the next section. The interaction of the perceived parameters of sound will follow the discussion of the individual parameters. These discussions provide critical information to understanding the breadth of the "artistic elements of sound" in audio recording, presented in the next chapter.

Physical Dimensions of Sound

Five physical dimensions of sound are central to the audio recording process. These physical dimensions are: the characteristics of the sound waveform as (1) *frequency* and (2) *amplitude* displacements, occurring within the continuum of (3) *time*; the fusion of the many frequency and amplitude anomalies of the single sound to create a global, complex

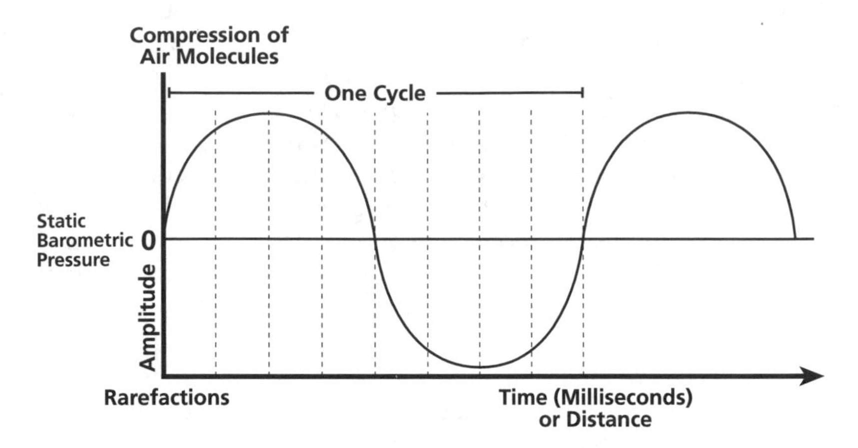

Figure 1-2 Dimensions of the waveform.

waveform as (4) *timbre*; and, the interaction of the sound source (timbre) and the environment in which it exists, create alterations to the waveform according to variables of (5) *space*.

Frequency is the number of similar, cyclical displacements in the medium, air, per time unit (measured in cycles of the waveform per second, or Hz). Each similar compression/rarefaction combination creates a single cycle of the waveform. Amplitude is the amount of displacement of the medium at any moment, within each cycle of the waveform (measured as the magnitude of displacement in relation to a reference level, or decibels).

Timbre

Timbre is a composite of a multitude of functions of frequency and amplitude displacements; it is the global result of all the amplitude and frequency components that create the individual sound. Timbre is the overall quality of a sound. Its primary component parts are the dynamic envelope, spectrum, and spectral envelope.

The *dynamic envelope* of a sound is the contour of the changes in the overall dynamic level of the sound throughout its existence. Dynamic envelopes of individual acoustic instruments and voices vary greatly in content and contours. The dynamic envelope is often thought of as being divided into a number of component parts. These component parts may or may not be present in any individual sound. The widely accepted components of the dynamic envelope are: attack (time), initial decay (time), initial sustain level, secondary decay (time), primary sustain level, and final decay (release time).

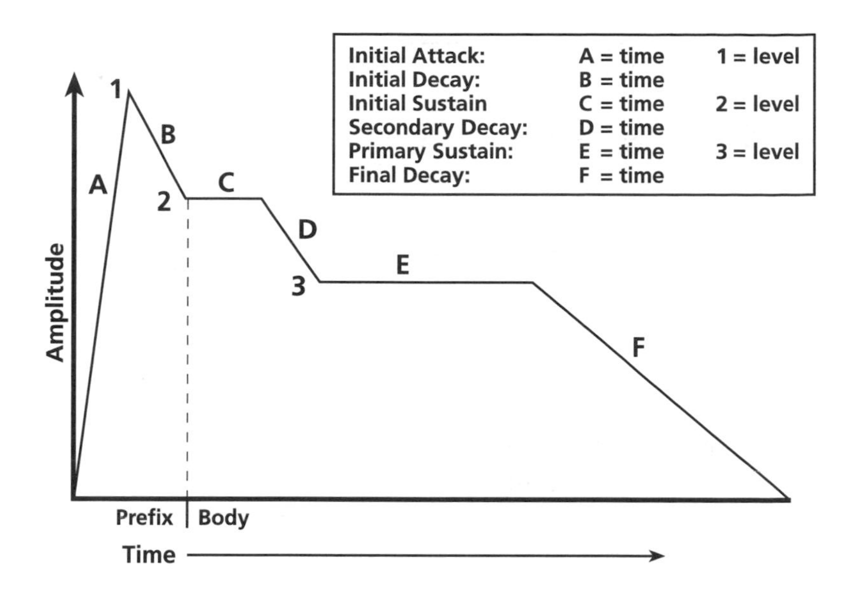

Figure 1-3 Dynamic envelope.

Dynamic envelope shapes other than those created by the above outline are common. Many musical instruments have more or fewer parts to their characteristic dynamic envelope. Further, vocalists and the performers of many instruments have great control over the sustaining portions of the envelope, providing internal dynamic changes to sounds. Musical sounds that do not have some variation of level during the sustain portion of the envelope are rare; the organ is one such exception.

The *spectrum* of a sound is the composite of all of the frequency components of the sound. It is comprised of the fundamental frequency, harmonics, and overtones.

The periodic vibration of the waveform produces the sensation of a dominant frequency. The number of periodic vibrations, or cycles of the waveform is the *fundamental frequency*. The fundamental frequency is also that frequency at which the sounding body resonates along its entire length. The fundamental frequency is often the most prominent frequency in the spectrum, and will often have the greatest amplitude of any component of the spectrum.

In all sounds except the pure sine wave, frequencies other than the fundamental are present in the spectrum. These frequencies are usually higher than the fundamental frequency. They may or may not be in a whole-number relationship to the fundamental. Frequency components of the spectrum that are whole-number multiples of the fundamental are *harmonics*; these frequencies reinforce the prominence of the fundamental

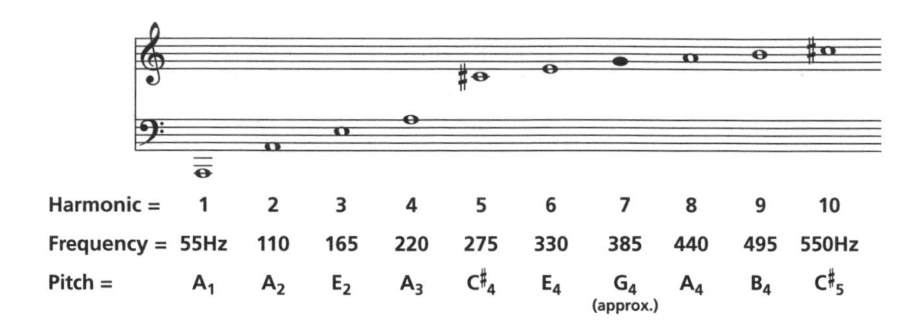

Figure 1-4 Harmonic series.

frequency (and the pitched quality of the sound). Those components of the spectrum that are not proportionally related to the fundamental we will refer to as *overtones*. Traditional musical acoustics studies define overtones as being proportional to the fundamental, but with a different sequence than harmonics (first overtone = second harmonic, etc.); this traditional definition is herein replaced by a differentiation between *partials* that are proportional to the fundamental (harmonics), and those that are not (overtones). This distinction will prove important in the evaluation of timbre and sound quality in later chapters. All of the individual components of the spectrum are partials. Partials (overtones and harmonics) can exist below the fundamental frequency as well as above; they are accordingly referred to as subharmonics and subtones.

For each individual instrument or voice, certain ranges of frequencies within the spectrum will be emphasized consistently, no matter the fundamental frequency. Instruments and voices will have resonances that will strengthen those spectral components that fall within these definable frequency ranges. These areas are called *formants, formant regions*, or *resonance peaks*. Formants remain largely constant, and modify the same frequency areas no matter the fundamental frequency. Spectral modifications will be present in all occurrences of the sound source with harmonics or overtones in the formant regions. Formants can appear as increases in the amplitudes of partials that appear in certain frequency bands, or as spectral components in themselves (such as noise transients caused by a hammer striking a string). They can also be associated with resonances of the particular mechanism that produced the source sound. Formants are largely responsible for shaping the characteristic sounds of specific instruments; they allow us to differentiate between the instruments of different manufacturers, or even to tell the difference between two instruments of the same model/make.

A sound's spectrum is comprised primarily of partials that create a characteristic pattern, which is recognizable as being characteristic of a particular instrument or voice. This pattern of spectrum will transpose

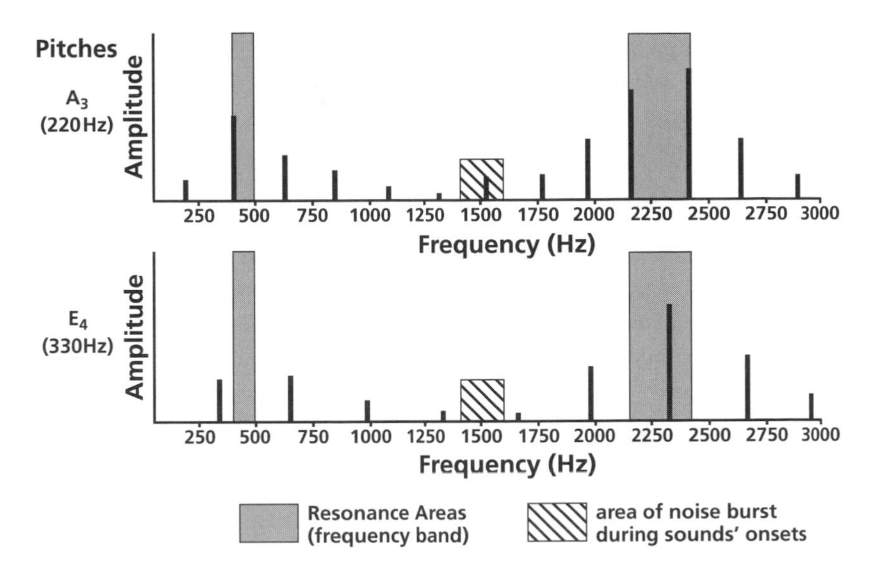

Figure 1-5 Formant regions of two pitches from a hypothetical instrument. Vertical lines represent the partials of the two pitches, placed at specific frequencies and at specific amplitudes.

(change level) with every new fundamental frequency of the same sound source and remain mostly unchanged. This consistent pattern will form a similar timbre at different pitch levels. Formants establish frequency areas that will be emphasized for a particular instrument or voice. These areas will not change with varied fundamental frequency, as they are fixed characteristics (such as resonant frequencies) of the device that created the sound. Formants may also take the form of spectral information that is present in all sounds produced by the instrument or voice.

The frequencies that comprise the spectrum (fundamental frequency, harmonics, overtones, subharmonics and subtones) all have different amplitudes that change independently over the sound's duration. Thus, each partial has a different dynamic envelope. Altogether these dynamic envelopes of all the partials make up the *spectral envelope*. The spectral envelope is the composite of each individual dynamic level and dynamic envelope of all of the components of the spectrum.

The component parts of timbre (dynamic envelope, spectrum, and spectral envelope) display strikingly different characteristics during different parts of the duration of the sound. The duration of a sound is commonly divided into two time units: the *prefix* or *onset*, and the *body*. The initial portion of the sound is the prefix or onset; it is markedly different from the remainder of the sound, the body. The time length of the prefix is usually determined by the way a sound is initiated, and is often the same time unit as the initial attack. The actual time increment of the prefix may be anywhere from a few microseconds to 20–30 milliseconds.

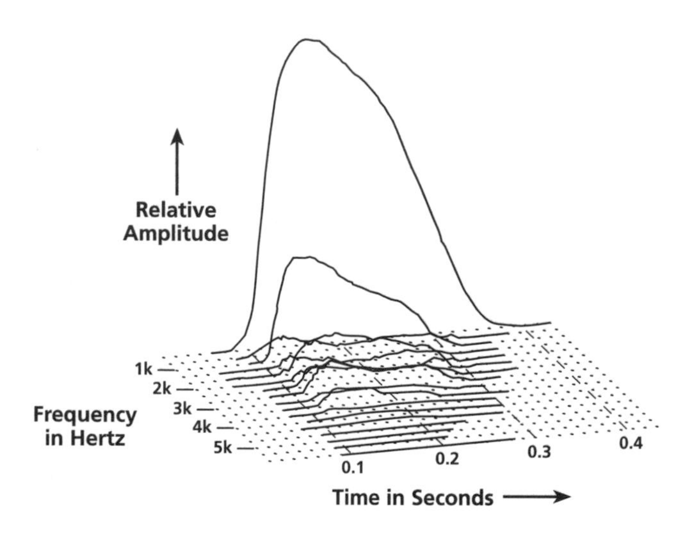

Figure 1-6 Spectral envelope.

The prefix is defined as the initial portion of the sound that has markedly different characteristics of dynamic envelope, spectrum, and spectral envelope, than the remainder of the sound. The body of the sound is usually much longer in duration than the prefix. (See Figure 1-3.)

Space

The interaction of the sound source (timbre) and the environment, in which it is produced, will create alterations to sound. These changes to the sound source's sound quality are created by the acoustic space. The nature of these alterations are directly related to (1) the characteristics of the acoustic space in which the sound is produced and (2) the location of the sound source within the environment.

Space-related sound measurements must be performed at a specific, physical location. The measurements are calculated from the point in space where a receptor (perhaps a microphone or a listener) will capture the composite sound (the sound source within the acoustic space). The location of the listener (or other receptor) becomes a reference in the measurement of the acoustic properties of space.

The aspects of space that influence sound in audio recording are: (1) the *distance* of the sound source to the listener, (2) the *angle* of the sound source to the listener, (3) the *geometry of the environment* in which the sound source is sounding, and (4) the *location* of the sound source *within the host environment*.

The environment in which the sound source is sounding is often referred to as the *host environment*. Within the host environment, sound

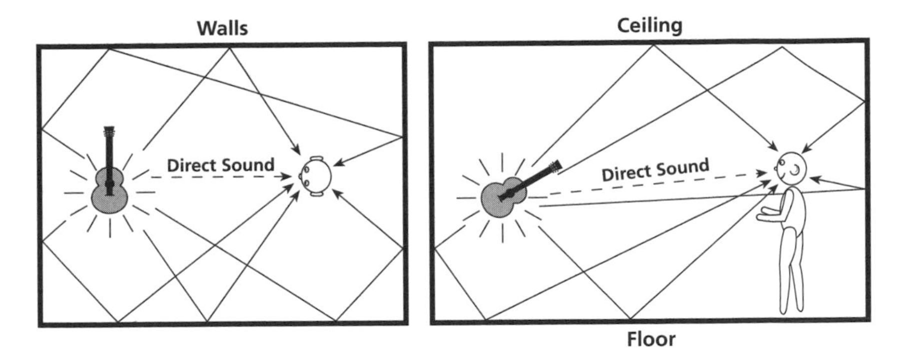

Figure 1-7 Paths of reflected sound within an enclosed space.

will travel on a direct path to the listener (as *direct sound*) and sound will bounce off reflective surfaces before arriving at the listener (as *reflected sound*).

Reverberant sound is a composite of many reflections of the sound arriving at the listener (or microphone) in close succession. The many reflections that comprise the reverberant sound are spaced so closely that the individual reflections cannot be measured; these many reflections are therefore considered as a single entity. As time progresses, these closely spaced reflections become more closely spaced and of diminishing amplitude, until they are no longer of consequence. *Reverberation time* (often referred to as RT60) is the length of time required for the reflections to

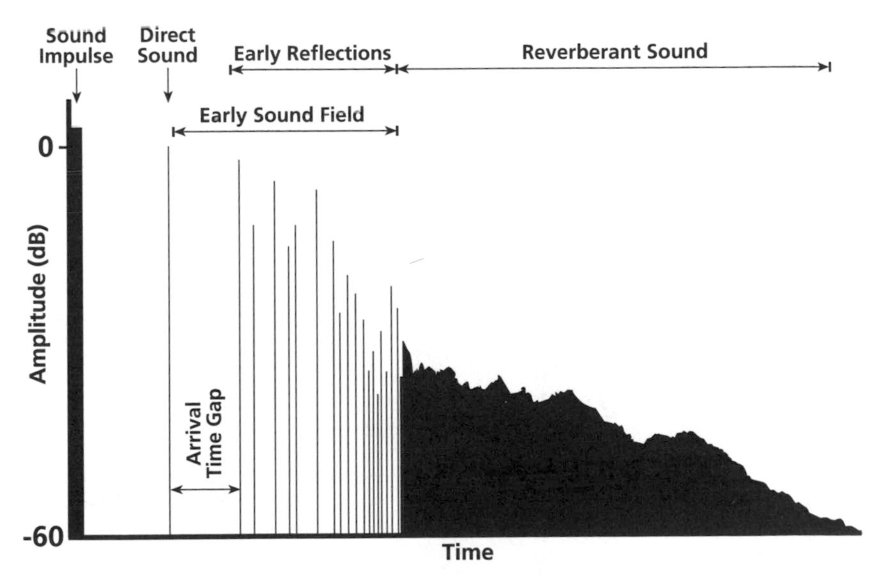

Figure 1-8 Reflected sound.

reach an amplitude level of 60 dB lower than that of the original sound source.

Early reflections are those reflections that arrive at the ear or microphone within around 50 milliseconds of the direct sound. As a collection, the reflections that arrive at the receptor within the first 50 milliseconds after the arrival of the direct sound comprise the *early sound field.*

Varying the *distance* of the sound source from the receptor (ear or microphone) alters the sound at the receptor. The sound at the receptor will be a composite of the direct sound and the reflected sounds (reverberation and early sound field). The composite sound at the receptor is affected by the distance of the sound source from the receptor in two ways: (1) low amplitude portions of the sound's spectrum (usually high frequencies) are lost with increasing distance of the sound source to the receptor and (2) reflected sound increases in prominence to the direct sound as distance increases. Figure 1-9 illustrates the loss of *timbral detail* (the subtle aspects and changes in the content of a sound's timbre, also called *definition of timbre*) with increasing distance as well as the change of the proportion of direct to reflected sound.

The characteristic changes to the composite sound caused by the geometry of the host environment and by the location of the sound source within the host environment, are also influenced by the changes caused by distance.

These two dimensions of the relationship of the sound source to its acoustic space may alter the composite sound in four, additional ways: (1) timbre differences between the direct and reflected sounds; (2) time differences between the arrivals of the direct sound, the initial reflections, and the reverberant sound; (3) spacing in time of the early reflections; and (4) amplitude differences between direct and reflected sounds.

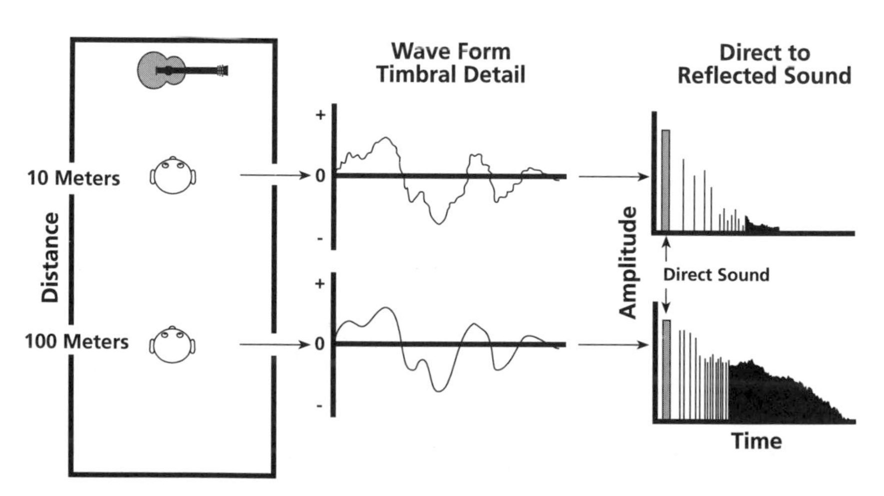

Figure 1-9 Changes in sound with distance.

The geometry of the host environment greatly influences the content of the composite sound. The dimensions and volume of the space, the angles of boundaries (walls, floors, ceilings), materials of construction, and the presence of openings (such as windows) and large objects within the space will all alter the composite sound. Host environments cover the gamut of all the physical spaces and open areas that create our reality (from small room, to a large concert hall; from the corridor of a city street, to an open field, etc.).

Unique sequences of reflected sound are created when a sound is produced within an environment, and sequences are shaped by the location of the sound source within the host environment. These unique sequences contain patterns of reflections that are defined by the spacing of reflections over time and the amplitudes of the reflections. A "rhythm of reflections" exists, and will form the basis of important observations in later chapters. By altering the early time field and reverberant sound, the location of the sound source within the host environment may cause significant alterations to the composite sound at the receptor (ear or microphone).

The location of the sound source within the host environment may strongly influence the composite sound. The amount will be directly related to the proximity of the sound source to the walls, ceiling, floor, openings (such as windows and doors), and large objects reflecting sound within the host environment.

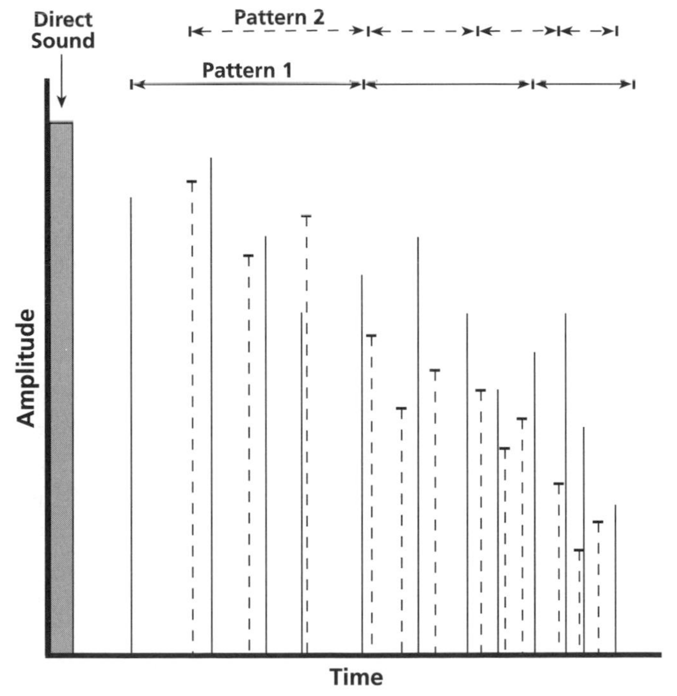

Figure 1-10 Patterns of reflections.

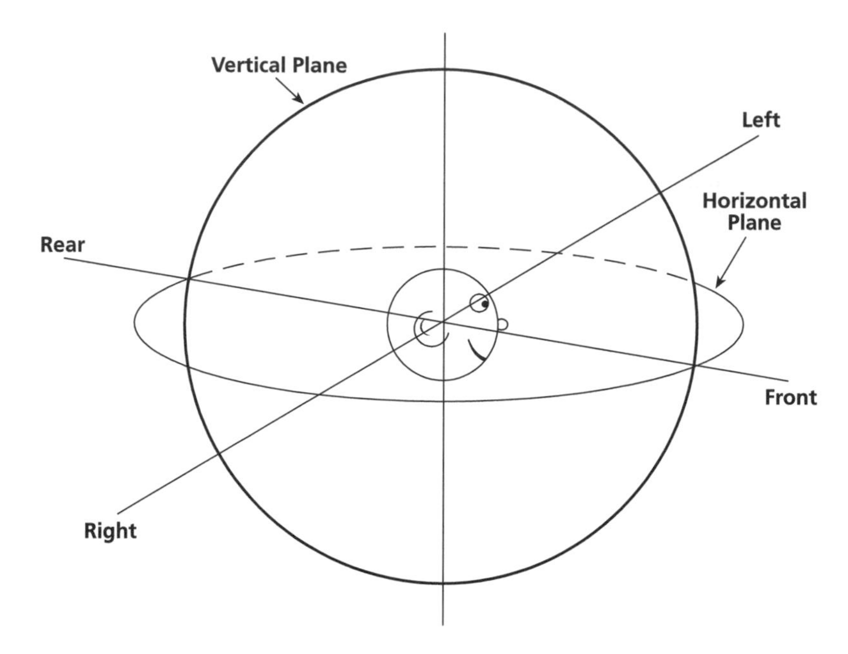

Figure 1-11 Horizontal and vertical planes.

In audio production, the spatial properties of host environments and the location of a sound source within the environment can be generated artificially. It is common to use reverberation units and delays to create environmental cues. These cues may be very realistic representations of natural spaces or they may be environmental characteristics that cannot occur in our physical reality.

The angle of the sound source to the receptor is an important influence in audio recording. The sound source may be at any angle from the receptor (listener or omnidirectional microphone) and be detected. The sound source may be present at any location in the sphere surrounding the receptor. The location is calculated with reference to the 360° vertical and horizontal planes that encompass the receptor.

The angle of the sound source to the receptor may be calculated against the horizontal plane (parallel to the floor), the vertical plane (height), or by combining the two (in a way very similar to positioning locations on a globe). Defining elevation (vertical plane) and direction (left, right, front, rear) can determine the precise location of the sound source within our three-dimensional space by precise increments of degrees.

Angles of source locations on the horizontal plane are captured or generated in audio recording to provide stereo and surround sound. To date, the vertical plane has received little attention in audio because of

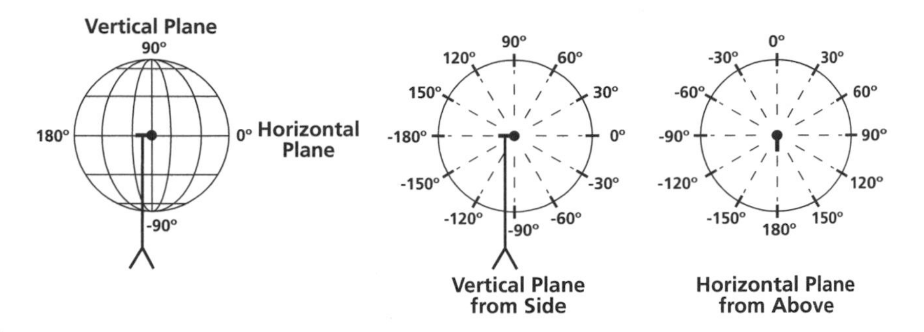

Figure 1-12 Defining sound source angle from a microphone.

playback format difficulties. Recent surround sound advances have produced formats that provide these cues in ways that can strikingly enhance programs.

Perceived Parameters of Sound

The five physical dimensions of sound translate into respective perceived parameters of sound. Sound as it exists in human perception is quite different from sound in its physical state, in air. Our perception of sound is a result of the physical dimensions being transformed by the ear and interpreted by the mind. The perceived parameters of sound are human perceptions of the physical dimensions of sound.

This translation process from the physical dimensions to the perceived parameters is nonlinear, and differs between individuals. The hearing mechanism does not directly transfer acoustic energy into equivalent nerve impulses. The human ear is not equally sensitive in all frequency ranges, nor is it equally sensitive to sound at all amplitude levels. This nonlinearity in transferring acoustic energy to neural impulses causes sound to be in a different state in our perception, than what exists in air. Thus, the physical states of sound captured by recording equipment will be heard by the recordist in ways that may be unexpected, without knowledge of these differences.

Complicating this further, there is no reason to believe any two people actually hear the characteristics of sound in exactly the same way. If it were possible for all conditions for two sounds to be identically sent to two listeners, the two people likely would hear slightly (or strikingly) different characteristics. We only need notice the different ear shapes around us to recognize no two people will pick up acoustic energy in precisely the same way.

Table 1-1
The Physical Dimensions and the Perceived Parameters of Sound

Physical Dimensions	*Perceived Parameters*
Frequency	Pitch
Amplitude	Loudness
Time	Duration
Timbre (physical components)	Timbre (perceived overall quality)
Space (physical components)	Space (perceived characteristics)

Pitch

Pitch is the perception of the frequency of the waveform. The frequency area most widely accepted as encompassing the hearing range of the normal human spans the boundaries of 20–20,000 Hz (20 kHz), though recent research has proven that humans are sensitive to (if not actually hear) frequencies below and above this range.

Most humans cannot identify specific pitch-levels. Some people have been blessed with, or have developed the ability to recognize specific pitch levels (in relation to specific tuning systems). These people are said to have "absolute" or "perfect pitch." The ability to accurately recognize pitch-levels is not common even among well-trained musicians.

It is commonly within human ability, however, to determine the relative placement of a pitch within the hearing range. A *register* is a specific portion of the *range*. It is entirely possible to determine, within certain consistent limits of accuracy, the relative register of a perceived pitch-level. This skill can be developed, and accuracy improved significantly.

Humans are able to consistently perform the estimation of the approximate level of a pitch, associating pitch-level with register. With practice, this consistency can be accurate to within a minor third (within three semi-tones). This skill in the "estimation of pitch-level" will be an important part of the method for evaluating sound presented later.

Humans perceive pitch most accurately as relationships. We perceive pitch as the relationship between two or more soundings of the same or related sound sources. We do not perceive pitch as identifiable, discrete increments; we do not listen to pitch material to define the letter-names (increments) of pitches. Instead, humans calculate the distance (or interval) between pitches by gauging the distance between the perceived levels of the two (or more) pitches.

The interval between pitches becomes the basis for all judgments that define and relate the sounds. Thus, melody is the perception of successively sounded pitches (creating linear intervals), and chords are the perceptions of simultaneously sounded pitches (creating harmonic intervals). We often perceive pitch in relation to a reference level (one predominating pitch that acts as the key or pitch-center of a piece of music), or to a system of organization to which pitches can be related (a tonal system, such as major or minor).

Our ability to recognize the interval between two pitches is not consistent throughout the hearing range. Most listeners have the ability to accurately judge the size of the semi-tone (or minor second, the smallest musical interval of the equal tempered system) within the range of 60 Hz and 4 kHz. As pitch material moves below 60 Hz, a typical listener will have increased difficulty in accurately judging interval size. As pitch material moves above 4 kHz, the typical listener will also experience increased difficulty in accurately judging interval size.

The smallest interval humans can accurately perceive changes with the register and placement of the two pitches creating the interval. The size of the minimum audible interval varies from about 1/12 of a semi-tone between 1–4 kHz, to about 1/2 of a semi-tone (a quarter-tone) at approximately 65 Hz. These figures are dependent upon optimum duration and loudness levels of the pitches; sudden changes of pitch-level are up to 30 times easier to detect than gradual changes. It is possible for humans to distinguish up to 1,500 individual pitch levels by spacing out the appropriate minimum-audible intervals, throughout the hearing range.

With all factors being equal, the perception of harmonic intervals (simultaneously sounding pitches) is more accurate than the perception of melodic intervals (successively sounded pitches). Up to approximately 500 Hz, melodic and harmonic intervals are perceived equally well. Above 1 kHz, humans begin to be able to judge harmonic intervals with greater accuracy than melodic intervals; above 3,500 Hz, this difference becomes pronounced.

Loudness

Loudness is the perception of the overall excursion of the waveform (amplitude). Amplitude can be physically measured as a sound pressure level. In perception, loudness level cannot be accurately perceived in discrete levels.

Loudness is referred to in relative values, not as having separate and distinct levels of value. Traditionally loudness levels have been described by analogy ("louder than," "softer than," etc.) or by relative values ("soft," "medium loud," "very soft," "extremely loud," etc.). Humans compare loudness levels and conceive loudness levels as being "louder than" or "softer than" the previous, succeeding, or remembered loudness level(s).

A great difference exists between loudness as perceived by humans and the physical amplitude of the sound wave. This difference can be quite large at certain frequencies. In order for a sound of 20 Hz to be audible, a sound pressure level of 75 dB must be present. At 1 kHz, the human ear

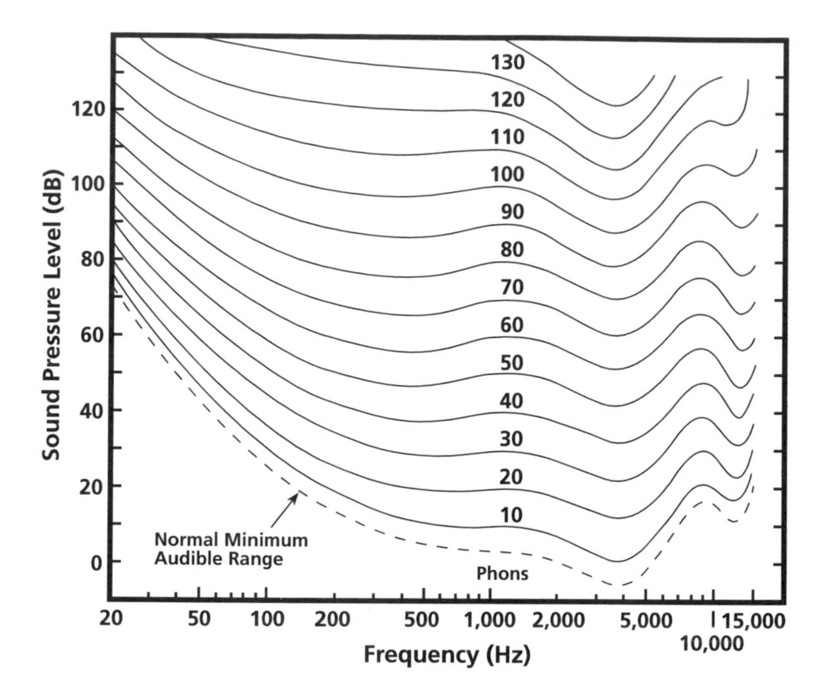

Figure 1-13 Equal loudness contour.

will perceive the sound with a minute amount of sound pressure level, and at 10 kHz a sound pressure level of approximately 18 dB is required for audibility. The unit *phon* is the measure of perceived loudness established at 1 kHz, based on subjective listening tests.

The nonlinear frequency response of the ear and the fatigue of the hearing mechanism over time play an important role in further inaccuracies of the human perception of loudness. With sounds of long durations and steady loudness level, loudness will be perceived as increasing with the progression of the sound until approximately 0.2 seconds of duration. At that time, the gradual fatigue of the ear (and possibly shifts of attention by the listener) will cause perceived loudness to diminish.

As loudness level of the sound is increased, the ear requires increasingly more time between soundings, before it can accurately judge the loudness level of a succeeding sound. We are unable to accurately judge the individual loudness levels of a sequence of high-intensity sounds as accurately as we can judge the individual loudness levels of mid- to low-intensity sounds; the inner ear itself requires time to reestablish a state of normalcy, from which it can accurately track the next sound level.

As a sound of long duration is being sustained, its perceived loudness level will gradually diminish. This is especially true for sounds with high sound pressure levels. The ear gradually becomes desensitized to the loudness level. The physical masking (covering) of softer sounds and an inability to accurately judge changes in loudness levels will result from the fatigue. When the listener is hearing under listening fatigue, slight changes of loudness may be judged as being large. Listening fatigue may desensitize the ear's ability to detect new sounds at frequencies within the frequency band (frequency area) where the high sound pressure level was formerly present.

Duration

Humans perceive time as *duration*. Sound durations are not perceived individually. We cannot accurately judge time increments without a reference time unit. Regular reference time units are found in musical contexts, and rarely in other types of human experiences. Even the human heartbeat is rarely consistent enough to act as a reliable reference. The underlying metric pulse of a piece of music does, however, allow for accurate duration perception. This accuracy cannot be achieved in any other context of the human experience.

In music, the listener remembers the relative duration values of successive sounds, in a similar process to that of perceiving melodic pitch intervals. These successive durations create musical rhythm. The listener calculates the length of time between when a sound starts and when it ends, in relation to what precedes it, what follows it, what occurs simultaneously with it, and what is known (what has been remembered). Instead of calculating an interval of pitch, the listener proceeds to calculate a span of time, as a durational value.

Metric Grid

A *metric grid*, or an underlying pulse, is quickly established in the perception of the listener, as a piece of music unfolds. This creates a reference pulse against which all durations can be defined. The listener is thereby able to make rhythmic judgments in a precise and consistent manner. The equal divisions of the grid allow the listener to compare all durations, and to calculate the pulse-related values of the perceived sounds. Durations are calculated as being in proportion to the underlying pulse: at the pulse, half pulses, quarter of the pulse, double the pulse, etc.

In the absence of the metric grid, durational values cannot be accurately perceived as proportional ratios. Humans will not be able to perceive slight differences in duration when a metric grid is not established.

The listener is only able to establish a metric grid within certain limits. Humans will be able to accurately utilize the metric grid between 30 to 260 pulses per minute. Beyond these boundaries, the pulse is not perceived as the primary underlying division of the grid. The human mind

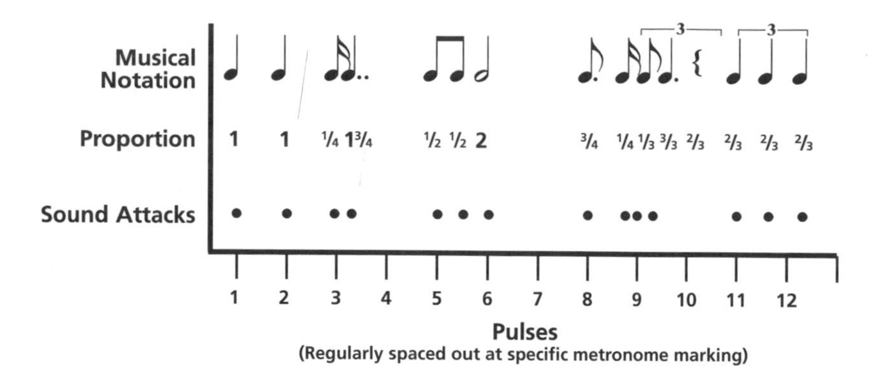

Figure 1-14 Metric grid.

will replace the pulse with a duration of either one-half or twice the value, or the listener might become confused and unable to make sense of the rhythmic activity.

The metric grid is the dominant factor in our perception of tempo, as well as musical rhythm. In most instances, the metric grid itself represents the steady pulsation of the tempo of a piece of music.

Time

The listener's *perception of time* plays a peripheral role in the perception of rhythm. Time perception is significant to the perception of the global qualities of a piece of music, and to the estimation of durations when a metric grid is not present in the music. The global qualities of aesthetic, communicative, and extra-musical ideas within a piece of music are largely dependent on the living experience of music; on the passage of musical materials across the listener's time perception of their existence.

Time perception is distinctly different from duration perception. The human mind makes judgments of elapsed time based on the perceived length of the present. The length of time humans perceive to be "the present" is normally two to three seconds, but might be extended to as much as five seconds and beyond.

The "present" is our window of consciousness, through which we perceive the world, and listen to sound. We are at once experiencing the moment of our existence, evaluating the immediate past of what has just happened and anticipating the future (projecting what will follow the present moment, given our experiences of the recently passed moments, and our knowledge of previous, similar events).

Human time judgments are imprecise. The speed at which events take place and the amount of information that takes place within the "present" greatly influences time judgments. The amount of time perceived to have passed will change to conform to the number of events experienced within the present; the listener will estimate the amount of time passed in relation to the number of experiences during the present, and make time judgments accordingly.

Time judgments are greatly influenced by the individual listener's attentiveness and interest in what is being heard. If the material stimulates thought within the listener, the event will seem shorter; if the listener finds the listening activity desirable in some way, the experience will seem to occupy less time than would an undesirable experience of the same (or even shorter) length. Expectations caused by, boredom with, interest in, contemplation of, and even pleasure caused by music, alter the listener's sense of elapsed time.

The time length of a piece of music (or any time-based art form, such as a motion picture) is separate and distinct from clock time. A lifetime can pass in a moment, through the experience of a work of art. A brief moment of sound might elevate the listener to extend the experience to an infinity of existence.

Timbre in Perception

The overall quality of a sound, its *timbre*, is the perception of the mixture of all of the physical aspects that comprise a sound. Timbre is the global form, or the overall character of a sound, which we can recognize as being unique.

The overall form (timbre) is perceived as the states and interactions of its component parts. The physical dimensions of sound discussed above, are perceived as dynamic envelope, spectral content, and spectral envelope (perceived values, not physical values). The perceived dimensions are interpreted, and shape an overall quality, or conception of the sound.

Humans remember timbres as entities, as single objects having an overall quality (that is comprised of many unique characteristics), and sometimes as having meaning in themselves (as a timbre can bring with it associations in the mind of the listener). We recognize the sounds of hundreds of human voices because we remember their timbres. We remember the timbres of a multitude of sounds from our living experiences. We remember the timbres of many musical instruments and their different timbres as they are performed in many different ways.

The global quality that is timbre allows us to remember and recognize specific timbres as unique and identifiable objects.

Humans have the ability to recognize and remember a large number of timbres. Further, listeners have the ability to scan timbres and relate unknown sounds to sounds stored in the listener's long-term memory. The listener is then able to make meaningful comparisons of the states and values of the component parts of those timbres. These skills will serve as meaningful points of departure for the method for evaluating timbre in Part 2.

Sufficient time is required for the mind to process the many characteristics of a sound in order to recognize and understand its overall image. The time required to perceive the component parts of timbre vary significantly with the complexity of the sound, and the listener's previous knowledge of the sound. For rather simple sounds, the time required for accurate perception is approximately 60 milliseconds. As the complexity of the sound is increased, the time needed to perceive the sound's component parts will also increase. All sounds lasting less than 50 milliseconds are perceived as noise-like, since a specific timbre cannot be identified at that short a duration; exceptions occur when the listener is well acquainted with the sound, and the timbre can be recognized from this small bit of information.

The partials of the timbre's spectrum fuse to create the impression of a single sound. Although many frequencies are present, the tendency of our perception is to combine them into one overall texture. We fuse partials that are harmonically related to the fundamental frequency, as well as overtones that are distantly related to the fundamental, into a single impression.

It is especially important for the recording professional to note, fusion can also occur between two separate timbres (two individual sound sources) if the proper conditions are present. Timbres that are attacked simultaneously, or are of a close harmonic relationship to each other, are most likely to fuse into the perception of a single sound. The more complex the individual sound, the more likely that fusion will not occur. Furthermore, if the listener recognizes one of the timbres, fusion will be far less likely to occur. Also related to recognizing timbre, synthesized sounds are more likely to fuse with other sounds than are known sounds of an acoustic origin.

Spatial Characteristics

The perception of the *spatial characteristics* of sound is the impression of the physical location of a sound source in an environment, together with the modifications the environment itself places on the sound source's timbre.

The perception of *space* in audio recording (reproduction) is not the same as the perception of space of an acoustic source in a physical environment. In an acoustic space, listeners perceive the location of sound in relation to the three-dimensional space around them: distance, vertical plane, and horizontal plane. Sound is perceived at any possible angle from the listener, and sound is perceived at a distance from the listener; both of these perceptions involve an evaluation of the characteristics of the sound source's host environment.

In audio recording, illusions of space are created. Sound sources are given spatial characteristics through the recording process and/or through signal processing. This spatial information is intended to complement the timbre of the sound source. The spatial characteristics may simulate particular known, physical environments or activities, or be intended to provide spatial cues that have no relation to our reality. In theory, all of the interactions of the sound with its host environment are captured with, or can be simulated and applied to, the sound source; upon playback through two or more loudspeakers, the spatial cues are reproduced.

Playback Environment

Recordings and their sound sources (combined with their spatial characteristics) are heard through two (or more) loudspeakers. The loudspeakers themselves are placed in and interact with, a playback environment—such as a living room or automobile. The playback environment is nearly always quite unrelated to the spaces on the recording. Thus, spatial characteristics applied to the sound source are ultimately perceived by the listener after they have been altered by the characteristics of loudspeakers, altered by the interaction of the loudspeakers and the environment (caused by placement of loudspeakers within the playback environment), and altered by the playback environment itself. The listener perceives the reproduced spatial characteristics of the sound source within the three-dimensional space of their listening environment (headphone monitoring is not a solution, as will be later discussed).

To accurately perceive the spatial information of an audio recording, the listening environment must be acoustically neutral, and the listener must be carefully positioned within the environment and in relation to the loudspeakers. The listening environment (including the loudspeakers) should not place additional spatial cues onto the reproduced sound.

Perceived Spatial Relationships and Current Sound Reproduction

Humans perceive spatial relationships (1) as the location of the sound source being at an angle to the listener (above, below, behind, to the left, to the right, in front, etc.), (2) as the location of the sound source being at distance from the listener, (3) as an impression of the type, size, and properties of the host environment, and at times (4) as the location of the sound source being within an environment.

These perceptions are transferred into the recording medium, to provide a realistic illusion of space, with one major exception. The angular location is severely restricted in audio reproduction, as compared to human perceptual abilities. Currently used audio playback formats can only accurately and consistently reproduce localization cues on the horizontal plane, and then only slightly beyond the loudspeaker array in

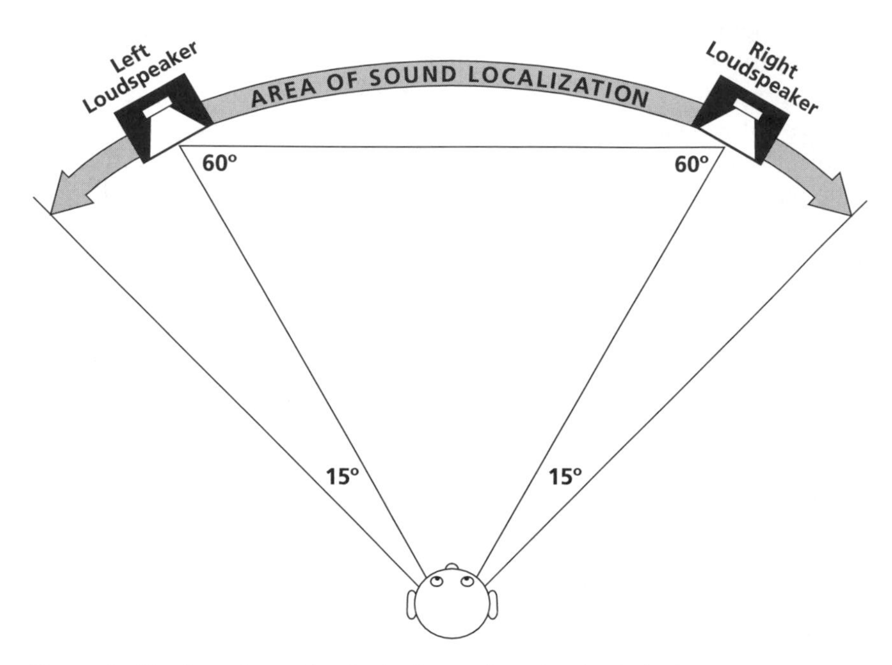

Figure 1-15 Area of sound localization in two-channel audio playback.

stereo recordings. The three-dimensional space of our reality is simulated (with dubious accuracy) in the two dimensions of audio recording.

Much research and development is taking place attempting to extend the localization of sound sources to behind the listener and to the vertical plane, as well as to provide a more realistic reproduction of distance and environmental cues. Significant advances are being made. Surround sound is now widely accepted and provides reproduction around the listener that can be mostly stable and accurate. Control of sound localization on the vertical plane and a more complete simulation of environmental characteristics are not, however, presently feasible, though the technology of the future will likely address these areas as well.

The following discussions of the perception of the spatial dimensions of reproduced sound refer to common two-channel, stereo and to surround sound systems. These concepts will transfer to other systems such as binaural recordings, holophonics, and others; but with different boundaries of the limits of perception inherent with each particular format.

Localization of Direction

The ability for humans to localize sounds is one of the survival mechanisms we have retained from our ancient past. This ability has been developed throughout our evolution; we have learned to perceive the direction of even those sounds that our hearing mechanism has difficulty processing.

Humans use differences in the same sound wave appearing at the two ears for the accurate *localization of direction*. Interaural time differences (ITD) are the result of the sound arriving at each ear at a different time. A sound that is not precisely in front or in back of the listener will arrive at the ear closest to the source before it reaches the furthest ear. These time differences are sometimes referred to as phase differences. The sounds arriving at each ear are almost identical during the initial moments of the sound, except the sound at each ear is at a different point in the waveform's cycle (and might contain minute spectral differences).

Interaural amplitude differences (IAD) work in conjunction with ITD in the localization of the direction of the sound source. IAD are also referred to as interaural spectral differences. IAD is the result of sound pressure level differences at high frequencies present at the two ears. The head of the listener, which blocks certain frequencies from the furthest ear (when the sound is not centered), causes the interaural spectral differences. This occurrence has been termed the "shadow effect." Interaural amplitude differences (IAD) will at times consist solely of amplitude differences between the two ears, with the spectral content of the waveform being the same at both ears.

The sound wave is almost always different at each ear. The differences between the sound waves may be time/phase-related, amplitude/spectrum-related, and/or spectrum differences. These differences in the waveforms are essential to the perception of the direction of the sound source. In addition, these same cues play a major role in perceiving the characteristics of host environments.

Up to approximately 800 Hz, humans rely on ITD for localization cues. Phase differences are utilized for localization perception up to about 800 Hz, as amplitude appears to be the same at both ears.

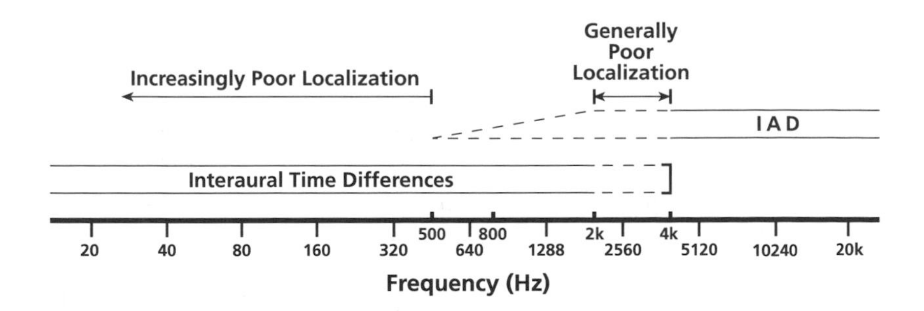

Figure 1-16 Frequency ranges of localization cues.

Between 800 Hz and about 2 kHz both phase and amplitude differences are present between the two ears. IAD and ITD are both used for the perception of direction in this frequency range, with amplitude differences becoming significant around 1,250 Hz.

In general, time/phase differences seem to dominate the perception of direction up to about 4 kHz. Although IAD are present, ITD dominates the perception of direction between 2 kHz and 4 kHz. Humans have poor localization ability for sounds in this frequency band.

Above 4 kHz, interaural amplitude differences (IAD) are the cues that determine the perception of location. Localization ability improves at 4 kHz and is quite accurate throughout the upper registers of our hearing range.

Recent studies have revealed the human body also generates physical cues for localization. The chest, head, shoulders, and outer ears all affect sounds of various wavelengths in different ways. Our body parts' different sizes and their different angles to the hearing canal create a very complex source of reflected and diffracted sound waves. These waves all lead to important interaural spectrum differences between the two ears. These differences are created by a comb-filter effect, comprised of minute cancellations and reinforcements of frequencies. The brain processes these subtle differences to aid in identifying a sound source's direction.

As we have seen, humans do not perceive direction accurately at all frequencies. Below approximately 500 Hz, our perception of the angle of the sound source becomes increasingly inaccurate, to the point where sounds seem to have no apparent, focused location. An area exists around 3 kHz where localization is also poor; wavelength similarities between the distance between the two ears and those of the frequencies around 3 kHz cause interaural time/phase differences to be unstable.

Humans have a well-refined ability to localize sounds in the approximate frequency areas: 500 Hz–2 kHz, and 4 kHz to upper threshold of hearing (whatever that might be). Within these areas, the minimum discernible angle is approximately one to two degrees, with less accuracy at the sides and back than in the front. Sounds that have fundamental frequencies outside of these frequency areas, but that have considerable spectral content within these bands, will also be localized quite accurately.

Interaural spectral differences occur throughout the frequency spectrum. While they may be subtle, it appears they are important for the localization of objects in frequency ranges where IAD and ITD are ineffective. The makeup of these interaural spectral differences will inherently be unique to individuals, as (obviously) no two people are the same size and shape, and no two outer ears are alike.

Pinna is the name of the outer ear. This part of our anatomy (an elaborately shaped piece of cartilage) plays several important roles in our perception of direction. The pinna gathers sound and funnels it into the ear canal. As the ridges of the outer ear reflect sound into the ear, the ridges introduce small time delays between their reflections and the direct sound that travels directly to the ear canal. These small time delays vary according to sound source location, and are important components of the interaural spectral differences described above. The pinna and the delays it generates aid us in differentiating between sounds arriving from the front and those arriving from the rear.

Resonances also appear to be excited in the outer ear. These also alter the frequency response of the sound source in predictable ways that vary between individuals. The brain learns these patterns of spectral changes to assist in localization. Pinna cues play significant roles in direction perception even though each individual has a unique ear-shape, and the resultant spectrum changes are equally unique. Location cues based on spectrum are thus not universal, but unique to the individual and are learned.

Pinnae serve a critical function in front to back localization. When sound arrives at the head from the rear, ridge reflections are not generated. The pinna actually blocks the direct sound from reaching the hearing canal and its ridges when sounds are generated beyond 130° from the front center. The pinna allows us to perceive the sound source as being generated to our rear because of the absence of spectral differences.

It is interesting to note, our distance and location judgments are not as accurate to the sides and the rear. The absence of this spectral information generated and collected by the outer ear may well play a role.

We actually move our heads involuntarily to assist in locating sounds—especially those sounds that are not in front. In moving our heads to remove location confusion we bring the source into our front listening field, and thus reintroduce the IAD, ITD and spectral differences of the pinnae. We also instinctively seek to bring the source into visual view, which eliminates all ambiguity for acoustic sources—but not for phantom images.

Front-back hearing is only partially understood. Little relevant research in spatial perception of sounds arriving from the rear and the sides is available. Certainly more will need to be accomplished before we are able to more thoroughly understand this area. It is increasingly important, however, that we understand how we perceive sound arrival from the rear and the sides, and the different qualities of those sounds, if we are to fully understand and control the differences between surround sound and stereo.

Distance Perception

Distance perception has not been studied thoroughly. The following information is well documented, and it is likely that numerous subtleties will be discovered in the future.

Two impressions lead to the perception of the distance of a sound source from the listener: (1) the ratio of the amount of direct sound to reverberant sound, and (2) the primary determinant, the loss of low amplitude (usually high frequency) partials from the sound's spectrum with increasing distance (*definition of timbre* or *timbral detail*). Both of these functions rely on the listener's knowledge of the sound's timbre for accurate perception of distance-location. While sound pressure decreases with distance, loudness itself does not factor into distance location perception.

Low energy spectral information is lost with the compressions and rarefactions of the waveform over distance. Some information is simply absorbed by the atmosphere due to air friction. This leads to the listener's determination of the level of timbral detail (definition of timbre) that is the major factor in distance perception.

Some timbre-related distance information results from waveform travel and the speed of sound. As high frequencies travel slightly faster than low frequencies, the spectrum of the sound is altered with increasing distance. The partials of complex sounds will become increasingly out of phase with the fundamental frequency, and between themselves, the longer the propagation of the sound.

The percentage of direct sound decreases while the percentage of reflected sound increases, as the source moves from the listener. This pertains to enclosed spaces only. This ratio of direct to reflected sound may play a significant role in distance perception when timbral detail cues are diminished or are unknown.

The listener must know the timbre of a sound in order to recognize missing timbral detail. If the sound is unknown or not recognized, the listener cannot identify the potential loss of low energy components from its spectrum. With knowledge of the sound source, the listener will be able to calculate how much low energy information is missing and thus be able to determine the general amount of distance between them and the object.

Knowledge of the timbre of the sound source will assist the listener in recognizing the absence of spectral information and/or perceiving the reiterations of the direct sound and the reverberant sound. These perceptions will provide the listener with the required information to judge distance. The previous experiences and listening skills of the listener will play a major role in the accuracy of judgments made.

Without prior knowledge of the timbre of a sound, perception of distance location is considerably less accurate, if not impossible.

Related to the ratio of direct to reflected sound, the time difference between the ceasing of the direct sound and the ceasing of the reverberant energy will increase with distance. Through *temporal fusion* we perceive the reverberant sound as being a part of the direct sound. This creates a single impression of the sound in its environment (referred to as the composite sound, above). As distance increases, temporal fusion begins to diminish and the ending of the direct sound and the continuance of the reverberant energy become more prominent.

Perception of Environmental Characteristics

The perceptions of the *characteristics of the host environment* and the *placement of the sound source within the host environment* are also dependent upon the ratio of direct to reflected sound and the loss of low-level spectral components with increasing distance. In addition, the characteristics of the host environment are perceived through (1) the time difference between the arrival of the direct sound and the arrival of the initial reflections, (2) the spacing in time of the early reflections, (3) amplitude differences between the direct sound and all reflected sound (the individual initial reflections and the reverberant sound), and (4) timbre differences between the direct sound, the initial reflections, and the reverberant sound.

The time delay between the direct and the reflected sounds is directly related to (1) the distance between the sound source and the listener, (2) the distance between the sound source and the reflective surfaces (which send the reflected sound to the listener), and (3) the distance of the reflective surfaces from the listener. These three physical distances also create the patterns of time relationships (the rhythms) of the early reflections.

Early reflections arrive at the listener within 50 milliseconds of the direct sound. These early reflections comprise the *early sound field*. The early sound field is composed of the first few reflections that reach the listener before the beginning of the diffused, reverberant sound (see Figure 1-8). Many of the characteristics of a host environment are disclosed during this initial portion of the sound. The early sound field contains information that provides clues as to the size of the environment, the type and angles of the reflective surfaces, even the construction materials and surface coverings of the space.

Humans have the ability to accurately judge the size and characteristics of the host environments of sound sources. This is accomplished by evaluating the sound qualities of the environment. Humans experience and remember the sound qualities of a great many natural environments in much the same way as we recognize and remember timbres. Further, we have the ability to compare the sound qualities of new environments we encounter to our memories of environments we have previously experienced.

The listening skill needed to evaluate and recognize environmental characteristics can be developed to a highly refined level. Some people who work regularly with acoustical environments develop these listening skills to a point where many can perceive the dimensions and volume of an environment, its surface coverings, or even openings within the space (doors, windows, etc.).

Interaction of the Perceived Parameters

The perception of any parameter of sound is always dependent upon the current states of the other parameters. Altering any of the perceived parameters of sound will cause a change in the perceived state of at least one other parameter.

The parameters of sound interact, causing the perception of the state of one parameter to be altered by the state of another. Certain occurrences of these interactions were noted under individual perceived parameters. The following are additional examples of note and are separated for clarity.

Duration for Pitch Perception

Sufficient duration is required for the ear to perceive pitch. If the duration is too short, the sound will be perceived as having indefinite pitch, as being noise-like. The time necessary for the mind to determine the pitch of a sound is dependent on the frequency of the sound. Sounds lower than 500 Hz and sounds higher than 4 kHz require more time to establish pitch quality, than sounds pitched between 2 kHz and 4 kHz where pitch perception is most acute. At the extremes of the hearing range, pitch quality may require as much as 60 milliseconds to become established.

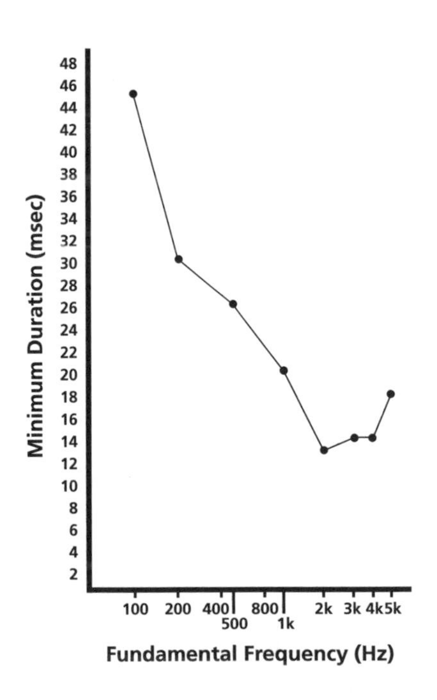

Figure 1-17 Minimum durations for pitch perception at select frequencies.

The length of time required to establish a perception of pitch will also depend on the sound's attack characteristics and its spectral content (its timbre). Sounds with complex (but mostly harmonic) spectra and sounds with short attack times will establish a perception of pitch sooner than other sounds.

Loudness and Pitch Perception

Loudness will influence the perception of pitch, as humans will perceive a change of pitch with a change of loudness (dynamic) level. The level 60 dB SPL (or about the loudness level of normal conversation) is considered to be a threshold where increases or decreases in loudness affect pitch perception oppositely. Above 60 dB, for sounds below 2 kHz a substantial increase in loudness level will cause an apparent lowering of the pitch level; the sound will appear to go flat, although no actual change of pitch level has occurred. Similarly, a substantial increase in the dynamic (loudness) level of a pitch above 2 kHz will cause the sound to appear to go sharp; an impression of the raising of the pitch level is created, although the actual pitch level of the sound has remained unaltered. Below 60 dB an increase in loudness will cause sounds below 2 kHz to be perceived as getting sharp and sounds below 2 kHz perceived as going flat.

Loudness and Time Perception

Loudness level can influence perceived time relationships. When two sounds begin simultaneously, they will appear to have staggered entrances if one of the two sounds is significantly louder than the other. The louder sound will be perceived as having been started first.

Perceived loudness level is often distorted by the speed at which information is processed. When a large number of sounds occur in a short period of time, the listener will perceive those sounds as having a higher loudness level than sounds of the same sound pressure level, but that are distributed over a longer period of time. This distortion of loudness level is caused by the amount of information being processed within a specific period of time (the time period is related to the perceived length of the present).

Loudness Perception Altered by Duration and Timbre

Duration can distort the perception of loudness. Humans tend to average loudness levels over a time period of about 2/10 of a second. Sounds of shorter durations will appear to be louder than sounds (of the same intensity) with durations longer than 2/10 of a second.

Timbre can also influence perceived loudness. Sounds with a complex spectrum will be perceived as being louder than sounds that contain fewer partials. Similarly, sounds with more complex spectra (including many overtones) will be perceived as louder than those that contain a greater number of proportionally-related partials (harmonics) when both are at the same sound pressure level. Following this principle, a change of timbre during the sustained duration of a sound will result in a perceived change in loudness.

Pitch Perception and Spectrum

As a product of the interaction of the harmonics and closely related overtones of a sound's spectrum, a timbre can create a perception of pitch where a fundamental frequency is not physically present. The harmonics of a sound reinforce its fundamental frequency to enhance the perception of pitch. This phenomenon is so capable of producing the perception of the fundamental frequency that a harmonic spectrum can provide the perception of pitch when the fundamental frequency is not physically present (missing fundamental). A perception of the periodicity of the fundamental frequency is created by the spectrum of the sound, although that frequency itself may not be present.

Amplitude, Time, and Location

The amplitude of two reiterated sounds (separated in time) can influence location perception. The precedence or Haas effect results when two loudspeakers reproduce the same sound in close succession. The effect works against the principle that when two loudspeakers reproduce a sound simultaneously, and at the same amplitude, the sound appears to be centered between the two loudspeakers.

When two loudspeakers reproduce the same sound source in close succession, normal perception would seem to be to localize a sound source at the earliest sounding loudspeaker, then to shift the image to center when the second loudspeaker is sounded. The Haas effect functions to continue the localization of the sound source at the first speaker location, while adding the second loudspeaker (to reinforce the sound intensity of the first speaker) without losing the localization of the sound source at the location of the first loudspeaker. The time difference between the sounding of the loudspeakers must be at least 3 milliseconds to keep the sound at the leading speaker, with 5 milliseconds being a more effective minimum; a maximum delay of approximately 25–30 milliseconds may be used before the delayed signal is perceived as an echo (echoes will be perceived at all frequencies at a delay of 50 milliseconds). If the leading channel is lowered by 10 dB, or the following channel increased by 10 dB, the sound source will again be centered.

Masking

Masking occurs when a sound (or a portion of a sound) is not perceived because of the qualities of another sound. The simultaneous sounding of two or more sounds can cause a sound of lower loudness level, or a sound of more simple spectral content, to be masked or hidden from the perception of the listener. The masking of sounds is a common problem for people beginning their studies or work in audio recording.

When two simultaneous sounds of relatively simple spectral content have close fundamental frequencies, the sounds will tend to mask each other and blend into a single, perceived sound. As the two sounds become separated in frequency, the masking will become less pronounced until both sounds are clearly distinguishable.

Sounds of relatively simple spectral content tend to mask sounds that are at higher frequencies. This masking becomes more pronounced as the loudness of the lower sound is increased, and is more likely to occur when a large interval separates the two pitch-levels of the sounds. This masking is especially prominent if the two pitch-levels are in a simple harmonic relationship (especially 2:1, 3:1, and 5:1). A higher pitched sound can mask a lower pitched sound if the higher sound is significantly higher in loudness level, and given the same conditions as above; the higher the loudness level, the broader the range of frequencies a sound can mask.

Masking can occur between successive sounds. With sounds separated in time by up to 20–30 milliseconds, the second sound may not be perceived if the initial sound is of sufficient loudness level to draw and retain listener attention, or to fatigue the ear. In a similar way, a sound may not be perceived if it is followed by another sound of great intensity within 10 milliseconds.

Audio equipment can produce "white" and other broadband noise that can mask sounds at all or many frequencies. An entire program might be masked by the noise of the sound system itself, should the loudness level of the noise be sufficiently higher than that of the program. This type of masking problem will first be noticed in the high frequencies, where low loudness levels exist in the upper components of the sound's spectrum.

Summary

The three states of sound that concern audio recording are sound as it exists in air (the physical dimensions of sound), sound as it exists in human perception (the perceived parameters of sound), the understanding of the meaning of a sound (sound as a resource for artistic expression). The physical dimensions of sound in air are transformed into neural impulses as the perceived parameters of the sound by the ear and brain. The perceived parameters of sound become understood as a resource of elements that allow for the communication and understanding of the meaning of sound (and artistic expression).

The two physical dimensions of the waveform are frequency and amplitude. They function in time, and form the basis for our understanding of timbre and spatial properties. As frequency becomes perceived as pitch, amplitude as loudness, time as duration, timbre as timbral characteristics, and space as perceived locations and environmental characteristics—the anomalies of human hearing transform acoustic energy into our perception with marked changes. What these changes actually are, and how these changes take place, are of great concern to the recording professional as they work in the many ways sound is captured, created, modified, and perceived.

Sound as it is perceived and understood by the human mind, becomes the resource for creative and artistic expression in sound. The perceived parameters of sound become the artistic elements of sound in creating musical material and in communicating other meaningful messages. The aesthetic and artistic elements of sound in audio recording are presented in the next chapter.

2 The Aesthetic and Artistic Elements of Sound in Audio Recordings

The audio recording process has given the creative artist the tools to very finely shape perceived sound (the perceived parameters of sound) through a direct control of the physical dimensions of sound. This control of sound is well beyond that which was available to composers and performers before the presence of modern recording technology. This controlling of sound in new ways has led to new artistic elements in music, and has led to a redefinition of the musician, to new characteristics of sound, and to new dimensions in music. While our discussion is focused on musical applications, it should be remembered that all aspects of these artistic elements of music also function as artistic elements in other areas of audio (such as broadcast media, film, multimedia, etc.).

A new creative artist has evolved. This person uses the tools of recording technology as sound resources for the creation (or recreation) of an artistic product. This person may be a performer or composer in the traditional sense, or this person may be one of the new musicians: a producer, or sound engineer, or any of the host of other, related job titles. Throughout this book, these people are referred to as recordists.

Through its detailed control of sound, the audio recording medium has resources for creative expression that are not possible acoustically. Sounds are created, altered, and combined in ways that are beyond the reality of live, acoustic performance. New creative ideas and new additions to our musical language have emerged as a result of recording techniques and technologies.

Creative ideas are defined by these aesthetic and artistic elements. The artistic elements are the aspects of sound that comprise or characterize creative ideas (or entire works of art, pieces of music). Study of the artistic elements will allow us to understand individual musical ideas and the larger musical event, and to recognize how those ideas and sound events contribute to the entire piece of music. Discussion will emphasize the artistic elements that are unique to recorded music, especially music created through the use of modern recording techniques and technologies.

As we have learned from the previous chapter, the artistic elements of sound are the mind/brain's interpretation of the perceived parameters of sound. Sound as it is perceived and understood by the human mind, becomes the resource for creative and artistic expression in sound. The perceived parameters of sound are utilized as the artistic elements of sound to create and ensure the communication of meaningful (musical) messages.

The Art of Recording occurs when the parameters of sound are perceived as a resource for artistic expression. Recording becomes an art when it is used to shape the substance of sound and music. These materials that allow for artistic expression will be understood through a study of their component parts: the artistic elements of sound.

The States of Sound and the Aesthetic/Artistic Elements

After the perception of sound, the recorded material is understood as being comprised of sound elements that are interpreted by the mind/brain, and thus communicate artistic ideas. The aesthetic/artistic elements are directly related to specific perceived parameters of sound, just as the perceived parameters of sound were directly related to specific physical dimensions of sound.

As will be remembered, sound in audio recording is in three states: physical dimensions, perceived parameters, and artistic elements.

The artistic elements are used by the recordist to shape music (sound), resulting in artistic expression. The perceived parameters translate into the artistic elements:

Table 2-1
The Perceived Parameters and the Aesthetic/Artistic Elements of Sound

Perceived Parameters	*Aesthetic/Artistic Elements*
Pitch	Pitch Levels and Relationships
Loudness	Dynamic Levels and Relationships
Duration	Rhythmic Patterns and Rate of Activity
Timbre (perceived overall quality)	Sound Sources and Sound Quality
Space (perceived characteristics)	Spatial Properties

The audio production process allows for considerable variation and a very refined control of ALL of the artistic elements of sound. All of the artistic elements of sound can be accurately and precisely controlled through many states of variation, in ways that were possible with ONLY pitch on traditional musical instruments.

Table 2-2
The States of Sound in Audio Recording

Physical Dimensions (Acoustic State)	*Perceived Parameters* (Psychoacoustic Conception)	*Artistic/Aesthetic Elements* (Resources for Artistic Expression)
Frequency	Pitch	Pitch Levels and Relationships—melodic lines, chords, register, range, tonal organization, pitch density, pitch areas, vibrato
Amplitude	Loudness	Dynamic Levels and Relationships—dynamic contour, accents, tremolo, musical balance
Time	Duration (time perception)	Rhythmic Patterns and Rates of Activities—tempo, time, patterns of durations
Timbre (comprised of physical components: dynamic envelope, spectrum and spectral envelope)	Timbre (perceived as overall quality)	Sound Sources and Sound Quality—sound sources, groupings of sound sources, instrumentation, performance intensity, performance techniques
Space (comprised of physical components created by the interaction of the sound source and the environment, and their relationship to a microphone)	Space (perception of the sound source as it interacts with the environment, and perception of the physical relationship of the sound source and the listener)	Spatial Properties—stereo location, surround location, phantom images, moving sources, distance location, sound stage dimensions, imaging, environmental characteristics, perceived performance environment, space within space

Pitch Levels and Relationships

Pitch level relationships present most of the significant information in music. The artistic message of most of today's music is communicated (to a large extent) by pitch relationships. The listener has been trained, by the music heard throughout their life, to focus on this element to obtain the most significant musical information. The other artistic elements often support pitch patterns and relationships.

Pitch is the most precisely controlled artistic element in traditional music. The use of pitch relationships and pitch levels in music is more sophisticated than the use of the other artistic elements. Complex relationships of pitch patterns and levels are common in music.

Information about the artistic element of pitch levels and relationships will be related to:

1. The relative dominance of certain pitch levels,
2. The relative register placement of pitch levels and patterns, or
3. Pitch relationships: patterns of successive intervals, relationships of those patterns, and relationships of simultaneous intervals.

Traditional Uses of Pitch

The aesthetic/artistic element of pitch levels and relationships is broken into the component parts: melodic lines, chords, tonal organization, register, range, pitch density, pitch areas, and tonal speech inflection.

A series of successive, related pitches creates *melodic lines*. Melodic lines are perceived as a sequence of intervals that appear in a specific ordering and that have rhythmic characteristics. The melodic line is often the primary carrier of the artistic message of a piece of music.

The ordering of intervals, coupled with or independent from rhythm, creates patterns. *Pattern perception* is central to how humans perceive objects and events. These basic principles relate to all of the components of the artistic elements. Melodic lines are organized by patterns of intervals (short melodic ideas, riffs, or motives), supported by corresponding rhythmic patterns. The complexity of the patterns, the ways in which the patterns are repeated, and the ways in which the patterns are modified provide the melodic line with its unique character.

Two or more simultaneously sounding pitches create *chords*. In much of our music, chords are based on superimposing, or stacking, the intervals of a third (intervals containing three and four semitones, most commonly). Chords comprised of three pitches, combining two intervals of a third, are called *triads*. Continued stacking of thirds results in seventh, ninth, eleventh, and thirteenth chords.

The movement from one chord to another, or *harmonic progression*, is the most stylized of all the components of the artistic elements. Harmonic progression is the pattern created by successive chords, as based on the lowest note (the root) of the triads (or more complex chords). These patterns of chord progressions have become established as having general principles that occur consistently in certain types of music. Certain types of music will have stylized chord progressions (progressions that occur most frequently), other types of music will have quite different movement between chords, and perhaps emphasize more complex chord types. The patterns of the harmonic progression create *harmony*.

Harmony is one of the primary components that support the melodic line. The chords in the harmonic progression reinforce pitches of the melody. The speed and direction of the melodic line is often supported by the speed at which chords are changed, and the patterns created by the changing chords: *harmonic rhythm*.

The expectations of harmonic progression create a sequence of chords, which will present areas of tension and areas of repose within the musical composition. The tendencies of *harmonic motion* do much to shape the momentum of a piece of music, and can greatly enhance the character of the melodic line and musical message. Performers utilize the psychological tendencies of harmonic progression, exploiting its directional and dramatic tendencies. The expectations of harmonic movement and the psychological characteristics of harmonic progression have become important aspects of musical expression and musical performance.

The melodic and harmonic pitch materials are related through *tonal organization*. Certain pitch materials are emphasized over others, in varying degrees, in nearly all music. This emphasis creates systems of tonal organization in which a hierarchy of pitch levels exist. A hierarchy will most often place one pitch in a predominant role, with all other pitches having functions of varying importance, in supporting the primary pitch. The primary pitch, or *tonal center*, becomes a reference level, to which all pitch material is related, and around which pitch patterns are organized.

Many tonal organization systems exist. These systems tend to vary significantly by cultures, with most cultures using several different, but related systems. The major and minor tonal organization systems of Western music are examples of different, but related systems, as are the whole-tone and pentatonic systems of Eastern Asia. The reader should consult appropriate music theory texts for more detailed information on tonal organization, as necessary.

The New Pitch Concerns of Audio Production

Certain components of pitch levels and relationships have become more prominent in musical contexts (and other areas of audio) because of the new treatments of pitch relationships in music recordings. The components of range, register, pitch density, and pitch area can be more closely controlled in recorded music, than in live (unamplified) performance. These components are more important in recorded music, because they are precisely controllable by the technology, and they have been controlled to support and enhance the musical material.

Range is the span of pitches of a sound source (any instrument or voice). Range is the area of pitches that encompasses the highest note possible (or present in a certain musical example) to the lowest note possible (or present) of a particular sound source.

A *register* is a portion of a sound source's range. A register will have a unique character (such as a unique timbre, or some other determining factor) that will differentiate it from all other areas of the same range. It is a small area within the source's range that is unique in some way. Ranges are often divided into many registers; registers may encompass a very small group of successive pitches, up to a considerable portion of the source's range.

A *pitch area* is a portion of any range (or of a register) that may or may not exhibit characteristics that are unique from other areas. Instead, it is a defined area between an upper and a lower pitch-level, in which a specific activity or sound exists.

Pitch density is the relative amount and register placement of simultaneously sounding pitch material, throughout the hearing range or within a specific pitch area. It is the amount and placement of pitch material in the composite musical texture (the overall sound of the piece of music), and is defined by its boundaries of highest and lowest sounding pitches.

With pitch density, sound sources are assigned (or perceived as occupying) a certain pitch area within the entire listening range (or the smaller pitch range used for a certain piece of music). Thus, certain pitch areas will have more activity than other pitch areas; certain sound sources will be present only in certain pitch areas, and other sources present only in other pitch areas; some sources may share pitch areas, and cause more activity to be present in those portions of the range; some pitch areas may be void of activity. Many possible variations exist.

Pitch density is a component of pitch-level relationships, and is directly related to traditional concerns of orchestration and instrumentation, with many new twists. Pitch density is a much more specific concern in recorded music because it is controllable in very fine increments. Traditional orchestration was concerned, basically, with the selection of instruments, and with the placement of the musical parts (performed by those assigned instruments and their sound qualities) against one another.

With the controls of signal processing (especially equalization), sound synthesis, and multitrack recording, the register placement of sound sources and their interaction with the other sound sources take on many more dimensions. Each sound source occupies a pitch area; the acoustic energy within the pitch area of a timbre's spectrum is distributed in ways that are unique to each sound source. The spectrum of each sound source is an individual pitch density, and the pitch density of the overall program (or musical texture) is the composite of all of the simultaneous pitch information from all sound sources.

Sound sources, and musical ideas, are often delineated by the pitch area they occupy within the composite pitch density. Sound sources are more easily perceived as being separate entities and individual ideas, when they occupy their own pitch area in the composite, pitch density of the musical texture. This area can be large or quite small, and still be effective.

Sounds that do not have well-defined pitch quality, occupy a *pitch area*. These types of sounds are noise-like, in that they cannot be perceived as being at a specific pitch. Such sounds may, however, have unique pitch characteristics.

Many sounds cannot be recognized as having a specific pitch, yet have a number of frequencies that dominate their spectrum. Cymbals and drums easily fall into this category. Cymbals are easily perceived as sounding higher or lower than one another. Yet a specific pitch cannot be assigned to the sound source.

We perceive these sounds as occupying a pitch area. We perceive a pitch-type quality based on (1) the register placement of the area of the highest concentration of pitch information (at the highest amplitude level) present in the sound, and (2) the relative density (closeness of the spacing of pitch levels) of the pitch information (spectral components). We are able to identify the approximate area of pitches in which this concentration of spectral energy occurs, and are thus able to relate that area to other sounds.

Pitch areas are defined as the range spanned by the lowest and highest dominant frequencies around the area of the spectral activity. This range is called the *bandwidth* of the pitch area. Many sounds will have several pitch areas where concentrated amounts of spectral energy occurs, with one range dominating and others less prominent. The size of the bandwidth and the density of spectral information (the number of frequencies within the bandwidth and the spacing of those frequencies) define the sound quality of the pitch area.

Dynamic Levels and Relationships

Dynamic levels and relationships have traditionally been used in musical contexts for expressive or dramatic purposes. Expressive changes in dynamic levels and the relationships of those changes have most often been used to support the motion of melodic lines, to enhance the sense of direction in harmonic motion, or to emphasize a particular musical idea. A change of dynamic level, in and of itself, can produce a dramatic musical event, and is a common musical occurrence. Changes in dynamic level can be gradual or sudden, subtle or extreme.

Dynamics have traditionally been described by analogy: louder than, softer than, very loud (fortissimo), soft (piano), medium loud (mezzo forte), etc. The artistic element of dynamics in a piece of music is judged in relation to context. Dynamic levels are gauged in relation to (1) the overall, reference dynamic level of the piece of music, (2) the sounds occurring simultaneously with a sound source in question, and (3) the sounds that immediately follow and precede a particular sound.

The components of dynamic levels and relationships in audio recording are dynamic contour (with gradual and abrupt changes in dynamic level), emphasis/deemphasis accents (abrupt changes in dynamic level), musical balance (gradual and abrupt changes in dynamic levels), and dynamic speech inflections.

Traditional Uses of Dynamics

It is common for the most important musical idea/sound source in a piece of music to be given prominence in one way or another. Making that sound the loudest is an easy way of achieving this prominence (though not always the most elegant). Arranging sounds by relating dynamic levels to the importance of the musical part is very common, and a very natural association of loudness and the center of one's attention.

Gradual changes in dynamic levels can be important. The crescendo (gradual increasing in loudness) can be used to support the motion of a melodic line (for instance), or it might be used on a sustained pitch as a musical gesture itself. Likewise a diminuendo or decrescendo (a gradual decrease in loudness) may be used in the same ways.

Rapid, slight alterations or changes in dynamic level for expressive purposes are often present in live performances. This is called *tremolo*, and is used primarily to add interest and substance to a sustained sound. *Tremolo* and *vibrato* are often confused. *Vibrato* is a rapid, slight variation of the pitch of a sound; it, also, is used to enhance the sound quality of the sound source. At times, performers may not be able to control their sound well enough to control tremolo and vibrato alterations; in these instances, tremolo and vibrato may detract from the source's sound quality, rather than contribute to it.

To support a musical idea or to create a sense of drama, musical ideas are often brought to the listener's attention by dynamic *emphasis accents* and *attenuation accents*. A shift in dynamic level that brings the listener's attention to a musical idea, is an accent. Accents are most often emphasis accents, making use of increasing the dynamic level of the sound to achieve the desired result. Much more difficult to successfully achieve, de-emphasis (or attenuation) accents draw the listener's attention to a musical idea, or a sound source, by a decrease in the dynamic level of the sound. Attenuation accents are often unsuccessful because the listener has a natural tendency to move attention away from softer sounds; these accents are most easily accomplished in sparse musical textures, where little else is going on to draw the listener's attention away from the material being accented.

New Concepts of Dynamic Levels and Relationships

Changes in dynamic levels over time comprise *dynamic contours*. Dynamic contours can be perceived for individual sounds, individual sound sources, individual musical ideas comprised of a number of sound sources, and the overall piece of music. Dynamic contours are perceived at many different *perspectives* (level of detail). At their extremes, they exist as the smallest changes within the spectral envelope of a single sound source, and as great changes in the overall dynamic level of a recording.

The interaction of the dynamic contours of all sound sources in a piece of music creates *musical balance*. Musical balance is the interrelationships of the dynamic levels of each sound source, to one another and to the entire musical texture. The dynamic level of a particular sound source in relation to another sound source is a comparison of two parts of the musical balance.

Dynamic contours and musical balance have been used in supportive roles in most traditional music. At times dynamic level changes have been used for their own dramatic impact on the music (as discussed with crescendo and diminuendo, above), but most often they are used to assist the effectiveness of another artistic element. The mixing process easily alters musical balance. Recordists exercise great control over this artistic element.

The dynamic levels and relationships of a performance may be significantly different in the final recording. The recording process has very precise control over the dynamic levels of a sound source in the musical balance of the final recording. An instrument may have an audible dynamic level in the musical balance of a recording that is very different from the dynamic level at which the instrument was originally performed. The timbre of the instrument will exhibit the dynamic levels at which it was performed (*perceived performance intensity*), but its relative dynamic level in relation to the other musical parts might be significantly altered by the mix. For example, an instrument may be recorded playing a passage loudly, and end up in the final musical balance (mix) at a very soft dynamic level; the timbre of the instrument will indicate that the passage was performed very loudly, yet the actual dynamic level will be quite soft in relation to the overall musical texture, and to the other instruments of the texture.

Many clear examples of this are found in The Beatles' recording of "Penny Lane." Listening carefully to the flutes, piccolo, and piccolo trumpet parts throughout the song, one will find many instances where the loudness levels of the performances are not reflected in the actual loudness levels of the instruments in the recording. Among many instances of conflicting levels and timbre cues, we hear moderately loud flutes that were performed softly; loudly played piccolo sounds at a soft level in the mix; and a piccolo trumpet appearing at a softer level in the performance. Other instruments and voices in the song also have inconsistent musical balance and performance intensity information.

The reader is encouraged to take the time now to perform the musical balance and performance intensity Exercise 2-1 at the end of this chapter.

The dynamic level of a sound source in relation to other sound sources, and musical balance, is quite different and distinct from the perceived distance of one sound source to another. Yet, these two occurrences are often confused, and are the source of much common, misleading terminology used by recordists. Significant differences are present between a softly generated sound that is close to the listener and a loudly performed sound that is at a great distance to the listener, even when the two sounds have precisely the same sound pressure level (SPL) or perceived loudness level. Loudness levels within the recording process are independently controllable from the loudness level at which the sound was performed, and are independently controllable from the distance of the sound source from the original receptor and from the perceived listening location of the final recording. Dynamics must not be confused with distance. Dynamic levels, themselves, do not define distance location.

Rhythmic Patterns and Rates of Activities

Durations of sounds (the length of time in which the sound exists) combine to create musical rhythm. Rhythm is based on the perception of a steadily recurring, underlying pulse. The pulse does not need to be strongly audible to be perceived. The underlying pulse (or *metric grid*) is easily recognized by humans as the strongest, common proportion of duration (note value) heard in the music.

The rate of the pulses of the metric grid is the *tempo* of a piece of music. Tempo is measured in metronome markings (pulses per minute, abbreviated "M.M."), or in some contexts as pulses per quarter note. Tempo, in a larger sense, can be the rate of activity of any large or small aspect of the piece of music (or of some other aspect of audio, for example the tempo of a dialogue).

Durations of sound are perceived proportionally in relation to the pulse of the metric grid. The human mind will organize the durations into groups of durations, or *rhythmic patterns*. In the same ways that we perceive patterns of pitches, we perceive patterns of durations. Pattern perception is transferable to all of the components of all of the artistic elements, and is the traditional way in which we perceive pitch and rhythmic relationships.

Rhythmic patterns are the durations of or between soundings of any artistic element. Rhythmic patterns might be created by the pulsing of a single percussion sound; in this way rhythmic patterns would be created by the durations between the occurrences of the starts of the same sound source. Rhythmic patterns comprised of the durations of successive, single pitches (perhaps including some silences) create melody. Rhythmic patterns of the durations of successive chords (groups of pitches) create harmonic rhythm. Extending this, in the same way rhythm can be transferred to ALL artistic elements. As examples, it is possible to have rhythms of sound location (as has become a common mixing technique for percussion sounds); it is likewise possible to have timbre melodies, or rhythms applied to patterns of identifiable timbres (this is often used for drum solos).

Sound Sources and Sound Quality

The selection, modification, or creation of *sound sources* is an important aesthetic and artistic element of audio recording. The *sound quality* of the sound sources (the timbre of the source), plays a central role in the presentation of musical ideas, and has become an increasingly significant form of musical expression.

The sound quality of a sound source may cause a musical part to stand out from others, or to blend into an ensemble. Sound quality alone can convey tension or repose, and give direction to a musical idea. Sound quality can add dramatic or extra-musical meaning or significance to a musical idea. Finally, the timbral quality of a sound source can, itself, be a primary musical idea, capable of conveying a meaningful musical message.

Until recently, composers used the sound quality of a sound source (1) to assist in delineating and differentiating musical ideas (making them easier to distinguish from one another), (2) to enhance the expression of a musical idea by the careful selection of the appropriate musical instrument to perform a particular musical idea, or (3) to create a composite timbre (or *texture*) of the ensemble, thereby forming a characteristic, overall sound quality.

Performers have always used the characteristic timbres of their instruments or voices to enhance musical interpretation. This activity has been greatly refined by the resources of recording and sound reinforcement technology. Performers now have greater flexibility in shaping the timbre of their instruments for creative expression. Of equally great importance, after the performance has been captured, the recording process allows for the opportunity to return to the performance for further (perhaps extensive) modifications of sound quality.

The selection of a sound source to represent (present) a particular musical idea is critical to the successful presentation of the idea. The act of selecting a sound source is among the most important decisions composers (and producers) make. The options for selecting sound sources are (1) to choose a particular instrumentation, (2) to modify the sound quality of an existing instrument or performance, or (3) to create, or synthesize, a sound source to meet the specific need of the musical idea.

The *selection of instrumentation* was once merely a matter of deciding which generic instrument of those available would perform a certain musical line. The selection of instrumentation has now become very specific and much more important. The performance that exists as a music recording may virtually live forever and be heard by countless people. This is very different from the typical, live music performance of the past that existed for only a passing moment and was heard by only those people present.

Today, the selection of instrumentation is often so specific, as to be a selection of a particular performer playing a particular model of an instrument. Generally, composers and producers are very much aware of the sound quality they want for a particular musical idea. The performer, the way the performer can develop a musical idea through their own personal performance techniques, and their ability to use sound quality for musical expression are all considerations in the selection of instrumentation.

Vocalists are commonly sought for the sound quality of their voice and their abilities to perform in particular singing styles. The vocal line of most songs is the focal point that carries the weight of musical expression. Vocalists make great use of performance techniques to enhance and develop their sound quality, as well as to support the drama and meaning of the text.

Performance techniques vary greatly between instruments, musical styles, performers, and functions of a musical idea. The most suitable performance techniques will be those that achieve the desired musical results, when the sound sources are finally combined. One performance technique consideration must be singled out for special attention—the intensity level of a performance.

As touched on in the above discussion with musical balance, a performance on a musical instrument will take place at a particular intensity level. This *perceived performance intensity* is comprised of loudness, energy exerted, performance technique, and the expressive qualities of the performance. Each performance at a different intensity level results in a different characteristic timbre of that instrument, at that loudness level.

The same sound source will thus have different timbres, at different loudness levels (and at different pitch-levels), through performance intensity.

Along with the timbre (sound quality) and loudness level, performance intensity can communicate a sense of drama and an artistically sensitive presentation of the music to the listener. Through performance intensity, louder sounds might be more urgent, more intense; softer sounds might be cause for relaxation of musical motion. The exact reverse is equally possible. The expressive qualities of music are contained in performance intensity cues.

Modifying a sound source is a common way of creating a desired sound quality. Instruments, voices, or any other sound may be modified (while being recorded, or afterwards) to achieve a desired sound quality. Most often, this takes the form of making detailed modifications to a particular instrument so it best presents the musical idea. The final sound quality will still have some (perhaps many, perhaps only a few) characteristic qualities of the original sound.

The extensive modification of an existing sound source, to the point where the characteristic qualities of the original sound are lost, is actually *the creation of a sound source*. The creation of new sound qualities (or inventing timbres) has become an important feature in many types or pieces of music. The recording process easily allows for the creation of new sound sources, with new sound qualities.

Sound qualities are created by either extensively modifying an existing sound through sound sampling technologies, or by synthesizing a waveform. Sound synthesis techniques allow precise control over these two processes, and are having a widespread impact on recording practice and musical styles. Many specific technologies and techniques exist for synthesizing and sampling sounds; all have unique sound qualities and unique ways of allowing the user to modify or synthesize a sound source.

A new sense of the importance of sound quality to communicate, as well as to enhance, the musical message has come from this increased emphasis on sound quality and timbre. Sound quality has become a central element in a number of the primary decisions of recording music, as well as in the creation of music through the recording process. In making these primary decisions, sound quality is conceptualized as an object. The sound is thought of as a complete and individual entity, capable of being pulled out of time and out of context.

In this way, sound quality is approached as a *sound object*. This important concept will be explored in detail later in Chapter 4, "Listening and Evaluating Sound for the Audio Professional."

The entire, composite sound of the music may also be conceptualized as a single entity, or overall quality comprised of any number of small, individualized sound sources and musical ideas. This sound quality of the overall sound, or entire program is called *texture*. Texture is perceived by the characteristics of its global sound quality.

Texture will nearly always be comprised of any number or types of individual sounds. Texture is perceived as an overall character, made up of the states and activities of all sounds and musical ideas. Pitch-register placements, rate of activities, dynamic contours, and spatial properties are all potentially important factors in defining a texture by the states or activities of its component parts.

Spatial Properties: Stereo and Surround Sound

The *spatial properties* of sound have traditionally not been used in musical contexts. The only exceptions are the location effects of antiphonal ensembles of certain Renaissance composers and in certain drama-related works of the nineteenth century, such as the 1837 *Requiem* by Hector Berlioz (with its brass ensembles stationed at the corners of the church, performing against the orchestra and choir on stage).

The spatial properties of sound play an important role in communicating the artistic message of recorded music. The roles of spatial properties of sound are many. Spatial properties may be used in supportive roles to enhance the character or effectiveness of musical ideas (large and small), to differentiate one sound source from another, to provide dramatic impact, to alter reality, or to reinforce reality by providing a performance space for the music. Further, spatial properties may be used as the primary idea of an artistic gesture. The spatial property of environmental characteristics even fuses with the timbre of the sound source to add a new dimension to its sound quality. Other possibilities certainly exist.

The number and types of roles that spatial location may play in music have yet to be exhausted or defined. The recent adoption of surround sound has further multiplied the possibilities.

All of the components of the spatial properties are under very precise and independent control. All of the spatial properties may be in many markedly different and fully audible states. Further, gradual and continuously variable change between those states is possible and common.

The spatial properties of sound that are of primary concern to recorded music (sound) are:

1. The *stereo location* of the sound source on the horizontal plane of the stereo array,
2. The *distance* of the sound source from the listener,
3. The perceived characteristics of the sound source's physical *environment*, and finally
4. The surround location of sound sources on the lateral plane 360° around the listener.

The perceived elevation of a sound source is not consistently reproducible in widely used playback systems, and has not yet become a resource for artistic expression.

Two-Channel Stereo

The first three spatial properties are realized through stereophonic sound reproduction. The spatial qualities of stereo are perceived as relationships of location and distance cues and relationships of sound sources. These create a perception of a *sound stage* contained within the *perceived performance environment* of the recording.

While surround sound is becoming more prevalent, two-channel sound reproduction remains the standard of the music recording industry, with monophonic capabilities still considered for AM broadcast and television sound applications. The two-channel array of *stereo sound* attempts to reproduce all spatial cues through two separate sound locations (loudspeakers), each with more-or-less independent content (channel). With the two channels, it is possible to create the illusion of sound location at a loudspeaker, in between the two loudspeakers, or slightly outside the boundaries of the loudspeaker array; location is limited to the area slightly beyond that covered by the stereo array, and to the horizontal plane. The characteristics of the sound source's environment and distance from the listener are created in much more subtle ways by stereo, but can be stunning nonetheless.

A setting is created by the two-channel playback format for the reproduction of a recorded or created performance (complete with spatial cues). This establishes a conceptual and physical environment within which the recording will be reproduced more-or-less accurately.

The reproduced recording presents an illusion of a live performance. This performance will be perceived as having existed in reality, in a real physical space; as the listener will conceive of this activity in relation to their own physical reality. The recording will appear to be contained in a single, *perceived physical environment*. Within this perceived space is an area that comprises the *sound stage*.

Sound Stage and Imaging

The sound stage is the perceived area within which all sound sources are located. It has an apparent physical size of width and depth. The sound sources of the recording will be grouped by the mind to occupy a single area. It is possible for different sound sources to occupy significantly different locations within the sound stage but still be grouped into the illusion of a single performance.

Imaging is the lateral location and distance placement of the individual sound sources within the sound stage. Imaging provides depth and

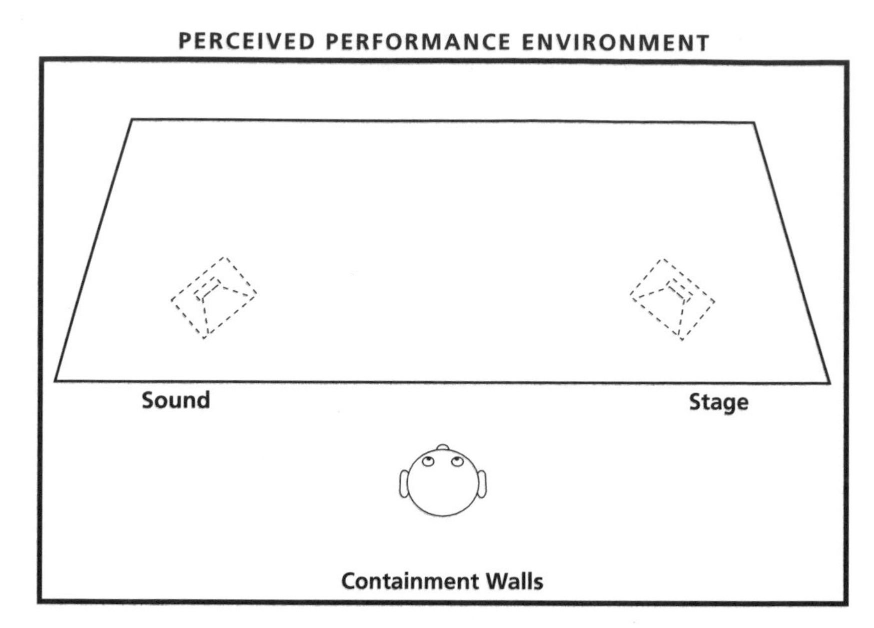

Figure 2-1 Sound stage and the perceived performance environment.

width to the sound stage. The perceived locations and relationships of the sound sources create imaging, as all sources appear to exist at a certain lateral and distance location within the stereo array.

Stereo Location

The *stereo* (lateral) *location* of a sound source is the perceived placement of the sound source in relation to the stereo array. Sound sources may be perceived at any lateral location within, or slightly beyond, the stereo array.

Phantom images are sound sources that are perceived to be sounding at locations where a physical sound source does not exist. Imaging relies on phantom imaging to create lateral localization cues for sound sources. Through the use of phantom images, sound sources may be perceived at any physical location within the stereo loudspeaker array, and up to 15° beyond the loudspeaker array. *Stage width* (sometimes called *stereo spread*) is the width of the entire sound stage. It is the area between the extreme left and right source images, and marks the sound stage boundaries.

Phantom images not only provide the illusion of the location of a sound source, they also create the illusion of the physical size (width) of the source. Two types of phantom images exist: the *spread image* and the *point source*.

A *point source* phantom image occupies a focused, precise point in the sound stage. The listener can close their eyes and point to a very precise

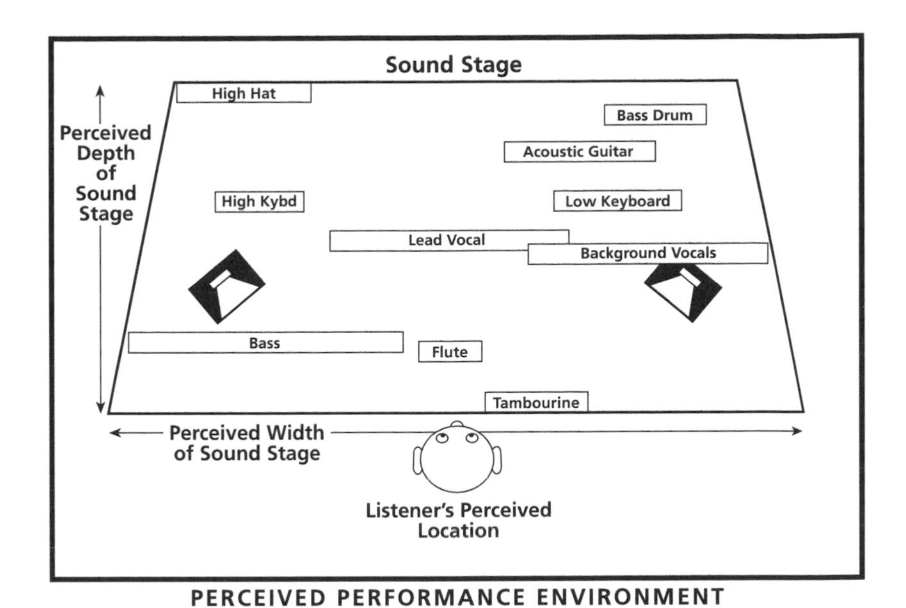

Figure 2-2 Sound stage and imaging, with phantom images of various sizes.

point of little area where the source is heard to originate. Point sources exist at a specific point in space; narrow in width, and precisely located in the sound stage.

The *spread image* appears to occupy an area. It is a phantom image that has a size that extends between two audible boundaries. The potential size of the spread image varies considerably; it might be slightly wider than a point source, or it may occupy the entire stereo array. The spread image is defined by its boundaries; it will be perceived to occupy an area between two points or edges. At times, a spread image may appear to have a hole in the middle, where it might occupy two more-or-less equal areas, one on either side on the stereo array.

The perceived lateral location of sound sources can be altered to provide the illusion of *moving sources*. Moving sound sources may be either point sources or spread images. Point sources and narrow spread images that change location most closely resemble our real life experiences of moving objects.

Many interesting examples of phantom images can be found on The Beatles' album *Abbey Road*. An apparent example of a spread image with a hole in the middle is the tambourine in the first chorus of "She Came in Through the Bathroom Window." The lead vocal in "You Never Give Me Your Money" begins the song as a point source. The image soon becomes a spread image that gradually grows wider, ultimately occupying a significant amount of the sound stage (this is partly due to the gradual addition and varying of environmental cues, which will be discussed shortly). In the second section of the work, the new lead vocal sound gradually moves from the right to the left side of the sound stage, while maintaining a spread image of moderate size.

Distance Location

Two categories of *distance* cues shape recorded music: (1) the distance of the listener to the sound stage, and (2) the distance of each sound source from the listener.

Both of these distances rely on a perception that the entire recording emanates from a single, global environment. This *perceived performance environment* establishes a reference location of the listener, from which all judgments of distance can be calculated.

The *stage-to-listener distance* establishes the front edge of the sound stage with respect to the listener and determines the level of intimacy of the music/recording. This is the distance between the grouped sources that make up the sound stage and the perceived position of the audience/listener. This stage-to-listener distance places the sound stage within the overall environment of the recording and provides a location for the listener.

The *depth of sound stage* is the area occupied by the distance of all sound sources. The boundaries of the depth of the sound stage are the perceived nearest and the perceived furthest sound sources (with the depths created by their environments, discussed below). The perceived distances of sound sources within the sound stage may be extreme; they may provide the illusion of great depth and a large area, or the exact opposite.

Stage-to-listener and depth of sound stage distance cues have different levels of importance in different applications. Depth of sound stage cues tend to be emphasized over stage-to-listener distance cues in many multitrack recordings; in those recordings, the cues of the distance of the source from the listener are often exploited for dramatic effect and/or to support musical ideas. In contrast, stage-to-listener distance cues are often carefully calculated in classical and some jazz recordings (especially those utilizing standardized stereo microphone techniques); in those recordings the stage-to-listener distance will not change and has been carefully selected to represent the most appropriate vantage point (the ideal seat) from which the music is to be heard.

Turning again to *Abbey Road*, the distance cues of the various instruments of “Golden Slumbers” gives the work and its companion “Carry That Weight” much space between the nearest and the furthest sources. The orchestral string and brass instruments are at some distance from the listener and give significant depth to the sound stage, while the piano brings the front edge of the sound stage very near the listener. Remembering that timbral detail is the primary determinant of distance location will help in accurately hearing these cues.

Environmental Characteristics

Matching a sound source to an environment with suitable sound and selecting the environment of the sound stage (the perceived performance environment) have become important parts of music recording. *Environmental characteristics* have the potential to significantly impact music and the quality of the recording.

Environmental characteristics fuse with the sound source to create a single sonic impression. Its host environment shapes the overall timbre/sound quality of each sound source; this is also true for the overall program (shaped by its perceived performance environment). Environmental characteristics contribute greatly to sound quality and also play an important role in the recording's sense of space. The characteristics provide a space for the sound sources to perform in, they supply some distance information that may be significant, and they contribute to the perceived depth of the sound stage.

The sound characteristics of the host environments of sound sources and the complete sound stage are precisely controllable. Each sound source has the potential to be assigned environmental characteristics that are different from the other sound sources. The recording process allows the potential for each sound source to be given a different environment, and for the characteristics of those environments to be varied as desired. Further, each source may occupy any distance from the listener within the applied host environment.

The *perceived performance environment* (or the environment of the sound stage) is the overall environment where the performance (recording) is heard as taking place. This environment binds all the individual spaces together into a single performance area.

The environment of the sound stage and an individual environment for each sound source (or groups of sound sources) often co-exist in the same music recording. This places the individual sound sources with their individual environments within the overall, perceived performance environment of the recording. The illusion of *space within space* is thus created, with the following potential perceptions:

1. That physical spaces may exist side-by-side,
2. That one physical space may exist within another physical space (where often a space with the sound qualities of a physically large room may be perceived to exist within a smaller physical space), and
3. That sounds may exist at various distances within the same host environments.

Any number of environments and associated stage-depth distance cues may occur simultaneously, and coexist within the same sound stage. The environments and associated distances are conceptually bound by the spatial impression of the perceived performance environment. These outer walls of the overall program establish a reference (subliminally, if not aurally) for the comparison of the sound sources.

Perhaps oddly, the overall space that serves as a reference, and that is perceived by the listener as being the space within which all activities occur, will often have the sound characteristics of an environment that is significantly smaller than the spaces it appears to contain. Such cues that send conflicting messages between our life experiences and the perceived musical occurrence are readily accepted by the listener and can be used to great artistic advantage. This is a very common space within space relationship.

Space within space will at times be coupled with distance cues to accentuate the different environments (spaces) of the sound sources, though often, this illusion is created solely by the environmental characteristics of the different spaces of each sound source.

"Here Comes the Sun," also from *Abbey Road*, provides some clear examples of space within space. Environments clearly exist side by side from the song's opening into the first verse. The guitar has an environment all to itself in the left channel, the electronic keyboard countermelody and Moog synthesizer glissando have similar environments distinctly different from the others, and the right channel voice has a very different third environment. The parts are held together by the notion that they all exist within a single performance space (perceived performance environment). The entry of additional instruments quickly adds numerous additional environments and enhances the sound stage. As the vocal lines are added, however, they appear to be within the same environment, though at distinctly different distances from the listener. The notion of spaces within spaces is also apparent in the drum parts; the trap set seems to occupy an area, with its characteristic environment, within which low toms in a larger space are contained.

Surround Sound

Music recordings are now being reasonably widely made in surround sound. Enough activity and interest is present that it is necessary for us to seriously explore this format now, but with some reservation. While some talented people have been working in this new format and some striking recordings have been made, few consistent uses of the unique sound qualities of surround have emerged. This section will discuss the most prevalent aesthetic and artistic elements currently found in surround music recordings, and will explore some potential applications. Without doubt, the artistic elements of surround will be further defined by recordists over the next few years; great changes and advances are likely, as the medium is just beginning to be explored in music production.

Listening to a stereo recording, we find ourselves observing a performance. We are viewing the activity as an outsider. And while we may get consumed by or immersed in the music, we are outside of the experience of the performance itself and are looking in. With surround sound, we can find ourselves enveloped by the music. We can be surrounded by the sound, and thereby contained within the space of the recording; we are no longer outside observers, but at least inside observers if not participants (at least in our perception of the experience). Now the listener can be enveloped by the sound (and become part of the space of the recording) or they might be oriented by the production techniques to observe a piece of music as a 360° panorama of sound. This aspect of surround sound has great potential of making a profound impact on music. Location and environmental characteristics will be approached differently for surround recordings, and distance cues will also take on new dimensions.

Surround Location

The sound stage of surround sound is vastly more complicated than stereo. Imaging takes on many strikingly new and different dimensions. With independent channels surrounding the listener, the potential exists for the sound stage to be extended enormously. This also places the listener in a listening position that is strikingly different from stereo.

As discussed, stereo is based on a single sound stage between two speakers. Five-channel surround (the format used for evaluations herein, discussed in detail in Chapter 9) provides the opportunity for as many as 26 possible combinations of speakers. This changes phantom image placement, width, and stability greatly.

The phantom images of stereo exist between two loudspeakers, and up to 15° beyond. Phantom imaging is more complex in surround. In surround there are five primary phantom image locations existing between adjacent pairs of speakers. These images tend to be the most stable and reliable between systems and playback environments.

Many secondary phantom images are possible as well. These can appear between speaker pairs that are not adjacent. These images contain inconsistencies in spectral information and are less stable. Implied are different distance locations for these images, as the trajectories between the pairs of speakers are closer to the listener position. These closer locations do not materialize in actual practice. The distance location of these images are actually pushed away somewhat by the diminished timbral clarity of these images.

When we consider locations caused by various groupings of three or four loudspeakers, placement options for phantom images get even more complex.

Phantom images can be of greatly different sizes in surround. They can range from completely surrounding the listener with a spread image

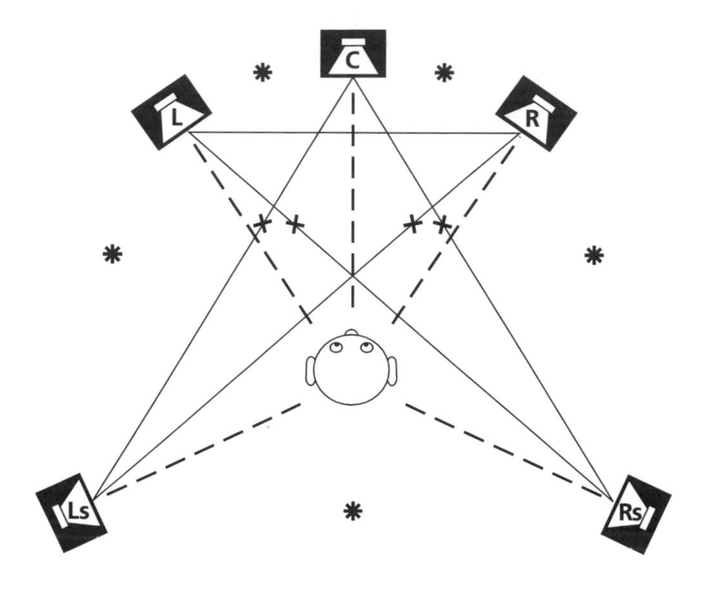

Figure 2-3 Phantom images in pairs of surround speakers.

of enormous size, to a small and precisely defined point source. Point sources and narrow spread images are common in current music productions, especially in the front sound field. The center channel has changed imaging on the traditional front sound stage tremendously. Images are often more defined in their locations and narrower in size.

Distance in Surround

Distance cues, of course, remain a product of timbral detail, with some reliance on environmental cues. The enhanced presence of ambience causes surround to more readily draw the listener into making inaccurate judgments of distance cues. Sounds are often perceived as further away than is accurate, largely because of an awareness of extra or enhanced environment information. The listener is drawn toward ambience and away from an awareness of timbral detail.

The depth of sound stage is extended all around the listener as well. The listener can perceive distance in all directions, and these cues can be present in surround recordings. This provides for creative opportunities not possible in stereo, but should be approached with reservation.

First, we know listeners accept sound stages of great depth in the front sound field. They easily imagine they are viewing something with proportions out of their physical confines. Listeners are not prepared to perceive sounds from the side and rear in the same way. When presented with similar materials from side and rear locations, listeners can be reluctant to place sounds in or behind walls (even after they have been observed and recognized at those locations). The same cues presented in a musically different way to envelop the listener in the space may allow this greater depth to be perceived.

Second, phantom images from the side and rear are inherently filled with phase anomalies of the listening room. This can cause a lack of timbral definition and detail, and distort distance cues. Further, it is also common for surround systems to give different timbral qualities to any instrument panned across its different speaker locations. These timbre changes often translate into distance changes and blur distance location imaging.

Finally, involuntary head movement contributes to our localization of sound sources. While this instinct has allowed our species to survive and evolve, it may well lead to a sense of apprehension in the listener. When presented with sounds from behind, the fight or flight instinct can be triggered, and thus distract the listener or create discomfort.

Environmental Characteristics and Surround Sound

Environmental characteristics can be directed to the listener from every direction in a very natural manner. This will immerse the listener in the cues, and provide the life-like experience of being present in the space where the recording was made. Spaciousness can be presented by both two-channel and five-channel systems to portray a sense of space, but only surround systems can provide the sensation of being there within the performance.

At present, environmental characteristics are mostly used in ways similar to stereo recordings. The characteristics of the perceived performance environment and of individual sound sources are crafted to shape the musicality of the recordings. One exception is fully or partially immersing the listener in the ambiance of a source's host environment, while localizing the source in a specific location elsewhere (usually in the front sound field).

The inherent qualities of environmental characteristics remain unchanged, except for changes in the direction(s) of the arrival of reflected sound. This itself is a great difference, as the fusing of environmental characteristics and direct sound can become challenged, with a variety of results such as enlarged images, unnatural effects (perhaps pleasing), distracting reflections, and many more. This is especially apparent when environment cues are sent to only a few channels and do not surround the listener; this can lead to many different illusions. The environmental cues of individual sound sources may be perceived as separated from the direct sound, may be used to enhance imaging and space within space illusions, and many other alternatives exist for this new dimension. This new set of illusions makes the perceived performance environment's tendency to bind all of the spaces of the individual sound sources together even more important. The perceived performance environment will continue to provide an important context for the recording and a critical point of reference.

The album *On Air* by Alan Parsons provides some clear examples of some of these concepts. The final track, "Blue Blue Sky," surrounds the listener with a gradually moving lead vocal. The vocalist moves from the left surround, behind the listener through the right surround and the right channel before arriving at the center. The source image remains largely consistent in size and allows us to appreciate how the movement of the most important sound source, and its temporary placement behind the head of the listener, can impact the musical idea. The listener is observing this activity around them more than being enveloped by it; the gentle nature of the musical line allows the listener to feel comfortable with this movement and the choices of location. The texture and sound stage change entirely at 1:43. Here instrumentation changes and a chorus of vocals enters, and the listener becomes very effectively enveloped by the surround sound stage. One can easily perceive themselves not only in the recording space, but also within the group of performers.

Conclusion

With the recording process, it is possible for any of the artistic elements of sound to be varied in considerable detail. In so doing, all artistic elements can be shaped for artistic purposes and used to create musical ideas. As all elements of sound can be varied by roughly equal amounts, it is possible for any element to play an important role in a piece of music. We commonly see this practice in today's music productions.

The artistic elements are used in very traditional roles in certain musical works and types of recording productions, and in very new ways in other works. These new ways the artistic elements are used tend to emphasize aspects of sound that can not be controlled in acoustic performances. The aesthetic/artistic elements unique to audio recording (especially sound quality and spatial properties) are commonly used to support and shape musical ideas. Different musical relationships and sound properties can exist in audio recordings rather than in acoustic music. Knowing and controlling these elements gives the recordist the opportunity to contribute to the creative process and the act of making music.

The potentials of the artistic elements to convey the musical message, the musical message itself, and the characteristics and limitations of the listener are explored in the following chapter.

Exercises

The following exercise should be practiced until you are comfortable with the material covered.

Exercise 2-1
General Musical Balance and Performance Intensity Observations.

1. Listen carefully to "Penny Lane" by The Beatles. Follow the flutes, piccolo, and piccolo trumpet parts to observe the conflicting levels/cues cited in the discussion above.
2. In succeeding hearings, find other instances where musical balance is at a different loudness than the performance intensity information of the instruments' sound qualities.
3. Listen again, while focusing attention on a specific instrument or voice you know well; follow that sound source carefully throughout the song, to make some general observations of performance intensity cues.
4. Listen again and note the actual loudness of that instrument/voice in relation to the other sound sources.
5. Finally, listen again for how these relationships change between major sections of the song (i.e., between verse and chorus).

3 The Musical Message and the Listener

This chapter will discuss the content of the musical message, and how the aesthetic and artistic elements function in communicating the message of a piece of music. The listener's ability to correctly interpret the sounds of the musical message largely determines their understanding of the intended artistic meaning of the music. The factors that limit the listener's ability to effectively interpret the artistic elements of sound into the intended musical message (or meaning of the music) will be explored; the listener as audience member, and as audio professional, will be contrasted.

Today's recordist can do much to shape music and is often a key person in the creative process. This chapter is intended to provide some insight into what is created and shaped when music is recorded.

The Musical Message

The message of a piece of music is related to the many purposes or functions of music. Each different purpose that music serves requires a different approach to listening. The approach will bring some aspect into the center of our attention or the center of our experience. As will be covered more thoroughly later, we listen to music with various levels of attention. This strongly impacts the listening experience. At the extremes, the listener will be focused and intent on extracting certain specific types of information (active listening); conversely their attention will be focused on some activity other than the music (eating, a conversation, a dentist's drill), with the listener perhaps not actually conscious of the music (passive listening). How the listener approaches the act of listening impacts the success of the music reaching the listener. The intended purpose and message of the music successfully reaches the listener only where the listener is appropriately receptive.

The purpose of a piece of music and associated characteristics of the musical message may take the forms of (1) conceptual communication, (2) portraying an emotive state, (3) aesthetic experience, and (4) utilitarian functions. The purposes are by no means exclusive; many pieces of music use different functions simultaneously, or at different points in time.

Music that includes a text, such as a song, will communicate other concepts. These works may tell a story, deliver the author's impressions of an experience, present a social commentary, etc. Music is used as a vehicle to deliver the tangible ideas of the author/composer. The interplay between the music and the drama of the text is often an important contributor to the total experience of these works.

It is difficult for music alone (without words) to communicate specific concepts, but it is possible. Often written works portray certain subjects without a text. The listener can associate sounds in music to their experiences with a subject if a connection can be made. The subjects of such works are often general in nature, such as Ludwig van Beethoven's "Pastoral" *Symphony No. 6*.

Certain concepts are associated with certain specific or types of musical materials (types of movie music, musical ideas associated with certain individuals or certain landscapes, etc.). These are exceptional cases where music alone can communicate specific concepts, with the aid of associations drawn from the listener's past experiences. It is easy to imagine a Western chase scene, or an impending shark attack, when listening to certain pieces of music—when one has heard the music and action together enough times.

Music communicates emotions easily. One of the reasons many people listen to music is for emotional escape, relief, or a journey to another place. Music may portray a specific mood, incite a specific emotional response from the listener, or create a more general and hard to define (yet convincing) feeling or emotive impression. The composer of the music draws from the past experiences of the listener to shape their emotive reactions to the material. This is found—at least to some extent—in all music. Works of this nature may include a text, or not.

Music may be an aesthetic experience. The perception of the relationships of the musical materials alone, without the associations of concepts or emotive states, may be the vehicle for the musical message. Music has the ability to communicate on a level that is separate and distinct from the verbal (conceptual) or the emotional. Music without words, without emphasis on the emotional level, can be tremendously successful in communicating a message of great substance. Music (as all of the arts) can reach beyond the human experience; ideas that cannot be verbally defined or represented as an emotional experience can be clearly communicated. Abstract concepts may be clearly communicated; the human spirit may reach beyond reality, to loftier ideals. Some people have been so moved as to compare the aesthetic appreciation of a substantial work of art to the impressions of religious experience. Works of Johann Sebastian Bach such as the *Brandenburg Concertos* or the *Cello Suites* are excellent examples of this type of music, from the multitude that encompass hundreds of years of history and nearly all of the world's cultures.

Music serves other functions. It is used to reinforce or accompany other art forms (motion pictures, musical theater, dance, video art), to enhance the audio and visual media (television, multimedia, radio, advertisements), and to fill dead air in every day experiences (supermarkets, elevators, etc.). In these instances, music is present to support dramatic or conceptual materials, to take the listener's attention away from some other sounds or activities (a dentist's office), or to make an environment more desirable (a restaurant, an automobile).

The complexity of the musical materials is often directly related to the function of the music. When music is the most important aspect of the listener's experience, the musical material may be more complex—as the listener will devote more effort to deciphering the materials. When the music is playing a supportive role, the materials are often less sophisticated and are directly related to the primary aspect of the listener's experience. When music is being used to cover undesirable noises or to fill a void of silence that would otherwise be ignored, the musical materials are often very simple, easily recognized, and easily heard without requiring the listener's attention.

Musical Form and Structure

Within the human experiences of time and space, nothing exists without shape or form. Music is no exception. Pieces of music have *form* as a global quality, as an overall concept and essence.

Pieces of music can be conceptualized as an overall quality. It is the human perception of form that provides the impression of a global quality that crystallizes the entire work into a single entity. Form is the piece of music as if perceived, in its entirety, in an instant; it is the substance and shape that is perceived from conceptualizing the whole.

Form is the global shape of a piece of music together with the fundamental concepts and expressions (emotions) it is communicating. It is the sum that is shaped from the interactions of its component parts. Form is a single concept and essence of the piece, given shape by structure; it is comprised of component parts that are the materials of the piece of music. The materials of the music and their interrelationships provide the *structure* of the work.

The structure of a work is the architecture of its musical materials. Structure includes the characteristics of the musical materials coupled with a *hierarchy* of the interrelationships of the musical materials, as they function to shape the work. The artistic elements of sound function to provide the musical materials with their unique character, as will be later discussed. All musical materials are related by structure.

The hierarchy of musical materials is the listener's perception of a general framework of the materials. The hierarchy provides an interrelationship of the materials (with their varying levels of importance) to one another, and to the musical message as a whole. Within the hierarchy of musical structure: all musical materials and artistic elements will have a greater importance to the musical message than other materials or elements (except the least significant); all sections or ideas in the music will have greater importance over others (except the least significant); all musical materials are subparts of other, more significant musical materials (except the most significant materials).

Further, the hierarchy of the musical structure organizes musical materials into patterns, and patterns of patterns. In this way, relationships are established between the subparts of a work and to the work as a whole. The hierarchy is such that any time span may contain any number of smaller time-spans, or be contained within any number of larger time spans; musical material at any level may be related to material at any other level of the hierarchy.

Form

Major Divisions: A B A

Structure

Major Divisions: Verse1 Chorus Verse2
tonal centers: I V V I
Subdivisions at / \ / / \ \
• intermediate levels:
• sub-levels:
phrases: a a' b c d c d' a' a b'
motives: / \ / \ / \ / \ / \\ / \ / \ / \\ / \
melodic
rhythmic
accomp patterns

harmonic progression:

instrumentation: #1. solo voice, guitar, bass, complete trap set | #2. solo voice, guitar, bass, kybd, background vocals, cymbals | Ensemble #1. | Ensemble #2.

Nontraditional artistic elements may function at any structural division and subdivision of the hierarchical level to create patterns & rhythms:
Dynamic Contour Pitch Density Sound Quality
Stereo Location Distance Location Environmental Characteristics

Primary and secondary elements are present throughout the structural hierarchy as Sound Events and Sound Objects.

Interrelationships of materials take place at all levels of the hierarchy and between artistic elements.

Figure 3-1 Form and structure in music.

A multitude of possibilities exists for unique musical structures, supporting very similar musical forms. The innumerable popular songs that have been written during the past few decades are evidence of this great potential for variation. Most of the songs have many similarities in their structures, but they also have many significant differences. The materials that comprise the works may be very different, but the materials work towards establishing an overall shape (form) that is quite similar between songs.

Many songs share similar forms. Their overall conceptions are very similar, although the materials of the music and the interrelationships of those materials may be strikingly different. Form is an overall design and conception that may be constructed of a multitude of materials and relationships.

Musical materials can be changed to dramatically alter the structure of a piece of music without altering its form. Many different structures can lead to the same overall design and portray the same basic artistic statement that creates form.

Musical Materials

As music moves and unfolds through time, the mind grasps the musical message through the act of understanding the meaning and significance of the progression of sounds. During this progression of sounds, the mind is drawn to certain artistic elements (that create the characteristics of the musical materials). We perceive the *musical materials* as small patterns (small musical ideas, often called motives or gestures), and group the small patterns into related larger patterns. The listener remembers the patterns, together with their associations to larger and smaller patterns (perceiving the structural hierarchy). In order for the listener to remember patterns, the listener must recognize some aspect of the organization of a pattern or some of the materials that comprised a pattern(s).

The use of contrast, repetition, and variation of patterns throughout the structure creates logic and coherence in the music. Several general ways in which musical materials are used and developed should make clear the multitude of possibilities. Materials are contrasted with other materials at the same and different hierarchical levels (above and below). Materials are repeated immediately or later in time, at the same or different hierarchical levels. Materials are varied by adding or deleting portions of an idea, by altering a portion of an idea, or perhaps by transposing an idea to different artistic elements (such as melodic ideas becoming rhythmic ideas; harmonic motion becoming dynamic motion).

A balance of similarities and differences within and between the musical materials are required for successfully engaging music. A musical work will not communicate the desired message, if this balance is not effectively presented to, or understood by, the listener.

The listener remembers the context in which the patterns were presented as well as the patterns themselves. In the presentation of materials, some patterns will draw the listener's attention and be perceived as being more important than other patterns—these are the *primary musical materials*. Other patterns will be perceived as being subordinate; these *secondary materials* will somehow enhance the presentation of the primary materials by their presence and activity in the music. The secondary materials that accompanied the patterns (or primary materials) are also remembered, as individual entities (capable of being recognized without the primary musical idea) and as being associated with the particular musical idea (patterns).

The primary materials are traditionally: melody (with related melodic fragments or motives) and (extra-musically) any text, or lyrics of the music. The secondary materials are traditionally: accompaniment passages, bass lines, percussion rhythms, harmonic progressions, and tonal centers. Secondary materials may also be dynamic contour, pitch density, timbre development, stereo/surround location, distance location, or environmental characteristics.

The secondary materials usually function to support the primary musical ideas. It is possible to have any number of equal, primary musical ideas. The potential groupings of primary and secondary musical ideas, in creating a single structural hierarchy, are limitless. Consider any number of secondary ideas (of varying degrees of importance in their support of the primary musical idea or ideas) that may coexist in a musical texture, with any number of related or unrelated primary musical ideas.

Musical materials are given their unique characters by the states and values of the aesthetic/artistic elements of sound. The artistic elements of sound function to shape and define the musical materials.

The Relationships of Artistic Elements and Musical Materials

Musical ideas are also comprised of *primary elements* and *secondary elements*. The primary elements are the aesthetic and artistic elements of sound that directly contribute to the basic shape or characteristics of a musical idea. The secondary elements are those aspects of the sound that assist, enhance, or support the primary elements.

It is possible (and in fact common) to have more than one primary element and more than one secondary element contributing to the basic character of a musical idea. Primary elements exhibit changes in states and values that provide the most significant characteristics of the musical material. The secondary elements provide support in defining or in providing movement to the primary elements.

At all levels of the structural hierarchy, musical materials (primary and secondary musical ideas) are comprised of primary and secondary artistic elements. Therefore, it is possible for a certain element of sound to be a primary element on one level of the hierarchy, and a secondary element on another level. This is not an uncommon situation. For example, a change in dynamic level of a drum roll might have primary significance at the hierarchical level of the individual sound source; at the same point in time but at the hierarchical level of the composite sound of the entire ensemble, changes in dynamics are insignificant to the communication of the musical message, with pitch changes being of primary importance.

All of the artistic elements of sound have the potential to function as the primary elements of the musical material. They have the potential to be the central carriers of the musical idea. Likewise, all of the artistic elements of sound have the potential to function as secondary elements of the musical material, and have the potential of functioning in supportive roles in relation to conveying the musical message. This concept of *equivalence* will be thoroughly explored.

In most music, pitch is the central element or the primary carrier of the musical message. The unique sound qualities of recording usually appear in the supportive roles of music, much more than as the primary elements. Most often, current production practice will use the unique artistic elements of recordings (such as the stereo location) to support or enhance the primary message (or perhaps to assist in defining an individual sound source). Rarely do the new sound resources function as the primary element of the primary musical idea, though this is entirely possible.

The new musical possibilities using all artistic elements can create convincing musical ideas when they are used as the primary carriers of the musical material. Current practice is likely to continue its gradual change towards further emphasis of these new artistic elements of sound unique to recording. It is important to recognize that the potential exists for any artistic element of sound to be the primary carrier of the musical material. The potential exists for any of the artistic elements of sound to function in support of any component of the musical idea. All of the artistic elements are equally capable of change, and that change can be perceived almost equally well in all of the artistic elements.

Traditionally, a break down of primary and secondary elements of a piece of music (with associated musical materials identified) would commonly appear similarly to Table 3-1. Pitch is the primary element, and is supported by rhythm and dynamics. The musical materials are differentiated by sound quality differences.

In many current recordings, a similar (and equally common) outline might appear as Table 3-2. While pitch remains a primary element, rhythm and sound quality changes are equally important in delivering the

Table 3-1
Traditional Hierarchy of Artistic Elements

Primary Elements	*Secondary Elements*
Pitch—melodic line #1	Pitch—harmony
Pitch—melodic line #2	Pitch—accompaniment patterns
	Dynamics—contour for expression
	Rhythm—supporting melody
	Sound Quality—instrument selection

musical message. Sound quality, in particular, has become more important with recording technology. Pitch, dynamics, and rhythm still play supportive roles, with spatial properties assisting sound quality in differentiating the musical ideas.

Equivalence and the Expression of Musical Ideas

Throughout Western music history, pitch (in its levels and relationships) has been the most important artistic element of music. The pitch relationships are utilized to create melodies, harmonies, accompaniment patterns, and tonal systems. Pitch has functioned as the central element in nearly all music that has descended from or has been significantly influenced by the European tradition. Pitch is the primary artistic element in much of the music we know and is the perceived parameter that contains

Table 3-2
Common Hierarchy in Current Music Productions

Primary Elements	*Secondary Elements*
Pitch—melodic line #1	Pitch—harmony
Pitch—melodic line #2	Pitch—accompaniment patterns
Rhythm—recurring patterns	Dynamics—contour changes without associated changes in timbre; accents; contour for expression
Sound Quality—changing texture	Rhythm—supporting melody Sound Quality—instrument selection; Expression changes without dynamic changes
	Spatial Properties—diverse host environments for each instrument; rhythmic pulses in different stereo locations; sound stage location of instruments widely varied

most of the information that is significant to the communication of the message of a piece of music.

Western music might have developed differently. While pitch relationships have been used as the primary generator of musical materials much more than other artistic elements, this need not have been so. Indeed, the music of other cultures uses the artistic elements of music in significantly different ways. Some cultures emphasize other artistic elements, such as rhythm, and many incorporate very different types of pitch relationships.

It is difficult to justify pitch's traditional prominence in the expression of musical ideas. While it is true that pitch is the perception of the primary attribute of the waveform (frequency, with the other attribute being amplitude), that of all the elements it is the most easily detected in many states and values, that pitch is the only artistic element that can be readily perceived as multiples of itself (the octave repetition of pitch levels, and the perceptions of real and tonal transposition of pitch-patterns)—these factors do not cause pitch to be a more prominent element than the others.

Our ability to perceive pitch is not significantly more refined (if at all) than our abilities to perceive the other parameters of sound. This is especially true of those parameters that utilize less precise pitch-related percepts (timbre, environmental characteristics, texture, and pitch density).

It follows that the artistic elements of sound other than pitch are equally capable of contributing to the communication of musical ideas.

This capability is being realized in the recordings of today's creative artists. In fact, this has been going on for quite some time, as the examples of recordings by The Beatles should indicate. This is occurring without conscious planning, but rather as a natural exploration of available sound relationships. Musicians (recordists, performers, composers, and producers) instinctively find roles for the unique artistic elements in recordings. The artistic elements that were not available, or that were underutilized, in traditional music performed in traditional contexts, are functioning in significant ways in modern music productions.

The concept that all of the artistic elements of sound have an equal potential to carry the most significant musical information is *equivalence*.

The states of the various components of the aesthetic and artistic elements of sound will make up the musical material. As such, they will function in primary or secondary roles of importance in the communication of the musical message. It is possible for any artistic element to function in any of the primary and secondary roles of shaping musical materials, and of generating the communication of the musical message. These are under the control of the recordist and have been commonly used to shape the musicality of music productions for years. Bringing these musical ideas into acceptance by the listener will require finesse and control of craft by the recordist.

Equivalence is also a framework for listening. It is a point of departure that reminds the recordist that any element or aspect of sound can change or can demand attention. Any change in the sound must be detected by the recordist and understood—no matter the significance to the music. Any aspect of sound might require the attention of the recordist at any moment during the listening experience. Equivalence provides this guidance and awareness.

Text as Song Lyrics

When a text is present in a piece of music, it is a significant addition to the musical experience. Through the text, language communicates a concept or describes a drama within the work. Further, the sound resources of the language will be exploited to enhance the aesthetic experience of the music. Songs are often relatively short musical pieces that contain a text (usually a single, rather short text). The song is the most common form of music today.

The text, or lyrics, of a song is a poem set to music. The text's elements are arranged in some sort of structure (as the structural construction of music), and the concepts of the text will create formal areas that are conceived as a single entity, as well as an overall idea and meaning of the text.

The lyrics of songs are constructed in many of the same ways as traditional poetry, written for its own sake and not intended to be set to music. The primary differences between the traditional poetry and poetry as song lyrics lie (1) in the repetitions of certain stanzas or phrases of the poem (unaltered or with slight changes), (2) in the careful crafting of the meters of the text, the rhythms of the lines, and the timing of the conceptual ideas of the text often found in song lyrics, and (3) in the sound qualities of the words that can be chosen to enhance the musical setting.

Literary Meaning

The *literary meaning* of the text brings the dimensions of verbal communication of ideas and concepts to the musical experience. Songs have been written on a multitude of subjects from common, everyday small occurrences to the highest of human ideals. The lyrics of a song might present a story line, or it might be a description of an event or the author's feelings about some aspect of the world around them. The text might be a presentation of the social-political philosophies of the author, or it might be a love song. The potential subjects for a song are perhaps limitless.

The presentation of the text's literary meaning is often enhanced by subordinate phrases of text segments that create new dimensions in the text. These subordinate ideas provide the turns of phrase or concepts that enrich the meaning of the text as a whole. The turning of the phrase allows for different interpretations of the meaning of certain ideas, at times different meanings to different individuals (or groups of people) depending on the experiences of the audience.

The potential for different interpretations allows for some (or much) ambiguity and intrigue in the text. The ambiguity may be clarified with a study of the central concepts of the song lyrics. Reevaluating a well-crafted text will often allow the listener to find new relationships of ideas or meanings of materials that enhance the experience of the song, for the listener (or recordist). This is common in songs from many styles of music and lends a considerable dimension to the musical experience.

The concepts used to enhance the literary meaning of the work may or may not be directly related to the central ideas of the text. These ancillary concepts may take many forms and are important in shaping the presentation of the communicative aspects of poetry. A study of poetry, or of the setting of texts to music, may be very appropriate for the individual recordist, but is out of the scope of this writing. Some general observations are instead offered.

Structure and Form of Song Lyrics

The *structure* of the text exists on many levels, similar to the hierarchy of musical structure. The conceptual meanings of the text and the sounds and rhythms of the text, do not allow for a clear division between the structural aspects of the text and the form-related aspects of the text. The structure of the text should address the sound-qualities of the text and its organization of mechanical parts. The *form* of the text should address the conceptual, often with a recurring concept or theme, a refrain (as the song's chorus).

Some cross-over will occur between the two areas: (1) the structure of the text's presentation may alter the statement's meaning, and (2) concepts can, at times, function as structural subparts. These are the result of the ways we conceptualize in verbal communication, and the previous experiences and social-cultural conditioning of the individual.

The components of the structure of a text will be major divisions of the materials of the text, and the subdivisions they contain. The materials that comprise the components are words, with all of their associated meanings, and the thoughts and feelings they invoke from within the individual. Words will be related by their sound qualities, rhyme schemes, rhythms and meters of groups of words, repetitions of words and words sounds, and by tonal and dynamic vocal inflections. Meanings of the

Form

Major Divisions:	A			B				A		
Subdivisions										
phrases:	a	b	c	d	e	d	f	g	h	i
concepts:										

Structure

Major Divisions:	Stanza #1	Refrain	Stanza #2
Sub-grouping by function:	beginning of plot	author's impressions	plot continued

Subdivisions
- groupings of phrases by:
  - rhythm
  - rhyme scheme
  - word usage
  - sound quality
- lines by:
  - rhythm
  - rhyme scheme
  - word usage
  - sound quality
- words by:
  - repetition
  - varied meaning
  - tonal inflection
  - dynamic inflection

Figure 3-2 Form and structure of song lyrics.

words, repetitions of words with different associated meanings, phrases created by the concepts (sentences), and groupings of phrases by subject matter or concepts are also used.

This format will not necessarily be directly transferable to all text settings, but these concepts can provide a meaningful point of departure.

Texts and Music in Combination

The structures of the text and the music interact in the overall perception of the song. They are perceived as being interrelated. They serve to enhance each other. The structures may complement one another, or they may serve as areas of contrast, with the text and the music grouped in overlapping segments, unfolding over time.

Both complementary and contrasting relationships of the structural elements of the text and the music exist in most works. The two play off one another, creating a sense of drama between the text and the music.

The relationships of structures create our impression of form; our conceptualization of grasping the essence of the entire work in an instant of realization. Within our impression of form as the overall conception of the work, we conceptualize points of climax and points of repose; we conceptualize the characteristics of design and shape of the materials that create the movement from one important event, or moment, to the next.

We recognize the shape and design of the work as it is represented in our perception of the significant moments of the work, and in the movement between the moments as they unfold over time.

The relationships of the musical materials create structure in a piece of music. Our perception of the design of structure is our conception of form. The structure of a piece of music may be altered significantly without altering its form. Even when the primary musical materials and the structure of a work are significantly altered, two very different interpretations of the same piece of music will be perceived as being similar when the form (or overall conception) of both performances are similar.

Contrast, for example, two performances of the song, "Every Little Thing," the original by The Beatles (*Beatles for Sale,* 1964) and a cover by Yes (*Yes,* 1969). The overall shape of the piece is not dramatically altered, but the structures of the two performances are quite different. Great differences exist between the lengths of sections, as well as the treatments of the basic musical materials and how they are organized. Few people would argue that both performances are of the same piece of music. Few people could not perceive dramatic changes in the structure and materials of the two different versions.

The reader is encouraged to perform the exercise in identifying the structure of a song found at the end of this chapter, to compile a time line similar to those in Figure 3-3.

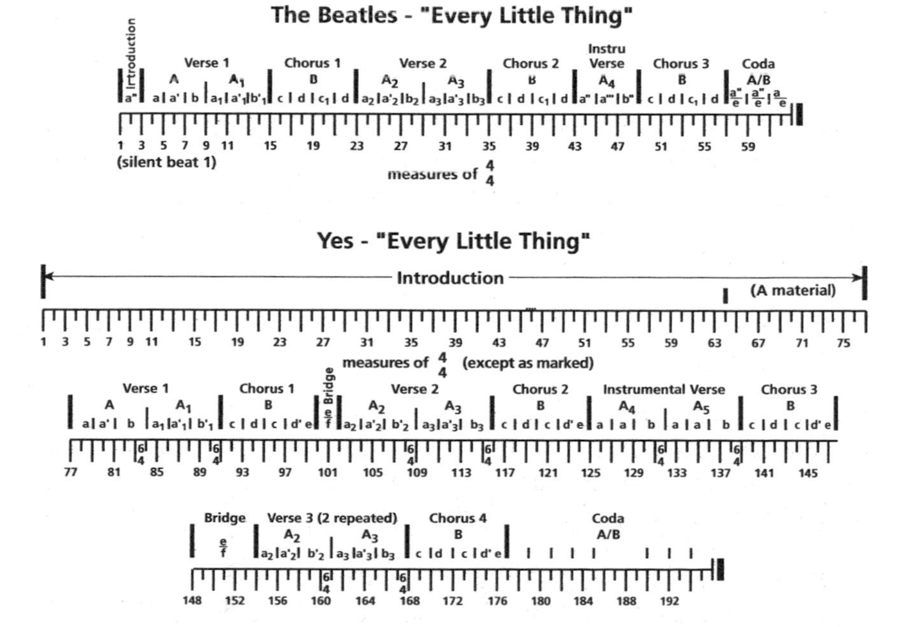

Figure 3-3 "Every Little Thing" as performed by The Beatles and by Yes.

The Listener

The audience member and the various audio professionals will have very different levels of listening expertise and usually dramatically different purposes for the listening process.

The recordist will have knowledge of the recording process (that which appears in Part 3, and more), the states of sound in audio recording, the materials of music, and of the hearing mechanism (previously discussed). Further, the recordist will have spent considerable time acquiring the listening skills for the evaluation of recorded and reproduced music, and sound that will be covered in Part 2. The recordist is often equally skilled at evaluating the technical integrity of the audio signal (perceived parameters of sound), and at evaluating the artistic elements of sound and the materials of the musical message, also to be covered in Part 2.

The *lay-listener* is the audience for a recording. The lay-listener will listen by relying primarily on their previous listening experiences. They will usually have little or no formal training. The lay-listener may be listening for some meaning in the music and be concerned with the relationship of the musical (or literary) message to their personal preferences of musical style, and musical and dramatic meanings. They may be listening for the sensual aspects of the music or be listening for the aesthetic experience.

The listener will likely be listening for pleasure and be concerned about enjoyment. People most often listen for entertainment, and perhaps for escape or enrichment.

While an album might be carefully crafted as a complete experience from beginning to end, most people will not sit in a position equidistant from two loudspeakers for about an hour listening to a CD. Most people do not dedicate time to focus their attention to music listening, or to sitting in one place while listening (unless they are driving, and then we hope they are not listening too intently). This is simply not the normal listening practice of people today.

Whatever the purpose for listening to music, the listener will not listen in the same way or for the same sound qualities as the recordist. Nor should they. The audience member should not be expected to listen in the same way as the recordist. Their purpose for listening is very different. It is necessary, however, for the recordist to deliver the recording in such a way that it can be understood.

The receiver and the quality of the communication limit the success of any communication. The receiver (listener) must be able to accurately process information (recording/music) for communication to occur. Humans are limited in their abilities to understand the content and/or meaning of what they perceive. These limitations are primarily the result of the listener's experience and knowledge, but are also dependent upon the listener's degree of interest in the material, intellect, and physical condition. The same material (music recording) will yield different information to different listeners (or to the same listener on different hearings), depending on knowledge, experience, analytic reasoning, social-cultural conditioning, expectations of context, attentiveness, and the condition of the hearing mechanism.

In crafting recordings for an audience, the recordist might need/wish to directly consider the listener (audience). An examination of the above factors will provide a realistic assessment of a target audience and perhaps allow the recordist to reach them more readily. These are the conditions that shape the listeners of recordings.

Knowledge

The listener's accumulated information related to what is being heard, as well as of all subjects related to their existence, plays a substantial role in the understanding of music and sound. *Knowledge* allows the listener to understand a sound, or a musical passage, by relating the experienced sound material to a body of known information. When the listener has a body of known information and/or possible circumstances, the music can be matched against those possibilities. With a match, listeners can then comprehend (and potentially reason) the meaning of the material.

Knowledge is the amassed body of learned information, or known truths. The listener can draw their knowledge to make evaluations and judgments on what is being heard. Knowledge areas related to the understanding of sound (and music) would include acoustics, psychoacoustics, music theory, music history and literature, language, audio recording theory and practice, mathematics, physics, engineering, computer science, communications, and more. The listener can formally and consciously know these subjects, or they might be more or less intuitively learned through sensitivity to life's experiences.

Experience

The listener's past life *experiences* are directly related to knowledge. Sound is experienced. In its conceptualized state, sound becomes experienced information. A personal knowledge, or experience of the sound, is the result of the listening process. Prior listening experiences are a resource that can be drawn from to recognize certain sound events or relationships. Sounds are mapped into the memory. The listener is better able to retain sound events in memory when a sound is the same as, or similar to, a sound that has been previously experienced. The act of listening is itself an experience; involving the learning of new information from what is going on in the listener's "present." New information is recognized and understood by comparing it with what has been previously experienced.

The type and quantity of listening experiences, and the personal knowledge gained will vary significantly between individuals. These listening experiences are significant factors in understanding the messages of music. Different types of music will communicate different messages and may communicate the message through different musical styles. Difficulties people experience in understanding or appreciating different types of music can often be attributed to a limited experience with a certain type of music. An individual's listening experiences may have limited their ability to understand the materials (language) of the music, or to appreciate what the music is trying to communicate. Increased knowledge of a type of music, and/or an increased number of experiences in listening to the type of music, will increase the listener's ability to understand or appreciate the type of music.

Listening experiences are greatly influenced by the life environment of the individual. The social and cultural environment(s) in which the individual lives, and has experienced, provide opportunities for a certain finite number of listening experiences. Within any environment, certain experiences will occur much more frequently than others. Certain types of listening experiences will be very common, and certain types of listening experiences will never occur or occur only rarely.

Social-cultural conditioning will predispose the listener to a certain set of available previous experiences. People are conditioned by their environment (social and cultural) to apply meanings to sounds, and to understand stylized musical relationships. We learn to listen for certain relationships in musical materials and the artistic elements of sound. For example, the music of India uses pitch and rhythm in significantly different ways than American popular music. Individuals from either culture will not readily understand the meaning or appreciate the subtleties of the music of the other culture, upon initial hearings.

The application of meanings to sounds is the basis for language. Sounds have meaning, and can represent ideas. In this manner, a series of short sounds as narrowly defined, isolated ideas can combine (in a prescribed ordering) to create a complex concept. Communication of simple ideas to complex thoughts is thus accomplished by language. We well know, different cultures have strikingly different languages. Some languages have common elements to other languages, and certain languages have elements that are largely unique.

Social context also plays a significant role in defining language sounds and meaning. Quite different meanings may be associated with a single sound, in the same language, by people of the same culture/society. This most often occurs between different social groups (ethnic origins, religious beliefs, age groups, etc.), groups of different economic status, and between geographic locations.

Sounds have meanings associated with their source. A sound produced by a car horn will invoke in the listener the thought of an automobile, not of the horn itself. Such referential listening only occurs when the listener has a certain set of life/listening experiences. Associations between sounds and their sources are largely dependent upon the listener's set of life/listening experiences, as provided by social-cultural environments. One can imagine living conditions under which an individual might never have experienced the sound of a car horn (perhaps the nineteenth century). The sound would not elicit the same response from this person as it would from a modern urbanite.

The meanings of musical sounds transfer between cultural and social groups in very similar ways to language sounds.

Social-cultural conditioning creates expectations as to the function of music. People are conditioned to relate various functions and applications to certain types of music. Dance, celebration, worship, ceremony, accompaniment to visual media, and aesthetic listening are but a few of the functions that music serves in various societies. Each function carries with it certain expectations for musical style. These expectations are defined differently in different cultures and societies.

The life/listening experiences of the listener shape the available resources from which the individual can draw on to understand any sound information. Social-cultural environment conditions the listener through (1) providing a predominance of certain listening experiences, (2) providing certain expectations as to the content of musical materials (the applications of the artistic elements), (3) providing certain expectations as to the context within which certain types of music will be heard (in church, in a club, in the street, etc.), (4) providing meanings of association for certain significant sounds (significant sounds being perhaps a siren, perhaps a falling tree), and (5) providing associations of group activity for certain types of music (ceremony, dance, group experiences).

While broadcast media have broadened the number of common elements between social and cultural groups throughout the world, great diversity still exists among human cultures and societal groups. Social-cultural conditioning must remain a significant factor in our realization of the limitations of the listener. For example, it might be unrealistic to expect the lay person from China to understand the musical nuance and message of rap music, just as it might be unrealistic to expect the typical American, suburban 16-year-old to understand the meaning and significance, or to appreciate the aesthetic qualities of Tibetan chant.

Expectation

Knowledge, experience, and social-cultural conditioning create *expectations* for the listener. The listener will expect to hear certain sounds (or sequences of sounds) under certain circumstances. They will expect certain types of sounds to follow what has already been heard. They can expect to hear materials in certain relationships (melody with certain harmony), and to hear certain sounds within a given physical environment (one would not expect to hear a lion sound on a city street). The listener will likewise expect certain sounds in a given musical context (an operatic vocal technique would be unexpected in a reggae work) and expect to hear certain kinds of music in certain social-cultural contexts (the listener will expect to hear different music in church, movies, dance clubs, etc.).

When the listener is presented with something that is not expected, they may be surprised if they are able recognize the material enough to understand it and its context, or may be confused if they cannot recognize the sound or relate the sound to its context. An unexpected sound might intrigue the listener as a unique turn of a musical idea or as a sound slightly out of context. Conversely, if unexpected sounds that are also unfamiliar to the listener are present, they will not be able to understand the sound, they will not receive the message of the material, and may likely be dissatisfied or frustrated by the listening experience. Among other possibilities, the listener might have a dislike for the original context of this unexpected sound, and thus cause this new experience to be unenjoyable.

Expected and unexpected sounds and relationships are balanced within all musical styles. A musical style is a set of expectations. Certain types of musical events and relationships are present that provide a musical style with consistency and a unique character.

Analytical Reasoning

The listener's knowledge, experience, and *analytical reasoning* play important roles in the understanding of musical messages within various musical styles. Too many unexpected sounds or situations will result in confusion and frustration on the part of the listener. If expectations are filled in predictable ways, the listener will become bored with the material. They perceive logic and coherence of the musical materials through a fulfillment of expectations in the characteristics and functions of the musical materials, coupled with enough unexpected activity to maintain interest.

Listeners use analytical reasoning to extract the meaning of musical materials, when they are unable to identify the material. Analytical reasoning, in music listening, is the ability to relate immediate listening experience to knowledge, in a manner capable of deducing meaningful observations and information. The ability to perform this type of listening activity is dependent upon intellect, the amount of knowledge the listener is able to draw from, the listener's previous experience in performing analytic reasoning exercises, and the listener's knowledge of the types of information to extract from the listening experience. This method of listening works similar skills as the critical and analytical listening skills addressed in Part 2.

Active and Passive Listening

The level of attentiveness of the listener plays an important role in their understanding of the musical message. Listener attentiveness and musical understanding are related to active and passive listening, and to the listener's interest in the music.

The difference between active and passive listening is the listener's attention and involvement. *Passive listening* occurs when the listener is not focused on the listening process, or on the music itself within the listening experience. Passive listening might find listeners otherwise occupied, and listening to music as a background activity (reading a book, for example). Alternatively, listeners might be listening to music for reasons other than understanding. They may be listening for relaxation purposes. Other types of passive listening include approaching music for its emotive state, or feeling, or listening to music for its pulse only, such as an accompaniment to dancing. In all of these cases, the listener is not listening to the musical materials themselves, and they might not be aware of the music during certain periods of time. In passive listening the music itself is not the center of the listener's attention.

Music is at the center of the listener's attention in *active listening*. Various levels of detail can be extracted during the active listening process. Among many possible states, active listening might take the form of listening to the text and primary melodic lines of a work. It may take the form of following the intricacies of motivic development in a Beethoven string quartet, or of evaluating the characteristics of a sound system. In all cases, the state of active listening has the listener's attention aware of musical materials or sound quality.

The listener is most likely to be an active listener if they are interested in the music or have a specific reason to be listening carefully. The listener's interest in the music may be determined by their mood or energy level at the time of the listening experience but is most often associated with listening preferences (and the previous experiences that have shaped those preferences). These preferences lead to the types of music the listener listens to most often and what they prefer to hear.

Hearing Mechanism Condition

The final variable between individual listeners is the *condition of the hearing mechanism*. Some individuals have impaired hearing—some have knowledge of their condition, and others do not. The hearing of the individual might vary from the norm because of a defect at birth, from accidental damage from physical trauma or prolonged exposure to high sound pressure levels, or from the natural deterioration caused by the aging process. The recordist cannot anticipate hearing impairment of the listener, nor create recordings that can be heard well by those so unfortunate. It is quite important, however, that the recordist know the condition of thier own hearing.

Variation of the individual's hearing characteristics from normal human hearing is of great importance to the recordist. The recordist must have knowledge of their own hearing and make use of that information in evaluating sound in their job function. Significant hearing problems may make a person poorly suited for certain positions in the recording industry. Normally, a recordist might find they are less sensitive to sound in certain frequency ranges or that the two ears have different frequency and amplitude sensitivities. This information will serve the recording professional well in evaluating sound, as it will allow them to make adjustments in their work by knowing how their personal perception is different from the existing sound.

Target Audience

The typical listener envisioned for a specific recording project, or piece of music, is often called a *target audience*. The target audience for a piece of music is often determined to help focus a project and to seek a way of predicting the success of the music in communicating its message. The target audience is defined by identifying the knowledge, musical and sociological expectations, and the listening experience of the typical audience member. The music can then be shaped to conform to the abilities and expectations of the typical member, thereby increasing the chance it will successfully communicate its musical message. The goal is to create a recording/song that this defined audience will find engaging and that will be commercially successful.

Conclusion

Recording professionals should not expect people to listen to their recordings with undivided attention or with the same level of accomplishment they have attained. At the same time the recordist must not underestimate or undervalue the listener. Listeners are often passionate about the music we record and the music they listen to.

Music audiences feel strongly about "their" music. They are often very possessive about the type of music they enjoy and the performers they follow. Music can speak deeply to people and bring people to identify with music on a very fundamental and personal level. Commentaries about their music can be perceived as reflections on themselves. While the listener may not know much about music or recording, they know what they like—and usually are willing to tell you about it.

Similarly, listeners are often quick to identify the quality of productions. Well-crafted and successful recordings are easily identified by listeners, not for their quality but because they present the musical material in a way that communicates well and directly to the listener. Further, sound qualities of the recordings of one type of music will be different from others, and will draw the listener, or not. The listener may not recognize technical integrity, but any signal problems will detract from the recording and the listening experience. The listener will not miss this. They may not be able to tell what is wrong but they will recognize that it is not right.

The recordist is in the position to play a central role in the creation of music. They may use the recording process to shape music performances and the music itself. This new role for the recordist has been widely recognized since the early 1960s. While perhaps it is new when we think back over hundreds of years of music, recordists are currently very much a part of the creative process of nearly all recordings. With over 35 years of sophisticated practice in crafting sounds through multitrack production and stereo reproduction, the recordist as an artist is no longer something new.

The more the recordist understands music and the listener, the more likely it is that they will be in a position to assist the artist in delivering a performance and recording of a piece of music that will be successful. To bring the reader to appreciate some of what is involved with this feat was the goal of this chapter. It is wished that the recordist would want the listener to find enjoyment in what was recorded—for the artist's sake and for the music.

Exercises

The following exercise should be practiced on a variety of pieces of music until you are comfortable with the material covered.

Exercise 3-1
Structure Exercise.

The purpose of this exercise is to create a time line of a song, divided into major structural divisions and phrases.

1. Select a recording of a song you know reasonably well and prepare a time line with measures numbered, up to perhaps 100.
2. Listen to the recording to identify where the major sections fall against the time line. Try following the time line while tapping the pulse of the song, or conducting. When a major section begins/ends, make a mark on the time line.
3. After listening to the song, write down the names of those divisions. Now try filling in additional information, such as other verse or chorus beginning/ending points and phrase lengths.
4. Repeat listening to the recording and writing down the information recognized.
5. The graph is completed when it includes all of the major structural divisions, the mid-level structural divisions and the smallest uniform phrase. Incorporating text information is also helpful.
6. Following the time line while listening may prove helpful in initial studies in identifying structural divisions. You are encouraged to wait until the music is stopped before writing observations. Clearly separating the listening and writing activities will assist you in improving listening skills and in learning to evaluate sound. This will become increasingly important.

Part Two

Developing Listening and Sound Evaluation Skills

4 Listening and Evaluating Sound for the Audio Professional

People in the audio industry need to listen to and evaluate sound. Carefully evaluating sound, for one reason or another, is an integral part of most positions in the audio industry. Sound must be evaluated in all areas of audio production, manufacturing, and support. These areas are very diverse. They may be equipment performance or microphone placements, music mixes or the technical quality of a sequence of test tones, or any one of many other possibilities.

Sound is being evaluated by the audio professional in all these cases and more.

Saying "sound is central to audio" is obvious to the point of sounding trivial. It is equally ironic that the audio and music community has not developed a way to clearly communicate meaningful information about sound. No language or vocabulary exists for qualities of sound. Part 2 begins the creation of a means and vocabulary to communicate about sound.

While this book is focused on the artistic roles of the recording professional, sound evaluation is important to everyone in the industry who listens to, evaluates, and talks about sound. Part 2 of the book can and should be used by anyone in need of developing the ability to understand, evaluate, and communicate about sound. It should be a primary objective of all people in the audio industry to be more sensitive and reliable in their evaluations of sound. While the term "recordist" will still be used in Part 2, it should be interpreted to mean "any audio professional" during discussions of sound evaluation. The sequence of chapters in this Part will present a system for evaluating sound that will substantially develop the reader's ability when mastered.

It is necessary for all people related to the audio industry to be accurate and consistent in their evaluations of the quality and content of sound and audio. As we have seen, the previous experience, knowledge, cultural conditioning, and expectations of the listener (in this case the audio professional) have a direct impact on the level of proficiency at which the listener is able to evaluate sound. With increased experience in evaluating sound comes increased skill and accuracy.

The act of listening and the process of evaluating sound can be learned and greatly refined. The following is a presentation of the need for sound evaluation and the listening process, leading to a discussion about how we talk about sound, and the development of listening and sound evaluation skills.

Why Audio Professionals Need to Evaluate Sound

Audio professionals need to evaluate sound to define what they hear, to understand what they hear, and to communicate with one another about sound. These are important aspects of the job functions for almost all people in audio.

Recording engineers and producers, obviously, must have well-developed listening skills because evaluating sound is one of the most important things they do in their work. The need for highly refined skills obviously holds true for composers and performing musicians, especially those involved in the audio recording processes. All audio professionals who listen to sound share a similar need for these skills. The technical people of the industry, those involved in artistic roles, and those in manufacturing or facility design, or product sales and many others, all must share observations and information about sound.

There are other reasons audio professionals need to evaluate sound in addition to talking about sound in precise and meaningful terms. The recording's sound qualities need to be observed, recognized, and understood to perform a great many jobs in the industry. Nearly all positions approach sound evaluation in a somewhat unique way. In fact, there might be as many reasons (significantly or slightly unique) for evaluating sound, as there are job functions within the multitude of positions in the audio industry.

For the recordist, there are additional benefits to sound evaluation, and some will be discussed in detail in later chapters. These include ways to (1) keep track of one's work so that the audio professional can return to those thoughts/activities in the future, (2) plan recording projects out of the studio, (3) understand the work and ideas of others, (4) recreate sounds and musical styles, and many more.

Nearly all people in audio work directly with some aspect of sound. The aspects of sound that people work with might be vastly different, yet they must communicate directly and accurately to share information. In order to share information, sound must first be evaluated and understood by the listener.

Understanding sound begins with perceiving the sound through active attention. One can then recognize what is happening in the sound or recognize the nature of the sound, provided the listener has sufficient knowledge and experience. The listener must know what to listen for (i.e., the artistic elements of sound) and where to find that information (perhaps a particular musical part). This recognition can lead to understanding, given sufficient information. What is understood can be communicated, given a vocabulary to exchange meaningful information that is based on a common experience.

Talking About Sound

People in the audio industry, as in all industries, work together towards common goals. In order to achieve those goals, people must communicate clearly and effectively. A vocabulary for communicating specific, pertinent information about sound quality does not currently exist. People have been talking about sound for hundreds of years without a vocabulary to describe the actual perceptions and experiences. Instead people have used terms to associate other perceptions and experiences to sound—unsuccessfully and inaccurately.

Describing the characteristics of sound quality through associations with the other senses (through terminology such as "dark," "crisp," or "bright" sounds) is of little use in communicating precise and meaningful information about the sound source. "Bright" to one person may be associated with a narrow, prominent band of spectral activity around 15 kHz throughout the sound source's duration. To another person the term may be associated with fast transient response in a broader frequency band around 8 kHz, and present only for the initial third of the sound's duration. A third person might easily provide a different, yet an equally valid definition of "bright" within the context of the same sound. The three people would be using different criteria of evaluation and would be identifying markedly different characteristics of the sound source, yet the three people would be calling three potentially quite different sounds the same thing—bright. This terminology will not communicate specific information about the sound and will not be universally understood. It will not have the same meaning to all people.

Analogies such as "metallic," "violin-like," "buzzing," or "percussive" might appear to supply more useful information about the sound than the inter-sensory approach. This is not so. Analogies are, by nature, imprecise. They compare a given sound quality to a sound the individuals already know. A common reference between the individuals attempting to communicate is often absent. Sounds have many possible states of sound quality.

"Violin-like" to one person may actually be quite different to another person. One person's reference experience of a "violin" sound may be an historic instrument built by Stradivarius and performed by a leading artist at Carnegie Hall. Another person may use the sound of a Bluegrass fiddler, performing a locally crafted instrument in the open air, as their reference for defining the sound quality of a "violin." The sound references are equally valid for the individuals involved, but the references are far from consistent and will not generate much common ground for communication. The sound qualities of the two sounds are strikingly different. The two people will be referencing different sound characteristics, while using the same term. An accurate exchange of information will not occur.

The imprecision of terminology related to sound quality is at its most extreme when sounds are categorized by mood connotations. Sound qualities are sometimes described in relation to the emotive response they invoke in the listener. The communication of sound quality through terminology such as "somber," for example, will mean very different things to different people. Such terminology is so imprecise it is useless in communicating meaningful information about sound.

People can only communicate effectively through the use of common experiences or knowledge. The sound source itself, as it exists in its physical dimensions in air, is presently the only common experience between two or more humans.

As we hear sounds, we make many individualized interpretations and personal experiences. These individual interpretations and impressions are present within the human perceptual functions of hearing and evaluating sound. They cause individualized changes of the meaning and content of the sound. Therefore, our interpretations and impressions are of little use in communicating about sound. Humans have few listening experiences that are common between individuals and that are available to function as the reference necessary for a meaningful exchange of information (communication).

This absence of reference experiences and knowledge makes it necessary for the sound source itself to be described. Meaningful communication about sound will not be precise and relevant without such a description.

The states and activities of the physical characteristics of the sound will be described in our communications about sound. This approach to evaluating sound requires knowledge of the physical dimensions of sound and how they are transformed by perception. Meaningful communication between individuals is possible when the actual, physical dimensions of sound are described through defining the activities of its component parts.

By describing the states and activities of the physical components of a sound, people may communicate precise, detailed, and meaningful information. The information must be communicated clearly and objectively. All of the listener's subjective impressions about the sound, and all subjective descriptions in relation to comparing the sound to other sounds, must be avoided for meaningful communication to occur.

Subjective information does not transfer to another individual. As people attempt to exchange their unique, personal impressions, the lack of a common reference does not allow for the ideas to be accurately exchanged.

Meaningful communication about sound can be accomplished through describing the values and activities of the physical states of sound. Sounds will be described by the characteristics that make them unique. Meaningful information about sound can be communicated through verbally describing the values and activities of the physical states of sound in a general way. Information is communicated in a more detailed and precise manner through graphing the activity, as will be described in the following chapters.

A vocabulary for sound is essential for audio professionals to recognize and understand their perceptions, as well as to convey to others what they hear.

The Listening Process

Recording engineers and other industry professionals must learn to listen in very exacting ways. The profile of the listener discussed in Part 1 assisted us in identifying how the recordist has different purposes for listening and needs a much higher required skill level. It is necessary for audio professionals to be accurate and consistent in the listening process and its observations.

An approach to listening will need to be developed by the recordist for them to function in their job. This will ultimately become a systematic process for hearing detail in sound. The recordist will not be listening passively but will rather be actively engaged in seeking out information with each passing sound. They will be concerned about a multitude of things, from the quality of a performance, to its technical accuracy; from the quality of a microphone selection, to its appropriate placement; from the quality of the signal path, to the inherent sound quality of a signal processor. All of these things and many more might pass through the thoughts of the recordist frequently and regularly throughout any work session. The listening experience of the audio professional will be multidimensional in many ways.

The recordist must acquire a systematic approach to listening that will involve quickly switching between critical and analytical listening information. It will involve quickly switching between levels of detail, or perspective, and focus on various artistic elements and musical materials.

In many ways the recordist's listening process is like a scanner—always moving between types of information and between levels of detail.

Critical Listening versus Analytical Listening

Audio professionals evaluate sound in two ways: *critical listening* and *analytical listening*. Critical listening and analytical listening seek different information from the same sound. Analytical listening evaluates the artistic elements of sound, and critical listening evaluates the perceived parameters. A different understanding of the sound is achieved in each case.

The artistic elements are the functions of the physical dimensions of sound, applied to the artistic message of the recording. We recognize the physical dimensions of sound through our perception, as perceived parameters. This allows understanding of the technical integrity of sound quality to be contrasted with musical meaning and relationships of the artistic elements.

The same aspects of sound quality may provide two different sets of information. This is entirely dependent upon the way we listen to the sound material—evaluating the sound for its own content (critical listening) or evaluating sound for its relationships to context (analytical listening). The recordist must understand how the components of sound function in relation to the musical ideas of a piece of music and the message of the piece itself. These are analytical listening tasks. The audio professional must also understand how the components of sound function to create the impression of a single sound quality, and how they function in relation to the technical quality of the audio signal. These aspects are critical listening tasks.

Analytical listening is the evaluation of the content and the function of the sound in relation to the musical or communication context in which it exists. Analytical listening seeks to define the function (or significance) of the musical material (or sound) to the other musical materials in the structural hierarchy. This type of listening is a detailed observation of the interrelationships of all musical materials, and of any text (lyrics). It will enhance the recordist's understanding of the music being recorded, and will allow the recordist to conceive of the artistic elements as musical materials that interact with traditional aspects of music.

Critical listening is the evaluation of the characteristics of the sound itself. It is the evaluation of the quality of the audio signal (technical integrity) through human perception, and it can be used for the evaluation of sound quality out of the context of a piece of music. Critical listening is the process of evaluating the dimensions of the artistic elements of sound as perceived parameters—out of the context of the music. In critical listening, the states and values of the artistic elements function as subparts of the perceived parameters of sound. These aspects of sound are perceived in relation to their contribution to the characteristics of the sound, or sound quality.

Critical listening seeks to define the perception of the physical dimensions of sound, as the dimensions appear throughout the recording process. It is concerned with making evaluations of the characteristics of the sound itself, without relation to the material surrounding the sound, or to the meaning of the sound. Critical listening must take place at all levels of listening *perspective* (see below), from the overall program to the minutest aspects of sound.

The Sound Event and Sound Object

The concepts of the *sound event* and the *sound object* assist in understanding how the musical materials (analytical listening) and sound quality (critical listening) are shaped by the artistic elements. A sound event is the shape or design of the musical idea (or abstract sound) as it is experienced over time. The sound object is the perception of the whole musical idea (or abstract sound) at an instant, out of time.

The sound event is a complete musical idea (at any hierarchical level) that is perceived by the states and values of the artistic elements of sound. The term designates a musical event that is perceived as being extended over time, and has significance to the meaning of the work. The sound event is a musical idea perceived by its various dimensions, as shaped by the artistic elements of sound. It is a perception of how the artistic elements of sound are used to provide the musical section with its unique character. The sound event is understood as unfolding and evolving over time, and is used in analytical listening observations.

Sound object refers to sound material out of its original musical context. For example, in a discussion of the sound quality of George Harrison's Gibson J-200 on "Here Comes The Sun" compared to its sound on "While My Guitar Gently Weeps," the two sound qualities of the instrument would be thought of as sound objects during that evaluation and comparison process. A sound object is a conceptualization of a sound as existing out of time, and without relationship to another sound (except its possible direct comparison with another sound object).

The concepts, sound object and sound event, are contrasted at any hierarchical level. They allow analytical listening and critical listening evaluations to be performed, interchangeably and/or simultaneously, on the same sound materials.

These concepts are able to provide an evaluation of the music's use of the artistic elements of sound, in ways that are not necessarily related to the importance or function of the musical materials. Rather, these concepts seek to determine information on the artistic elements (or perceived parameters) themselves, as they exist as singular and unique entities (sound objects), and as they change over time (sound events).

Perspective and Focus

For sound evaluation purposes, the audio professional must be able to understand the artistic elements of sound, how those elements relate to the perceived parameters of sound, and how those two conceptions of sound are used with *perspective* and with *focus*. The concepts of perspective and focus are central to the listening process and evaluating sound. The audience will go through this process in a general and intuitive manner. The audio professional must be thorough and systematized in approaching the listening process.

In order for the message carried by the artistic elements to be perceived, the listener (audience or audio professional) must recognize that important information is being communicated in a certain artistic element. The listener must then decipher the information to understand the message, or recognize the qualities of the sound. The listener will identify the artistic elements that are conveying the important information by scanning the sound material at different perspectives, while focusing attention on the various artistic elements at the various levels of perspective.

Focus is the act of bringing some aspect of sound to the center of one's attention. The listener is required to identify the appropriate, perceived parameter of sound that will become the center (focus) of attention in deciphering the sound information. Further, the listener needs to determine a specific level of detail on which to focus attention.

The *perspective* of the listener determines the level of detail at which the sound material will be perceived. Perspective is the perception of the piece of music (or of sound quality) at a specific level of the structural hierarchy. The content of a hierarchy is entirely dependent upon the nature of the individual work, or portion of a musical work.

In a musical context, the detail might break down as in Table 4-1. Each level of detail represents a unique perspective from which the material can be perceived. Each perspective will allow the listener to observe different characteristics and attributes of the sound material. A perspective might be thought of as a conceptual distance of the listener from the sound material; the nearer the listener to the material, the more detail the listener is able to perceive.

The listener may approach any perspective to extract analytical listening information (pertaining to the function of the musical materials

Table 4-1
Example Hierarchical Levels of Perspective

Level 1	Overall musical texture
Level 2	Text (lyrics)
Level 3	Individual musical parts (melody, harmony, etc.)
Level 4	Individual sound sources (instruments)
Level 5	Dynamic relationships of instruments
Level 6	Composite sound of individual sources (timbre and space)
Level 7	Pitch, duration, loudness, timbre, space, and duration elements of a particular sound source
Level 8	Dynamic contour; definition of important components of timbre and space
Level 9	Definition of prominent harmonics and overtones of the sound source
Level 10	Dynamic envelopes of prominent overtones and harmonics

and artistic elements at that level of the structural hierarchy) or to extract critical listening information (pertaining to defining the characteristics of the sound itself). Focus, again, is the act of bringing one's attention to the activity and information occurring at a specific perspective of the structural hierarchy, and/or within a particular artistic element.

Attention to focus and perspective are needed in both critical listening and analytical listening activities, and should be considered before starting any listening session. It is important the recording professional define the focus and level of perspective of the listening experience before the sound material begins, as they can shape the listening experience in strikingly different ways for different situations. In many listening situations, all parameters of sound will need to be continually scanned to determine their influence on the integrity of the audio signal, and all artistic elements will need to be scanned to determine their importance as carriers and shapers of the musical message. In other listening situations, the recordist might need to carefully follow a specific artistic element at a specific level of perspective throughout the listening experience. Different situations will require a different approach to listening. It is important that the recording professional have a clear idea of what needs to be the focus of their attention and the level of detail required (perspective) before beginning to listen—or of the need for continually shifting focus and levels of perspective.

Multidimensional Listening Skills

Equal attention must be given to all aspects of sound as, depending on the sound material and purpose of the listening, any perceived parameter of sound or any artistic element may be the correct focus of the listener's attention. An incorrect focus will cause important information to go unperceived and will cause unimportant information to incorrectly skew the listener's perception of the material. The recording professional will often face the possibility that a change might happen in any of the dimensions of sound, at any point in time, at any level of perspective. It is necessary that recording professionals hear, recognize, and understand the character of the sound and any changes that might occur. This awareness needs to be cultivated, as it is counter to our learned listening tendencies.

Audio professionals must develop their listening skills to be multidimensional. The listening process involves the potential need to listen to many things simultaneously. Though on one hand impossible, this is in practice often necessary. To accurately evaluate sound, they must learn to:

1. Shift perspective between all levels of detail,
2. Focus on appropriate elements and parameters at all levels of perspective (and not allow their attention to be pulled away to activity in another element or level of perspective), and
3. Shift between analytical listening (for the importance of the musical material) and critical listening (for the characteristics of the sound itself) to allow the evaluation of sound.

Distractions

It is often difficult for the recordist to keep from being distracted. Maintaining focus on the purpose and intent of the particular listening experience is very important. Common distractions are becoming preoccupied with the music, being drawn to sounds and sound qualities other than those under evaluation, and being curious about how a sound quality was created (as opposed to character of the sound).

Most of us are drawn to a career in recording because of a love of music. When working on a recording, we can lose our focus by becoming engaged with the musicality of the material. This focus is similar to listening for entertainment. However, there is a time and place to listen for entertainment. Most often recordists listen to qualities that are more precise and exacting. Even when listening within musical contexts, working directly with musical materials, and thinking about the musicality of the recording, the audio professional will be working at a level of perspective that is far removed from the passive music listening experience enjoyed by most people.

While focused on listening to the characteristics of one element of sound, the sound qualities of another element can draw the listener's attention. It is very important that the listener remain mindful of the purpose of the listening experience. For example, if the listening activity is intended to determine the musical balance of the snare drum against the toms, one should not allow oneself to get distracted by the sound quality of the piano.

In evaluating sound, the audio professional must remember that they are seeking to understand the sound that is present. It is possible for the listener to become distracted from listening by their own knowledge of the recording process or by their wanting to learn more about the recording process. At times people are drawn to thinking about how sound qualities were created—equipment, recording techniques, etc. Bringing production concerns into the process of evaluating sound is counterproductive, unless listening skills are specifically used to identify equipment choices and production techniques, but this is a different matter.

Listening sessions should have a clearly defined function. If the recordist is listening to determine equipment that may have been used in a recording, then that is the purpose of the session. If the recordist is listening to understand the sound quality of a certain environmental characteristic, then they should be listening to the various components of that sound and not be concerned about identifying the manufacturer or model number of the device that created the environment.

Personal Development for Listening and Sound Evaluation

The skills and thought processes required for listening and sound evaluation must be learned. The development of any skill requires regular, focused, and attentive practice. Patience is required to work through the many repetitions that will be needed to master all of the skills necessary to accurately evaluate sound. Each individual will develop at a separate pace, as with any other learning.

Memory Development

The recordist will evaluate sound more quickly and accurately with the development of their auditory memory. This will often be accomplished through their ability to recognize patterns in the various aspects of sound. The listener must be conscious of the memory of the sound event, and they must seek to develop their memory to sustain an impression of the sound long enough to describe, annotate, or graph certain characteristics about the sound event.

Auditory memory can be developed. As one learns what to listen for, and as one understands more about sound and how it is used, the listener's ability to remember material increases proportionally. This is similar to the process of learning to perform pieces of music through listening to recordings of performances and mimicking the performances. With repetition, this seemingly impossible task becomes a skill that is much easier to perform. Listeners often remember more than their confidence allows them to recognize. The listener must learn to explore their memory and immediately check their evaluations to confirm the information.

The human mind seeks to organize objects into patterns. Sound events have states or levels of activity of their component parts that will often tend to fall into an organized pattern. The listener must be sensitive to the possibility of patterns forming in all aspects of the sound event, to allow greater ease in the process of evaluating sound. Recognizing patterns will assist in understanding sound and sequences of sounds, and will make remembering them more possible.

Developing memory is very possible and very important. Considering sound takes place over time and can only exist by atmospheric changes over time, it should be understood that sound is a memory. Sound is an experience that is understood backwards in time. Sound is perceived after it is past, using memory. Sound does not happen now (at a specific moment) but rather it happened then. It can start or stop now, but it exists over a stretch of time (duration).

The reader is encouraged to work through the exercise at the end of this chapter and to return to that exercise regularly during the course of their work in listening skill development.

Success and Improvement

With increased experience in evaluating sound comes increased skill and accuracy. The act of listening and the process of evaluating sound can be learned and become greatly refined.

The reader will continue to become more accurate and consistent in evaluating sound the more they practice the skills and follow the exercises in the following chapters. The development of these skills must be viewed as a long-term undertaking. Some of the skills might seem difficult, or impossible, during the first attempts. The reader must remember their previous experiences might not have prepared them for certain tasks. The skills are, however, very obtainable. Further, the skills are desirable, as the individual will function at a much higher level of proficiency in the audio industry after they have obtained these evaluation and listening skills.

The mastery of the skills of sound evaluation is a lifelong process; one that should be consistently practiced and itself evaluated. New controls of sound are continually being developed by the audio industry. These new controls create new challenges on the listening abilities of those in the audio industry.

Discovering Sound

Things are present in recorded music that are difficult to identify. When something has never been experienced or perceived, one does not know it exists. It is possible for people to simply not hear some aspect of sound, simply because they do not have an awareness or sensitivity for that area. Once that awareness and sensitivity is developed, those sounds are heard as easily as any other.

In *Personal Knowledge* Michael Polanyi conveys the experience of a medical student attending a course in the X-ray diagnosis of pulmonary diseases. The student watches dark shadows on a fluorescent screen against a patient's chest while listening to the radiologist describe the significance of those shadows in detailed and specific terms. At first the student is puzzled and can only see the shadows of the heart and ribs, with some spidery blotches between them. The student does not see what is being discussed. It appears to be a figment of the radiologist's imagination. As a few weeks progress, and the student continues to look carefully at the X-rays of new cases and listen to the radiologist, a tentative understanding begins to dawn on the student. Gradually the student begins to forget about the ribs and the heart, and starts to see the lungs. With perseverance in maintaining intellectual involvement, the student ultimately perceives many significant details, and a rich panorama is revealed. The student has entered a new world. The student may still see only a fraction of what the seasoned radiologist sees, but the pictures now make definite sense, as do most of the comments made by the instructor.

Many readers will likely discover a new world of sound. Dimensions of sound exist that are out of normal listening experience. We are not aware of those sounds until we learn to bring the focus of our attention to those elements and discover them. Only then can we begin to understand them.

We have learned to focus our attention to certain aspects of sound. In music, we have learned that pitch relationships will give us the most important information. In speech, we know that the sound qualities of words make up language, and the sound qualities of the speaker will inform us who is talking. We know dynamics will simply enhance the message of these two communications, and we listen to them in that way. We have been taught that where a musical instrument is playing is not important (and therefore not worth the effort of recognizing the sound characteristics of location and environment), but what pitches they are playing *is* important (and worthy of attention).

The reader will now be asked to perform listening exercises and to evaluate sound in ways that work against these learned (and perhaps natural) listening tendencies. This requires conscious effort, focused attention, patience, and diligence. With the knowledge that the listener is working against natural tendencies, it will make sense that certain things are difficult. This does not mean they are impossible; many people daily accomplish them. Nor does it mean that the way of listening should not take place. This way of listening is necessary to evaluate and understand many aspects of recorded sound that are simply not normally at the center of one's attention. As we know, the audio professional needs to listen in ways and for things that are not part of normal listening experiences.

The student took the leap of faith that is necessary in learning. The student believed in the radiologist and continued to try to understand. They initially perceived the material as an illusion, not really present, a figment of the radiologist's imagination, but continued trying to see and to understand. The student reached a moment of revelation when suddenly an image was perceived. It was always there. The student was now able to see it because of increased sensitivity to the possibility of its existence and an understanding of what that existence might be.

If the reader can commit to a similar leap of faith, they may be rewarded with the discovery of a new world of sound.

Summary

Understanding sound must begin with perceiving sound. This requires active attention, and sufficient knowledge and experience to know what to listen for. One can then recognize what is happening in the sound or recognize the nature of the sound. This recognition can lead to understanding. What is understood can be communicated, given a vocabulary to exchange meaningful information that is based on a common experience.

A system for evaluating sound has been devised and is presented in following chapters. It will provide a means for evaluating sound in its many forms and uses, and will provide a vocabulary that can communicate meaningful information about sound. The audio professional needs to evaluate sound for its aesthetic and artistic elements and its perceived parameters, as they exist in critical listening and analytical listening applications and at all levels of perspective. The system for evaluating sound addresses these concerns, and more.

Exercises

The following exercise should be practiced until you are comfortable with the material covered.

Exercise 4-1
Musical Memory Development Exercise.

1. Select a recording of a song you know reasonably well and prepare a time line with measures numbered, up to perhaps 100.
2. Before listening to the recording, sit quietly and try to remember as much detail of the song as you can.
3. Now, write down the song's meter. In your mind, listen to the piece and write down where the major sections begin and end. If you cannot come up with those divisions easily, you might well be able to deduce that information by thinking about the patterns of phrases in the introduction, verses, choruses, etc. Write down as much information as you can.
4. Think carefully about what you wrote and identify aspects you are not certain about—things that need to be determined when you listen to the recording.
5. Now you can listen to the recording, but listen intently for the information you have determined you need. Do not follow your graph. Listen with your eyes closed. Listen to remember what you hear. Do not write while you are listening and do not correct your graph while you are listening.
6. Write down what you heard in your one listening and correct what you previously wrote. Then repeat steps 4 and 5 until you have created a time line and structure of the song—in as few listening sessions as possible. Check your information one last time while following your graph. All of the information you wrote should be checked for accuracy; make corrections to your graph.
7. Do not get discouraged. Keep trying.
8. Select another piece of music you know more thoroughly and perform the exercise again.

This exercise can be performed whenever a time line needs to be created. If faced with a new song, listen intently to the song once immediately after sketching a time line. Remember not to write while listening. Listen when it is time to listen. Write what you have recognized and remembered.

People remember more than they believe they do. If you will trust your memory and use it, your memory will develop.

This exercise should be continually modified to incorporate any sound element you need to evaluate. For example, stereo location could replace structure. The purpose of this exercise is to improve your memory for the perceived parameters and the aesthetic and artistic elements of sound—any and all of them.

---

5 A System for Evaluating Sound

The many different positions of the audio industry and their unique needs for sound evaluation point to a need for a sound evaluation system that can be readily transferred to a variety of contexts. It must easily yield meaningful and significant information to people of diverse backgrounds and job functions. The method must transfer between musical contexts and abstract, critical listening applications.

The aspects of sound evaluated by people in audio cannot be described using our current vocabulary. No way to talk about sound is available. The system seeks to establish guidelines for talking about sound, by describing the physical dimensions of sound, as they have been perceived. Using knowledge of the physical dimensions and the transformations caused by perception, these descriptions can be objective and accurate.

Outside languages, scientific measurements, and music, sound has no written form. A written form for sound will greatly assist people in audio in discussing sound, studying music recordings, evaluating sound, and keeping records. The system incorporates ways of graphing and notating the perceived parameters and the artistic elements of sound. This will greatly aid the listener in understanding, recognizing, or evaluating sound.

System Overview

The system for sound evaluations was created to supply objective information on the listening experience. It seeks to give the listener the tools to define what is being heard. This will lead to a better understanding of the unique qualities of recorded/reproduced sound, better communication between people discussing sound, and enhanced control of the artistic aspects of making music recordings.

The elements of sound are all evaluated independently, using a variety of techniques. These isolated evaluations may then be related to evaluations of other elements, to observe how they interact. The standard X-Y graph used in so many different scientific contexts has been adapted for many of these evaluations, especially those that take place against time. Other evaluations use unique diagrams such as the sound stage.

The system seeks to describe and define the activities of the five physical dimensions of sound, as they are used in recording production/reproduction. The system examines the changes of state and value of those dimensions of sound, as they appear in perception and in artistic expression. Table 5-1 outlines how the various evaluations of the system relate to the aesthetic and artistic elements or perceived parameters of sound.

Table 5-1
Evaluation Techniques for the Elements of Sound

Element of Sound	*Evaluation Graphs and Processes*
Time	Time Line of Song; with structure, phrase, and text indications
	Sound Sources Against Time Line
Pitch	Melodic Contour
	Pitch Area
	Pitch Density
Dynamics	Dynamic Contour
	Musical Balance
	Performance Intensity versus Musical Balance
Sound Quality	Sound Quality Evaluation
Spatial Properties	Distance Location
	Stereo Location
	Surround Location
	Sound Stage / Perceived Performance Environment
	Environmental Characteristics

The system starts with basic skills and builds on them. Skill in recognizing musical materials and building a time line lead to the development of skills in pitch-related perception and dynamics. Interspersed throughout the system are skills preparing the reader to undertake sound quality evaluations. Finally, skills in recognizing spatial properties and environmental characteristics are addressed.

A complete listing of exercises appears after the Table of Contents. They are arranged in the most effective order for skill development, and the reader is encouraged to work through the exercises in the presented sequence. The exercises appear at the end of the chapters that contain explanations of the material. Some skills will take longer to learn than others, and the reader should be careful in assessing their progress. The assistance of someone that is already a skilled listener or teacher will at times be valuable. Any one exercise should be learned well before progressing too far ahead, though mastery of skills is not necessary before moving ahead. Indeed, mastery of some of these skills might take years, and the reader is encouraged to return to those exercises throughout an extended period of time. The first exercises of Part 1 should be reviewed before continuing with the exercises of Part 2.

Notating or writing down the characteristics of a sound can greatly assist the listener in understanding the sound. These notations (written representations of the sound) can also be used for communicating with others about the sound, for evaluating the sound, remembering the characteristics of the sound, and even for recreating the sound. While the reader will not seek to perform a written evaluation of all sounds, the process of performing a detailed evaluation of the sound event will provide information that would otherwise go unobserved. Notating sound material in graph form will be used for finely developing the reader's sound evaluation skills. It will also provide the reader with a useful resource to assist in evaluating sound.

The system for evaluating sound has much in common with traditional forms of music-related ear training. Some of the skills learned by musicians will transfer to this process. An ability to take traditional music dictation will be beneficial to learning the process of evaluating sound, but is not required. Traditional listening skills emphasize pitch relationships in musical contexts. This comprises a very small part of our concerns about sound in audio. The skills of making time judgments and an awareness of activities in pitch, dynamics, and timbre will need to be developed much further than traditional approaches allow.

Many musicians start their studies by mimicking or repeating music on recordings. Music is often learned by the person listening to recordings and trying to playback what was heard. Many people have even learned to play musical instruments almost solely by listening to recordings. Repeated listening to the same recording is something many people have previously done—whether to learn something or for enjoyment. This experience will be important in the many exercises in developing skill in evaluating sound.

It is important that all information extracted from sound evaluations be objective. Audio professionals need to communicate about the characteristics of sound. Communications about how the sound makes them feel or whether or not they like the sound may come from clients or non-industry people and need to be interpreted into the audio professional's work activities, but the information is not relevant or valuable in the evaluation of sound.

The reader must learn to never use subjective impressions or descriptions of the sound event in the evaluation process. Such impressions are unique to each individual and cannot be accurately communicated between individuals (they mean something different to all people). They do not contribute to an understanding and recognition of the characteristics of the sound event. Subjective impressions or descriptions do not contribute pertinent, meaningful information about the sound, and will not contribute to understanding the characteristics of sound. They have no place in the sound evaluation process.

Sound Evaluation Sequence

The sound evaluation process will follow a sequence of activities:

1. Perception of the element of sound or the activity of material to be analyzed at a defined focus or perspective,
2. Recognizing the material,
3. Defining material, and
4. Observing the states or characteristics present in the material or its activity, or between the element and the musical context.

The evaluation of sound begins with the perception of the sound event or sound object. A sound event can be any sound, aspect of sound, or sequence(s) of sounds that can be recognized as forming a single unit. The event may be at any hierarchical level of musical context or of sound quality analysis—from a distant perspective (such as the shape of the overall piece of music) to a close perspective with a focus on some nuance (such as a small change in the spectral content of timbre). The sound event must, however, have a specific and defined perspective. Each sound evaluation will have its own focus on perspective that must be well defined in the listener's mind.

Next, the listener must recognize and, in some way, identify the sound event. This act is necessary to differentiate the event from the material that precedes it, follows it, or is occurring simultaneously. The sound event will have dimensions within which it exists, and through which it is defined. It will have points in time where it begins and ends. It will be perceived within the musical/communications context or in isolation. The sound event will be defined through an understanding of the unique states and activities of the components of sound (artistic elements or perceived parameters) that comprise the sound.

Whatever the content of the sound event, the listener must perceive the sound event as a single unit that is in need of definition and is capable of being evaluated. This will be accomplished through identifying and recognizing the boundaries within which the sound event exists.

Third, the sound event is defined by the listener through scanning their perception of the sound event, in their immediate and short-term memories. This definition process will seek to compile information on the sound event. This information will be the activities or unique qualities of the materials of the sound event that make it separate and distinct from the materials that preceded it, succeeded it, and/or that occurred simultaneously with the sound event. Defining this activity (calculating what is happening to or in the various artistic elements or perceived parameters) is often the most difficult task of sound evaluation.

Any number of repeated hearings to the sound event will be needed to define all of the information contained in the sound event. The skills required to define the sound event will need to be developed. As the listener acquires greater evaluation and listening skills, the number of hearings required will reduce significantly.

The final step is to seek to make sense of the information that accumulated in defining the sound event. In comparing the information of the components of sound that defined the sound event, meaningful observations can be made. The listener will use both long- and short-term memory to compare sound events recently experienced and events that are well known to the listener. These other sound events are evaluated for their relationship to the defined sound event. The listener will be looking for same, similar, and dissimilar states of activity and other attributes in the other known sound sources, as those that defined the sound source, to assist in making pertinent observations about the sound event. The process of evaluating the sound event will be completed when the listener has compiled enough detailed information to make all necessary observations of the event and its context.

The sound evaluation system is a clear set of routines. It is directly related to the listening experience. The routines follow the order:

1. Identification of desired perspective, with suitable alteration of the listener's sense of focus,
2. Definition of the boundaries of the sound event,
3. Gathering of detailed information on the material and activity, and
4. Observations made from the compiled information.

The perception of the individual sound event allows for the identification of the proper perspective and focus of the listener. The listener consciously decides the level of detail at which the sound event will be evaluated—the perspective. The listener consciously decides to focus attention on a specific component of the sound event. The sound event and its component parts can then be identified, isolated from all other aspects of sound and piece of music. The perspective at which the listener has identified the sound event becomes the reference-level of the hierarchy, or framework, for the individual X-Y graph (discussed below) of the sound event.

Next, the sound event will be defined by its boundaries of states and activities. It is most often defined by (1) when it exists (its time line), (2) its most significant sound elements (providing the event with its unique characteristics—the levels or values of the elements of sound that comprise the sound event, both unchanging and transient), (3) the highest and lowest levels (boundaries) within those sound elements (the extremes of levels of activity to be mapped against the time line, or that do not change over time—levels), and (4) the relationships of how the sound event's characteristics change over time (amount of change and rate of change of levels mapped against the time line).

Defining the sound event will include:

1. Determining the time line: beginning and ending points in time of the event, and identifying the suitable time increment to allow the activity of the components of sound that characterize the event to be clearly presented.
2. Determining which of the elements of sound hold the significant information that characterize the sound event. These are the components that supply the information that defines the unique characteristics of the sound event. These are the components that must be thoroughly evaluated to understand the content of the sound event.
3. Determining the boundaries of the components/elements of sound that characterize the event. These will be maximum and minimum values or levels found in each of the components of sound.
4. Determining the speed at which the fastest change takes place in the components of sound that comprise the event. This will assist in defining the most suitable smallest time increment of the time line.

The third step compiles detailed information on the material and activity. It will add detail to the above step. A listing of the sources (items to be analyzed) will draw the listener into the evaluation process quickly and directly, and it should become one of the very first steps in collecting detailed information for the evaluation of sound.

The components of sound (that characterize the sound event) will be evaluated to determine the precise levels of the various elements (such as pitch, dynamic levels, etc.), and the placement of those levels against the time line (that is plotting or notating the activity of the component parts of the sound event). The components of the sound will be closely evaluated, with as much detail as possible, to determine their precise levels throughout the sound event.

Most often the components of the sound event are transitory (change over time) and must have their levels related to a time line. This information will be plotted on a two-dimensional graph (discussed below); thus allowing the information to be written/notated (so it can assist in this evaluation process and be available for future use). This process involves all of the skills of taking music dictation. In fact, this process is a type of music dictation for some new and some previously ignored aspects of sound.

A written form of the sound (notating the various levels and states of activities of the components of sound in the sound event) will be created through the process of following Steps 1 through 3, above. Graphing the sound event makes it much easier to compile the information that will allow the listener to recognize and understand the characteristics of sound. When sound has been notated, it is possible for the listener to check previous observations for accuracy, to focus on particular portions of the sound event, and for the listener to be able to continue examining information on the sound out of real time.

The final activity in evaluating sound is examining the compiled information to make observations about the sound event. The type of observations made will vary considerably depending on context—a music mix or a microphone technique.

For example, if the observations are being made concerning the functioning of a particular piece of audio equipment, the evaluations will center on the aspects of sound that the particular piece of equipment acts upon. Observations might be focused around the effectiveness of the piece of equipment, the integrity of the audio signal, any differences between the input and the output signals, and how the device acts on the various dimensions of sound.

The listener/evaluator will formulate questions and will use the data compiled in the above steps to answer those questions. Which questions to ask will be determined by their appropriateness to the purpose of the evaluation. The answers acquired through this process will be ones of substance and will be directly related to the sound event. The answers produced by this process will not produce subjective impressions or opinions.

The observations made in this final, evaluation process need not be profound to be significant. Often the simplest, most obvious observations offer the most significant and important information concerning a sound event.

Graphing the States and Activity of Sound Components

The traditional two-dimensional line graph is quickly understood, easily designed, and readily completed by most people. Therefore, it has been selected as the basis for notating (creating written representations of) sound events and sound objects.

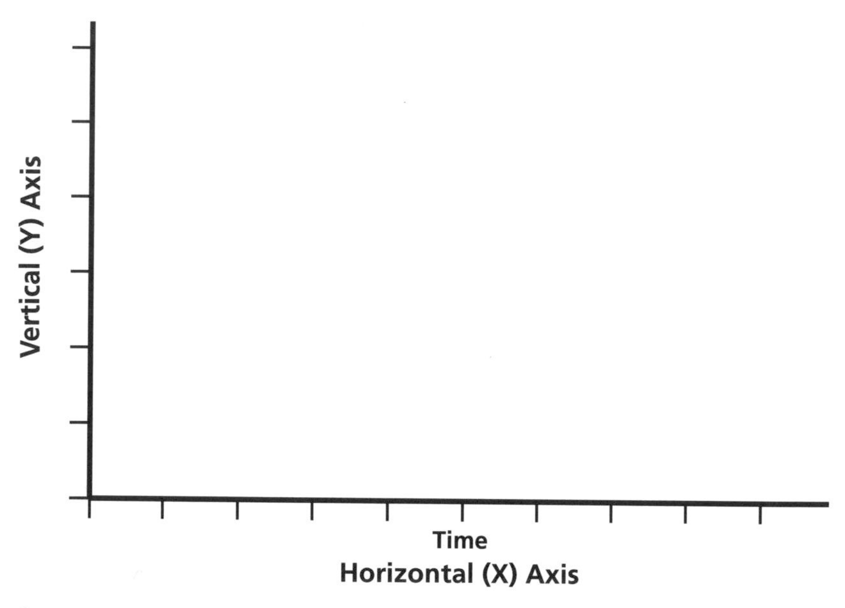

Figure 5-1 X-Y line graph.

The line graph will nearly always be used with time as the horizontal (X) axis. In this way, values of states (levels) of the component parts of the sound can be plotted with respect to time. This allows the sound to be observed from beginning to end at a glance, out of real time.

Time Line

The length of the sound event or sound object that can be plotted on a single graph is dependent upon the selected increments of the time axis, or *time line*. Events of great length (and little detail) may be plotted on a single graph, and events of short duration (and great detail) may be plotted on a single graph. A balance must be found in selecting the appropriate time increment for the time line. The sound should be easily observed in its totality (from beginning to end) and the graph should have sufficient detail to be of use in observing the qualities of the material.

Time increments will be selected for the X-axis that are appropriate for the sound. Time increments will take one of two forms: (1) units based on the second (millisecond, tenths of seconds, groups of seconds, etc.), and (2) units based on the metric grid (individual or subdivisions of pulses, measures, groups of measures). If the sound material is in a musical context, the metric grid will nearly always be the preferred unit for the time axis. Humans judge time increments most accurately with the recurring pulse of the metric grid acting as a reference.

In general, when the sound evaluation utilizes the metric grid, a process of analytical listening is occurring. Critical listening evaluations most often use real time increments and not the metric grid. The difference is one of context and focus.

If the sound material being evaluated is not in a musical context, increments based on the second must be used. It will be common to use increments based on the second in the evaluation of timbre relationships (including sound quality and environmental characteristics). While conceptualizing the pulse of MM:60 (or an integer or a multiple thereof) will provide some reference to the listener in making time judgments without a metric grid, this activity may not always be appropriate. It may distort the listener's perception of the material, and the reference may be unstable, as the listener's attention will rightly be focused elsewhere.

A stopwatch might assist in evaluating larger time units (to the tenths of seconds). The ability to judge time relationships can be developed. It is recommended the reader turn now to the Time Judgment Development Exercise at the end of this chapter. The exercise will, with practice and over time, allow the reader to refine their skills in accurately making time judgments, by learning to recognize the unique sound qualities (timbres) of various time units.

With practice, the listener will develop the ability to make accurate time judgments of a few milliseconds within the context of known, recognizable sound sources and materials. This skill will be invaluable in many of the advanced sound evaluation tasks regularly performed by audio professionals.

The time unit used in the line graph will be that which is most appropriate for the sound event or sound object. The time increment selected must allow the graph to depict the example accurately. The smallest perceivable change in the components of sound being analyzed must be readily apparent, and yet as much material as possible should be contained on a single graph.

Vertical (Y) Axis

The components of sound to be plotted and the boundaries of levels and activities of those components are next determined. In the initial two stages of the sound evaluation process, the listener determines those components of the sound event that provide it with its unique character. These components will be the ones most appropriately evaluated by plotting their activity on the line graph.

The component of the sound event to be evaluated will be placed on the vertical (Y) axis of the line graph. The second step of the sound evaluation process (above) is now followed. The listener will now determine the maximum and minimum levels reached in the sound event, in each of the components of sound to be graphed. These maximum and minimum levels will be slightly exceeded when establishing the upper and lower boundaries of the Y-axis.

Exceeding these perceived boundaries allows for errors that may have been made during initial judgments of the boundaries and allows for greater visual clarity of the graph. Boundaries should be exceeded by 5 to 15 percent, depending on context of the material and the space available on the line graph.

Next, the minimum changes of activity and levels are determined. Through Step 3 of the sound evaluation sequence described above, the listener will determine the smallest increment of level change for the components of the sound event.

This smallest increment of levels will serve as the reference in determining the correct division of the Y-axis. It is necessary for the Y-axis to be divided to allow the smallest value of the component of sound to be clearly represented, just as the X (time) axis of the graph was divided previously so the fastest change of level would be clear.

The division of the vertical axis must allow the graph to depict the material accurately. The smallest significant change in the components of sound being evaluated must be immediately visible to the reader of the graph, and yet the vertical axis must not occupy so much space as to distort the material. The reader of the graph must be able to identify the overall shape of the activity, as well as the small details of the activity of the component the graph represents. A balance between limitations of space and clarity of presentation of the materials must always be sought.

Multitiered Graphs

It is not always desirable for each component of the sound event to have a separate line graph. Many times several components of a sound event can be included on the same graph and plotted against the same time line. *Multitier graphs* allow several components to be represented against the same time line. With the advantage that all characteristics can be more easily related to one another, which will lead to greater understanding of the sound.

The vertical (Y) axis of the line graph is divided into segments. Each segment is dedicated to a different component of sound. Each segment will have its own boundaries and increments.

Plotting a number of components of sound against the same time line not only makes efficient use of space on the graph, it also allows a number of the characteristics of the sound (perhaps the entire sound) to be viewed simultaneously. By placing a number of the components of the sound against the same time line, it is possible to give a more complete and a more easily understood representation of the sound event or sound object.

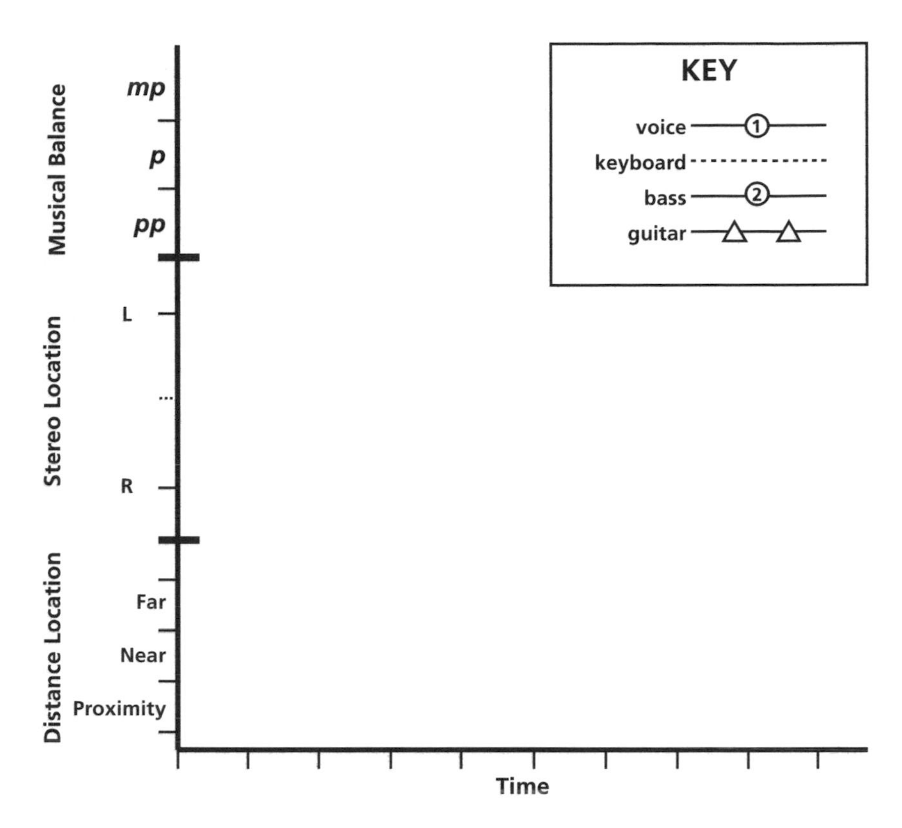

Figure 5-2 Multitier graph.

The person reading the graph will be able to extract information more quickly from a multitier graph than from a series of individual graphs. In addition, when several components of sound appear against the same time line the states and activities of the various components of the sound event/object can be compared in ways that would be difficult (if not impossible) were these components separated.

Specific multitier graphs will be used for certain evaluations in later chapters. In those cases, the graphs will always appear in a predetermined format, and greatly assist evaluations of components such as sound quality, musical balance versus performance intensity, and environmental characteristics.

Graphing Multiple Sound Sources

Multiple sound sources within the same component of the sound will also need to be graphed. It is quite common for more than one aspect within a component of the sound to be taking place at any one time (such as the sound of harmonics and overtones within the spectrum of a sound). This activity would require a separate tier of a multitier graph for each sound source in each component of the sound event/object being evaluated. The line graph would quickly become large and unclear.

As long as the segment of the graph can remain clear, it is possible for any number of sound sources to appear on any graph. When more than one sound source appears against the same two axes, the activities of each sound source must be clearly differentiated from the others. Sound sources may be differentiated in a number of ways. Each of these ways may be useful depending on the situation—what is available to the reader, the nature of the sound, or the context of the sound event.

The lines that denote each sound source may be labeled. The labeling of lines is accomplished by placing a number or the name of each sound source in or near the appropriate line on the graph. This type of differentiation is useful for graphs that contain relatively few sound sources.

Providing a different line configuration for each sound source is sometimes a suitable way of differentiating a number of sources on the same tier of an X-Y graph. Combinations of dots and dashes, or the insertion of geometric shapes into the source-lines may be useful for differentiating sound sources on the same graph—again for graphs with relatively few sources.

When sound sources are assigned lines of different colors, the graph can clearly display the largest number of sources. Only the number of easily recognized colors available then limits the number of sources that can be placed on the same graph.

The use of different colors has the further advantage of being able to define groups of sound sources by assigning a color to the group and assigning a different line configuration (combination of dots and dashes) to the individual sound source.

Using lines of various thickness to differentiate sound sources is not an option. This approach will obscure the information of the graph. Varying line thickness will cause the sound to visually appear to occupy an area of the vertical axis. This is a state that is only accurate for a few select components of sound.

The use of color is not always feasible, but it is the preferred method of placing a number of sound sources on the same graph. Using numbered lines or using varied and distinct line configurations for each sound source are the next, most flexible and clear methods of differentiating sound sources. Combinations of color and line configurations will produce the most organized and most useful graphs. Individual sound sources must always be easily distinguished on line graphs. Readily identifiable lines that have been precisely defined (by using a key, as described below) will ensure the clarity and usefulness of the graph.

The same sound sources may be depicted on a number of tiers of a multitier graph. In this case, care must be taken to define each sound source and to depict the sound sources in the same way on each tier (either by the same number, color, or line configuration). This will allow someone reading the graph to quickly and accurately determine the states and activities of all of the sound sources (or aspects of the sound sources) over time. A *key* of the sound sources plotted should be created to ensure this clarity.

A key is the listing of sound sources of the sound event, coupled with a chart of how the individual sound sources are represented on the line graph (see Figure 5-2). This listing of sound sources with their designations must be included in each line graph that contains more than one sound source (unless the lines are labeled).

The listing of sound sources is one of the first activities undertaken in the entire evaluation process. *Sound sources* are the individual elements of activity within the level of perspective that is the focus of the sound evaluation. A listing of the sources (elements to be analyzed) will draw the listener into the evaluation process quickly and directly, and it should become one of the very first steps in evaluating sound.

Plotting Sources Against a Time Line

Plotting the individual sources against the time line, without concern for levels and rates of activity of the component parts of a sound, will allow the listener to compile preliminary information on the material without getting overwhelmed by detail. This process is also an excellent first step in getting acquainted with the activity of writing down material that is being heard (the taking of dictation). It may become a common initial activity each time the listener undertakes a detailed evaluation of a sound event. A reliable ability to place sources against a time line (and, of course, correctly identifying the time line) will be assumed throughout the remainder of the book. This process will be repeated, at least conceptually, before almost all future exercises. This is also an excellent exercise for learning to identify all the sound sources (instruments and vocals) present in a mix—something that sounds simple and often proves otherwise.

Listing sound sources is an important first step in many evaluations and will need to be undertaken as a first step toward plotting sources against a time line. It is important that all individual sources be identified and listed separately. These individual sources often act independently and were usually recorded with some degree of separation, giving the recordist an independent control of the sound that will be evaluated in many ways. Lists of sound sources should identify all independent vocal parts separately. Groups of background vocals presenting one musical idea should be listed as a single sound source; similarly, groups of stringed instruments playing one line or musical idea would also be labeled as "strings." Instruments should be listed by names. When more than one instrument of the same type is used the instruments should be numbered either by order of appearance, or by range—with the highest instrument usually the lowest number. Sounds should not be listed by descriptive terms (lush guitar, happy flute, etc.). If the listener is at a loss as to what to call a sound, using terms such as "unknown 1" would be appropriate. Performing a sound quality evaluation (even a general one) would allow the listener to further define the sound as "unknown 1, with long final decay." When the listener does not know the names of instruments, or the sound sources are very unique, listing sound sources must be undertaken with care, and will take effort. For example, it would be a great undertaking to list the sound sources from "Tomorrow Never Knows" by The Beatles. Many of the song's sound sources would need to be described in terms that addressed the sound source or the sound quality.

As a common activity, the reader should (1) create a listing of the sound sources in the sound event, (2) create a time line of the event, and (3) plot the listed sound sources against the time line. The reader should practice Exercise 5-2 (at the end of the chapter), and become comfortable with the process.

Figure 5-3 provides an example of sound sources plotted against a time line. The listener will be able to follow the figure while listening to The Beatles' "She Said She Said." The recordist must be able to quickly recognize sound sources and focus on the activity of each. As an additional activity, complete the graph by plotting the presence of the high hat part against the time line.

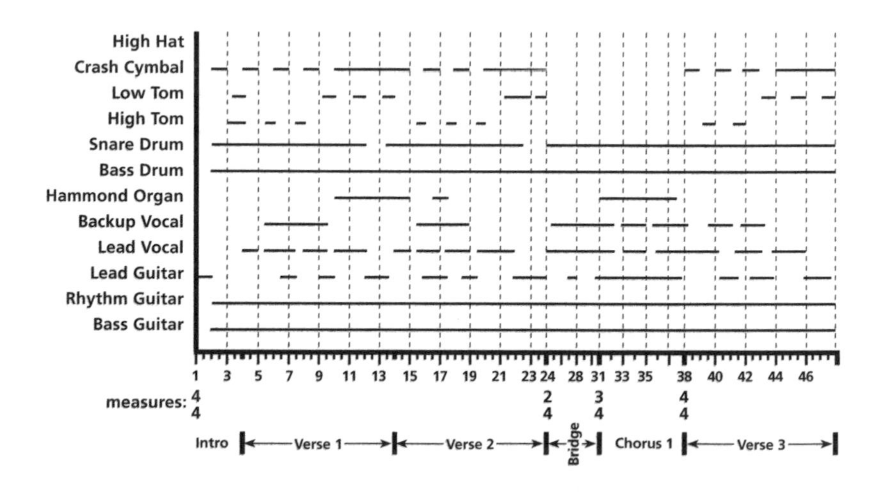

Figure 5-3 Sound sources placed against time line, The Beatles: "She Said She Said.".

Notating Sounds in Snapshots of Time

Some important components of sound might not change over time. While they are static, their status may be a very important characteristic of the sound event/object. These sound components can be evaluated without a time line, as states that exist throughout a defined time period. Sounds are examined for the states of their component parts that remain unchanged over the time period, as if examining a snapshot of the sound's existence over the time period.

The location of instruments on a sound stage is one such possibility. While it is very possible for sound sources to change stereo location or distance location, sound sources often remain in the same location for extended periods of time (entire verses or choruses, for example). It is also common for some sound sources to remain in the same location throughout an entire song.

The graphs used for plotting the components can also be used to show the static, unchanging states. The X-axis can be used to provide a place for each sound source that is present, instead of a time line. Many sources can then be placed against the same vertical axis. In addition, the diagrams of the perceived performance environment can also be used to show placement of sound sources and the dimensions of the sound stage, when appropriate (as will be discussed in Chapter 9).

Summary

Graphs will often need to be supplemented by verbal descriptions to complete evaluations of sound events or sound objects. Simply graphing a sound does not complete the evaluation.

The contents of the graphs need to be reviewed, and then described through observations of the content and activity of the materials. In all instances, the language and concepts used to define and describe the sound must be completely objective in nature. All descriptions refer to the actual levels and any changes of levels, of the components of the sound. The descriptions lead to overall observations about the qualities of the sound itself, and its relationships to the other events/objects and the entire recording.

The listener must never use subjective impressions or descriptions of the sound in the evaluation. Such impressions are unique to each individual, cannot be accurately communicated between individuals (they mean something different to all people), and do not contribute to an understanding and recognition of the characteristics of the sound. Subjective impressions or descriptions do not contribute pertinent, meaningful information about the sound.

Evaluations or descriptions that are based on an individual's personal perceptions or impressions about a sound cannot communicate concepts and ideas that will contribute to an understanding of the characteristics of sound. While they may be important in another way to the individual, they have no place in the evaluation process that seeks to discover and define information about the characteristics of the sound. Subjective observations may, however, be important to artists or listeners. These are different concerns and must be approached as such. Often these observations are based on feelings, moods, emotions, or personal interpretations of any other sort. These matters will not be addressed in this book.

The next five chapters will take the reader through the artistic elements and perceived parameters of recordings, working through the system for evaluating sound. The sequence of activities outlined above, and the process of notating (graphing) sound will be core activities of those chapters.

In order to benefit the most from these chapters, the reader should take the time to work carefully through the examples and exercises. The many nuances of sound contained in recordings will gradually become more apparent, and the reader will gradually acquire an ability to talk about sound in ways that have clear meaning to others. Further, the reader will find they are better able to understand their own recordings, and better able to approach recording projects creatively.

Exercises

The following exercises should be practiced until you are comfortable with the material covered.

Exercise 5-1
Time Judgment Exercise.

Using a digital delay unit and a recording of a high-pitch drum (or a drum machine):

1. Route a nondelayed signal to one loudspeaker and a delayed signal to the other loudspeaker.
2. Delay the signal and perform many repetitions of the sound while changing time-increments that are easily recognized. Always move by the same time increment, such as 100 ms, during any given listening session.
3. As confidence is obtained in being able to accurately judge certain time units, move to other time units—both smaller and larger—and repeat the sequence in Step 2.

4. When control of time relationships is accurate within certain defined limits and you are confident in this ability, test that accuracy by routing both the direct and delayed signals to both loudspeakers (or to a single loudspeaker).
5. Continue to work through many repetitions of time increments in a systematic manner, comparing the qualities of the time relationships of each listening to previous and successive material, in a logical sequence (a suggested pattern/sequence: 150 ms, 125 ms, 100 ms, 75 ms, 125 ms, 150 ms).
6. Continue moving to smaller and smaller time units, until consistency has been achieved at being able to accurately judge time increments of 3 to 5 ms.

Being able to recognize small time units—such as several milliseconds—will take considerable practice. It is, however, quite possible to develop skill in recognizing the sound quality of these time units when they are presented as delay times of repeated sounds. The different time delays will have distinctly different sound qualities, once you are able to recognize and remember the individual qualities. These time delays will be heard as "timbres of time." Time units will actually have a characteristic sound, as they transform the sound quality of the reiterated drum sound. These unique sound qualities of sound reiterations will transform all other sounds in a similar way.

Exercise 5-2
Exercise for Plotting the Presence of Sound Sources Against the Time Line.

Graph the first few major sections (verse, chorus, etc.) of a piece of popular music, using the following steps:

1. Compile a list of all the sound sources of the song. Individual percussion sounds and vocal parts should be listed separately.
2. Create a suitable time line by:
   a. Determining the pulse (metric grid) of the song;
   b. Grouping the pulses into measures (weak and strong beats); and
   c. Plotting those measures on the horizontal (X) axis in increments that clearly show the material being graphed.
3. Plot the individual sound sources against the time line. Each sound will have its own location on the vertical axis, making it unnecessary

to make distinctions between the lines of each sound source on the graph. When an instrument is playing, place a line for that instrument in the appropriate location against the time line. If the instrument is playing in the measure, extend the line through the entire measure. Alternately, you can change this resolution to make note of instruments appearing every half measure. If still more detail is sought, a smaller time increment could be used.

4. After several days return to the graph. Check the time line for accuracy and listen several times again for sound sources. It is not unusual for sound sources to appear in the music that were not heard in previous hearings.
5. Listen several more times to check the entrances and exits of the instruments against the time line.

6 Evaluating Pitch in Audio and Music Recordings

Pitch relationships shape musical materials more than the other artistic elements. This is obvious when we consider melody as a succession of pitch relationships, and chords/harmony as simultaneously sounding pitch relationships.

With the exception of percussion instruments, nearly all musical instruments were specifically designed to produce many precise variations in pitch, far fewer variations in timbre, and a continuously variable range of loudness. Most Western music places great emphasis on pitch information for the communication of the musical message. Pitch is the central artistic element of most music, with the other artistic elements most often supporting the activities occurring in pitch relationships.

In evaluating pitch in audio and music recordings, the recordist will work well beyond traditional concepts of pitch as melody and harmony. An acute sense of pitch will bring the recording professional to recognize pitch levels, and to identify pitch areas and frequency bands.

Pitch evaluation is used in both analytical and critical listening. In critical listening perceived pitch is often transferred into frequency calculations. The analytical listening process relates pitch relationships to the musical ideas and message.

The reader will be developing ways of analyzing musical sounds that will also develop critical listening skills. The same sensations of sound are perceived in each process. How the sound is evaluated (critically versus analytically) will be the difference.

The pitch analysis concepts presented in this chapter are of particular importance to sound recordings and for developing the skills of the recording professional. The information gathered will help one to understand the piece of music (analytical listening), and to evaluate sound quality (critical listening), depending on how the information is applied.

Analytical Systems

Numerous analytical systems have been devised to explain pitch relationships in music. These systems are made up of evaluation criteria that vary considerably between systems and are more or less specific to certain types of music. Generally, then, these systems can only be useful for examining certain styles or types of music. Any single analytical system may or may not yield pertinent information to the music being evaluated.

The recordist will need to recognize and apply the appropriate analytical system to study the pitch relationships of a particular piece of music, if such an evaluation is expected of them. This is rarely required. Many recordists do, however, innately sense the relationships that are explained through musical analysis. Understanding pitch relationships, and especially harmonic progressions, can often greatly assist the recording professional in crafting a recording.

Information about the artistic element of pitch levels and relationships will be related to (1) the relative dominance of certain pitch levels (which relate to tonal/key centers and chord progressions), (2) the relative register placement of pitch levels and patterns (related to arranging and orchestration, and will be later discussed as pitch density), or (3) pitch relationships: patterns of successive intervals (motives and melodies), relationships of those patterns, and patterns and relationships of simultaneous intervals (chords and progressions).

Study of the many systems used to analyze pitch relationships is well beyond the scope of this book. Theories about music attempt to explain the analytical listening experience, and to extract basic information about the structure and form of the music. The recordist can use such insights to control and craft the sound qualities of the recording so they will support and directly enhance how the recording process can best deliver musical ideas to the listener.

Realizing a Sense of Pitch

Everyone has an *internal pitch reference*. This is a sense of pitch level that is present unconsciously within each individual. This reference can be brought to the consciousness of the recording professional. This will, however, require focused attention and concerted efforts over a period of time. The skill acquired will be well worth the time and effort involved.

Every individual is different and has a unique internal pitch reference. The things that make us unique human beings likely also contribute to our unique sense of pitch/frequency. What goes into making this reference is complex and varies between individuals. It is often related to the timbre of certain sounds and uses our exceptional ability of remembering the correct sounds of a particular sound source. Therefore, the reference itself is often related to an instrument one has played for a length of time or to the sound and/or feel of one's own voice. Some people identify strongly to certain pieces of music and can use their memory of the pitch level (tonal center) of that piece of music as a reliable reference.

The process of realizing and then developing one's unique sense of pitch can seem perplexing, daunting, or simply impossible. This is something that may at first seem beyond human capability, simply because it is beyond our own experience. This is the first of the skills we will work to acquire that requires the reader to make a leap of faith; to believe something is possible is the first step toward making it so. One must become confused and grapple with the confusion in order to learn.

The reader should work through the Pitch Reference Exercise at the end of the chapter over a period of several weeks. Daily attention with a number of 5-, 10- or 15-minute works sessions will yield results quickly. The reader should always try to find a quiet location where they will not be distracted.

This exercise will bring the reader to develop a consistent and reliable sense of *relative pitch*. This relative pitch may change by 5% to 10% depending on mood, energy level, distractions, or countless other factors. Still the core of the individual's sense of pitch will be present, and can be relied upon for specific tasks or for general use. For tasks that require precision, periodic checks of the reference level for accuracy may be necessary especially in the beginning. As with any skill, the more it is used the greater it will be refined. Once the skill is acquired, refining and maintaining the skill of remembering a reference pitch level can become intuitive.

Recordists have cultivated this sense of pitch to a reliable reference for many practical uses. As examples, it is commonly used to identify frequency levels (such as what is required to immediately determine an appropriate equalization [EQ] setting). It is also used to keep performers playing in tune or to keep tuning constant throughout a project, especially for an ensemble without a keyboard. All of the judgments the recording professional makes related to pitch and frequency will be enhanced with a stable sense of pitch/frequency gained through understanding one's own internal pitch reference.

Recognizing Pitch Levels

The internal pitch reference is an important aid in recognizing frequencies and pitches throughout the hearing range. This reference will now be used to help identify the general locations of frequencies and pitches in relation to carefully devised registers. Recognizing and understanding pitch and frequency information is central to many routine tasks of recording, such as evaluating the performance of a microphone on the sound of a voice.

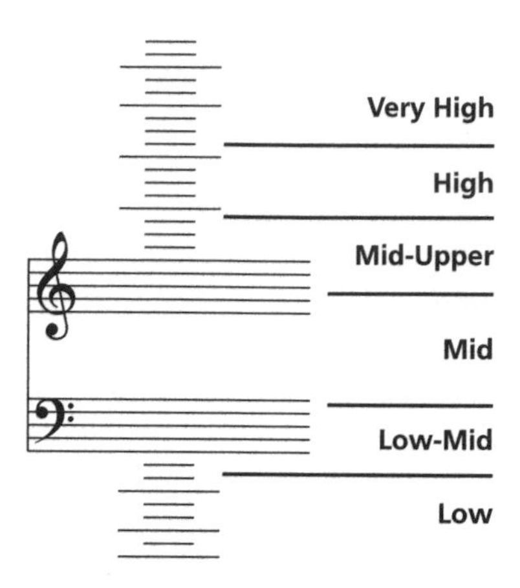

Figure 6-1 Pitch registers.

Critical listening and analytical listening define perceived pitch differently. Frequency estimation through pitch allows for critical listening observations, while the same sound may need to be translated into pitch relationships to understand analytical observations. The *pitch/frequency registers* (Figures 6-1 and 6-2) will be used to estimate the relative level of the pitch material and to allow the information to be directly transferred between these two contexts.

Frequency estimation is a fundamental skill that must be developed by the audio professional. The use of pitch/frequency registers will assist the reader in identifying perceived pitch and frequency levels. These registers will serve as reference areas and will provide a basis for a general description of perceived frequency and pitch levels. The registers will be used in many of the evaluation processes that follow and should be committed to memory. They will provide meaningful reference levels for many listening activities. Further, learning the frequency equivalents of pitches will also be invaluable to the reader, both in these studies and in the practice of recording.

Register	Pitch Range	Frequency Range
LOW	up to C_2	up to 65.41 Hz
LOW-MID	D_2 to G_3	73.42 to 196 Hz
MID	A_3 to A_4	220 to 440 Hz
MID-UPPER	B_4 to E_6	493.88 to 1,318.51 Hz
HIGH	F_6 to C_8	1,396.91 to 4,186.01 Hz
VERY HIGH	C_8 and above	4,186.01 and above

Figure 6-2 Pitch and frequency ranges of registers.

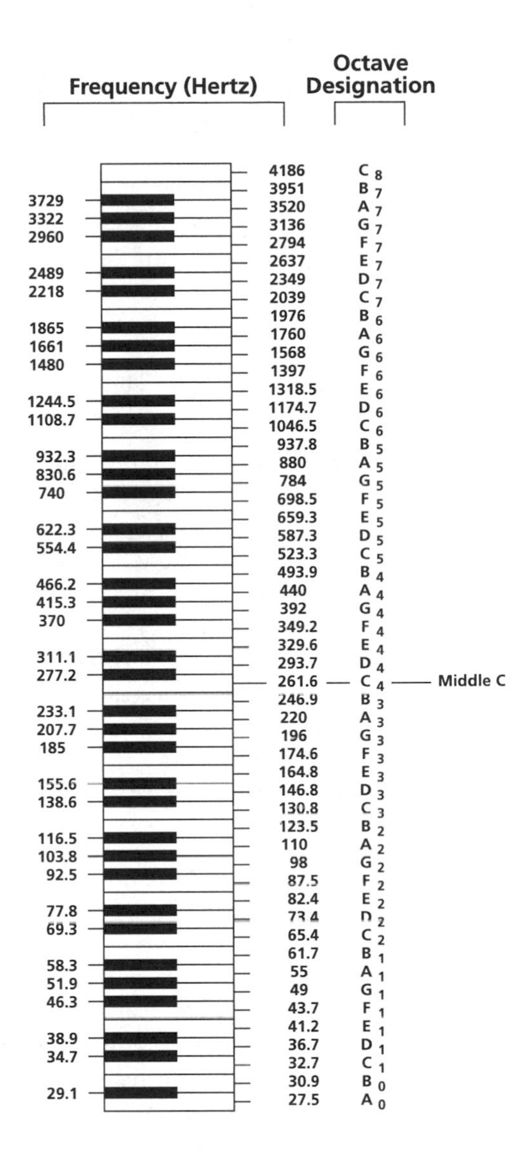

Figure 6-3 Keyboard with pitches, octave designations, and equivalent frequencies.

It will be helpful to relate this material to actual sound sources. The ranges of human singing voices stretch from the *low-mid* and *mid* registers (male voices) to the *mid* and *mid-upper* registers (female voices). Most musical activity occurs in the *mid* and *mid-upper* registers. This is where many instruments sound their fundamental frequencies, and where most melodic lines and most closely spaced chords are placed in musical practice. Take a moment to notice the frequency ranges spanned by these registers.

The sibilant sounds of the human voice occur primarily in the *high* register, typically around 2 to 3 kHz. Within the *very high* register, humans have the ability to hear nearly two and one-half octaves. While this register is not playable by acoustic instruments and by human voices, much spectral information is often in this register.

The reader should work through Exercise 6-2 (at the end of the chapter) to begin developing skill at estimating pitch-levels and octave placements. This is designed to take the reader from making general judgments to identifying precise pitch levels. The boundaries between registers are purposefully large to give the reader a sense of moving from the general to the specific and to acquire the skill with meaningful successes along the way.

This skill is central to many of the evaluations commonly performed by all people in audio and will be used throughout the remainder of this book. The listener can easily transfer perceived pitch-level into frequency estimation through using the above registers. The reader should practice transferring various pitch levels to frequency and the reverse.

It is possible for the experienced listener to consistently estimate pitch/frequency-level to within an interval of a minor third (one-quarter of an octave). After considerable practice and experience, and gaining a clear sense of an internal reference pitch, even greater accuracy is possible. Within several weeks of thoughtful effort, the reader should be consistent within a perfect fifth (a bit over one-half an octave), and accuracy will continue to increase at a rapid pace with regular study.

Pitch Area and Frequency Band Recognition

Recordists must often bring their attention to a specific range of frequencies, or frequency bands, to identify some aspect of sound quality or equipment performance. This section will present a rough equivalent in musical contexts that can be used to develop this skill and more.

Many percussion-related sounds occupy a *pitch area*, not a specific pitch. These sounds are perceived as existing in an area between two boundaries. The boundaries may, at times, be unstable and changing in pitch and dynamic level, and secondary pitch area(s) may also be present.

These sounds are evaluated and defined by (1) the density (amount) of pitch information within the pitch area, (2) the width of the pitch area (the distance between the two boundary pitches), and (3) the presence of secondary pitch areas and the dynamic relationships of those areas.

Some sounds will have several, separated pitch areas. The different pitch areas of the sound will be at different dynamic levels and have different densities (the amount and closeness of spacing of the pitch-information within the pitch areas). One pitch area will dominate the sound and be the primary pitch area. The other pitch areas will be secondary pitch areas.

Evaluating the pitch areas of several percussion sounds will develop a number of important listening skills. Frequency and pitch estimation (recognition) will be refined, and the listener will take the skills of the last section to a more detailed level of perspective. The focus will now be on identifying pitch/frequency levels within the spectrum of the source—using sound sources that allow us to approach this information in a more noticeable and higher level of perspective. This is a first step toward identifying subtle aspects of sound quality (that will be greatly refined later). Further, this study will be accomplished without the added tasks of a time line, as the graph will sum spectral information throughout the sound's duration. The reader will also make some rough observations on dynamic levels. This skill will also be greatly refined later.

It is possible that pitch area analysis will be unlike anything the reader has done in the past. Few reference points will be available for the listener to draw upon, other than the pitch estimation skills acquired above. Difficulties and frustrations are to be expected, along with great satisfaction with acquiring a very useful skill that will be used and improved throughout one's audio career.

To perform an analysis of pitch areas, sounds are plotted on a *pitch area analysis graph*. The graph incorporates:

1. The register designations for the Y-axis;
2. A space on the X-axis of the graph is dedicated to each sound instead of the passage of time (since this evaluation sums all information during the sound's duration);
3. The pitch areas are boxed off in relation to these two axes;
4. The density of each pitch area designated by a number within each box that relates to a relative scale from very dense to very sparse; and
5. Assigning a number to the relative dynamic levels of pitch areas, especially important to identifying the predominant (primary) pitch area (dynamics will receive detailed coverage in the next chapter).

Figure 6-4 presents the pitch areas of the prominent bass drum sound found in The Beatles' "Come Together" (*1* version) at 0:34.

The percussion sound is comprised of four pitch areas. The primary pitch area is the second from the bottom. It is moderately dense and is the loudest and dominant area. The density of the lowest pitch area is a bit more dense, but considerably softer. The area between 167 and 265 Hz is moderately dense and a bit louder than the lowest area, and not as loud as the primary pitch area. The highest pitch area (approximately 315–395 Hz) has a rather sparse density and is at the lowest loudness level of all four pitch areas. It is interesting to note that the lower boundaries of the

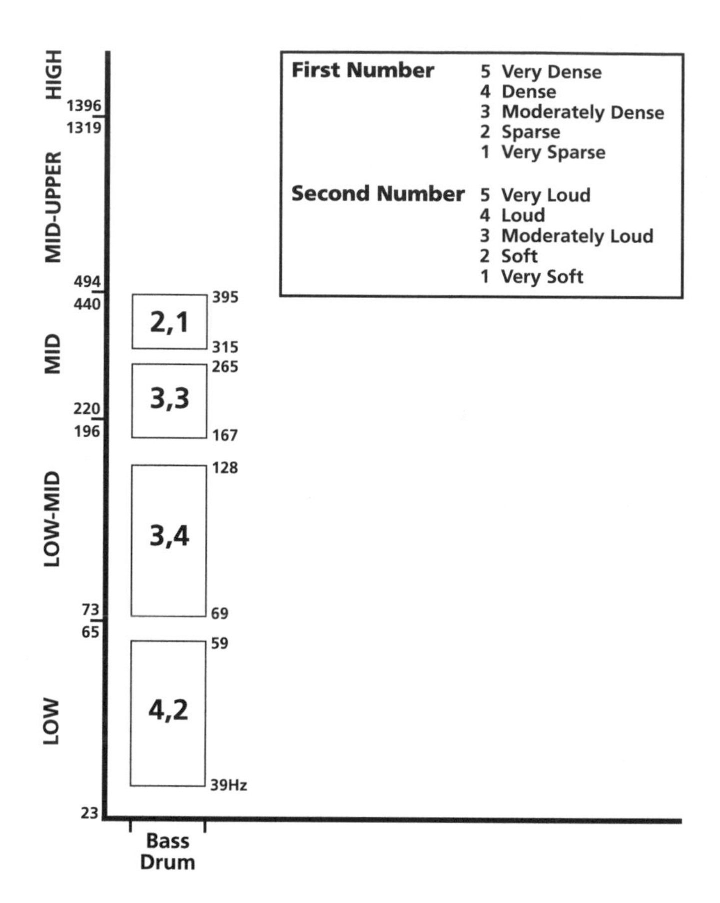

Figure 6-4 Pitch area analysis of the bass drum sound from The Beatles, "Come Together."

four pitch areas are nearly whole number multiples and harmonically related, but far enough away to create strong noise elements. These types of relationships are common for drum resonances and head vibrations.

The pitch area graph and the objective descriptions of density and dynamic relationships of the pitch areas, provide much useful and universally perceived information about the sound. Meaningful communication about this sound is possible with this information.

Next, the Pitch Area Analysis Exercise (Exercise 6-3 at the end of the chapter) should be performed until the material is learned. The reader is encouraged to evaluate drums and cymbals of a variety of sizes directly from music recordings.

A number of percussion sounds from the drum solo of The Beatles' "The End" appear in Figure 6-5. The reader can determine the dynamic relationships of the pitch areas and observe their densities. Study the example carefully, seeking to identify the boundaries of the pitch area,

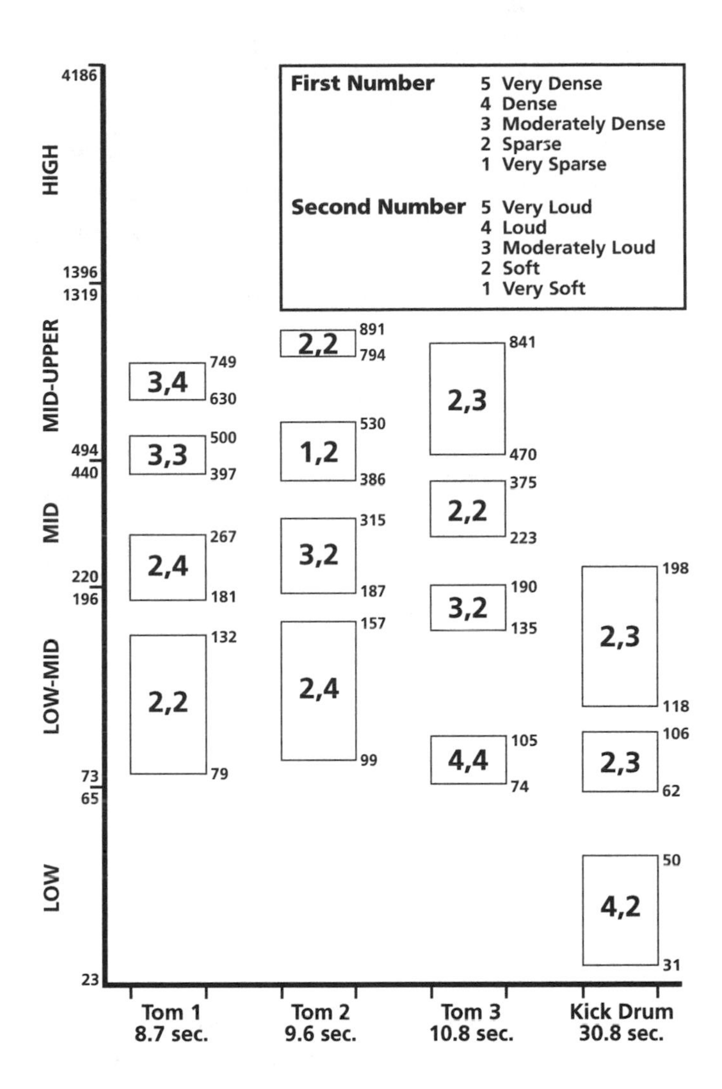

Figure 6-5 Drum sounds from The Beatles, "The End."

and confirm the density information presented. Once the pitch areas can be identified, the reader can observe the dynamic relationships of the pitch areas.

The reader will be able to apply the skills and concepts of pitch area to the recognition of *frequency bands* in critical listening applications. The information for evaluating the states and changes of frequency levels will be directly transferred from the perceived pitch information. The skills gained through the recognition of pitch areas and frequency bands will later be directly applied to the evaluation of timbre and sound quality. In fact, these pitch area analyses are rudimentary timbre analyses.

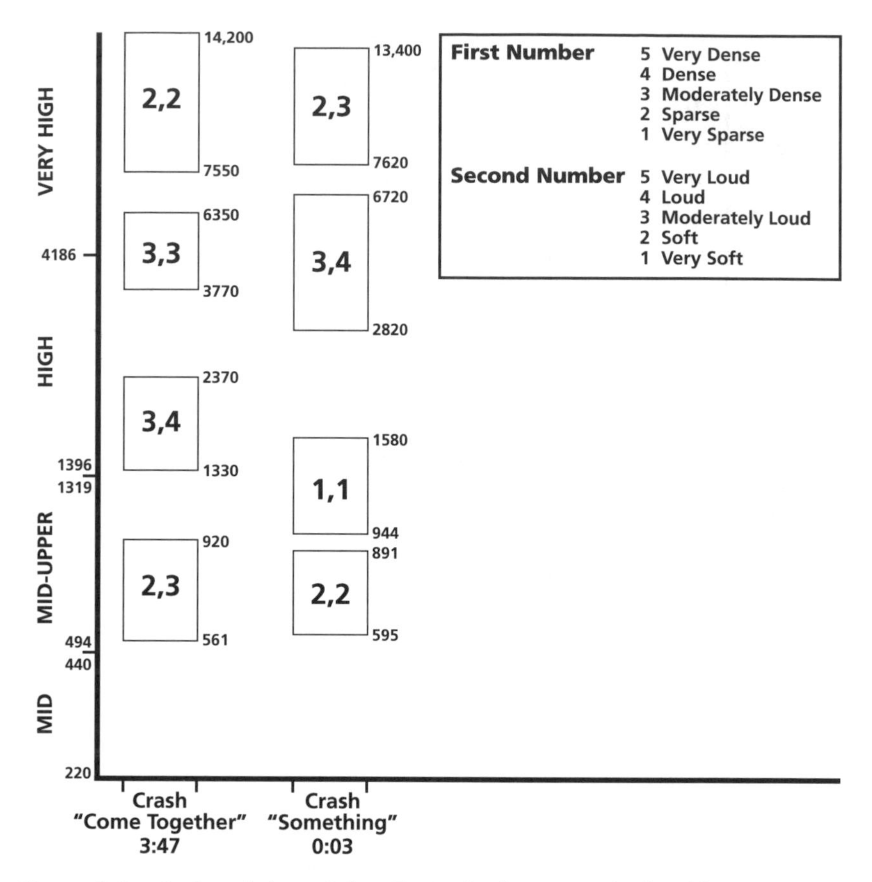

Figure 6-6 Crash cymbal sounds from The Beatles, "Come Together" and "Something" (both from *1*).

A common critical listening situation is to compare two sounds. Figure 6-6 compares the crash cymbal from two different songs. Notice the characteristics of the sounds through careful listening to identify pitch areas and their densities. From the graph we can see the areas are similar in location and nearly identical in their densities. Some areas have markedly different relative dynamic levels.

As skill increases, the listener will gradually come to understand and recognize slight differences present in the spectra of similar sounds. Eventually the reader will arrive at the point where evaluations of the high hat and crash cymbal sounds from three different "Let It Be" releases can be compared (listed in the Discography). The reader is encouraged to listen carefully to these sounds. The sounds change within each song and are slightly different in each recording. The reader will notice many striking differences and perhaps a few of the many subtleties. These differences will gradually seem more and more significant. Ultimately, the reader will recognize the sounds as very different and will perceive the details of their sound qualities.

Melodic Contour

Graphing melodic contour will assist the recordist in a number of ways. First, at the early stages of the listening skill development, graphing melodic contour will help develop skills in placing pitch/frequency changes and levels against time. These same skills will later be used in the much more detailed (and initially more difficult) task of evaluating pitch-related information in timbre/sound quality and environmental characteristics evaluations.

Learning to plot melodic contours against a time line will be productive in developing the skills of recognizing pitch levels, of perceiving metric units and rhythm of time, and of mapping pitch contours. These skills directly transfer into many of the listening functions of the audio professional.

Second, recognizing the contour (or shape) of the melodic line is important to understanding certain pieces of music or musical ideas. In certain pieces of music, the contour of a melodic line is perceived instead of the individual intervals. When the melodic line is performed very rapidly (as is easily accomplished with technology), the perception of the line fuses into an outline or shape. The series of intervals that comprise the melodic line are not perceived. The contour of the melodic line is instead perceived.

The melodic contour graph allows the contour of the musical idea to be recognized and evaluated for its unique qualities.

Graphing Material Against a Time Line

Nearly all exercises in the book graph material against a time line, and the recording professional is continually engaged in listening to material to notice changes in sound over time. The process for defining the activities of any artistic element against a time line follows. This sequence of activities will be applied to both musical (analytical) and critical listening contexts. It will be only slightly modified for each exercise in the following chapters, and the reader should become familiar with the order of activity:

1. During the first hearing(s) of the material, focus listening activity to establish the length of the time line. At the same time, notice prominent activity of the material (in this case melodic contour) and its placement against the time line.

2. Check the time line for accuracy, and make any alterations. Establish a complete list of sound sources (instruments and voices), and then place those sound sources against the completed time line.
3. Notice the activity of the material being graphed for boundaries of levels (here, the highest and lowest pitches of the example, and the smallest changes between pitch levels) and speed of activity (noting the fastest changes of levels). The boundary of speed will establish the smallest time unit required to clearly show the smallest significant change of the material (melodic contour). The boundaries of levels of activity will establish the smallest increment of the Y-axis required to plot the smallest change of the material. This step will establish the perspective of the graph, which is the graph's level of detail. The Y-axis should allow some space above and below the highest or lowest level for the material to be clearly observed, and for the possibility of adding new material or corrections in that area of the graph.
4. Begin plotting the activity of the material (melodic contour) on the graph. First, establish prominent points of reference within the activity; these reference points might be the highest or lowest levels, the beginning and ending levels, points immediately after or before silences, or any other points that stand out from the remainder of the activity; place these levels correctly against the time line. Use the points of reference to calculate the activity of the preceding and following material. Alternate focus on the contour, speed, and amounts of level changes to complete the plotting of the activity of the artistic element. The evaluation is complete when the smallest significant detail has been heard, understood, and added to the graph.

Listening and Writing

Many listenings will be involved for each of the above steps. Each listening should seek specific new information and should confirm what has already been noticed about the material. Before listening to the material, the listener must be prepared to extract certain information. Listeners must have a clear idea of what they will be listening to and/or for, and work to keep from being distracted. Attention should be focused at a specific level of perspective and on a specific task. The listener must both confirm their previous observations and be receptive to new discoveries about the example. Previous observations will be checked often, while seeking new information.

Listening to only small, specific portions of the example may assist certain observations at certain points in the evaluation process. In these situations, the listener should intersperse listenings to the entire example with the rehearings of small sections, to be certain consistency is being maintained throughout the evaluation and to maintain proper perspective.

The reader should very rarely write observations while listening. Instead, the reader should concentrate on the material, and attempt to memorize their observations. This activity will develop auditory memory and will ultimately greatly reduce the number of hearings required to evaluate sounds. When the recording/sound has stopped, the reader should recall what was heard and only then write. If the material is not clearly remembered, listen again—perhaps to a shorter segment.

One must first recognize what has been heard before it can be written down. This process is about recognizing what is heard and making a written record of the experience. The reader should try to make the sequence "hear, recognize, remember, write" automatic.

Melodic Contour Graph

Melodic contour information is plotted on a *melodic contour graph*. The graph incorporates:

1. Pitch area register designations for the Y-axis, using the registers needed to clearly show the activity of the musical example;
2. The X-axis of the graph is dedicated to a time line divided into appropriate time units of a metric grid or real time (depending on the material and context);
3. Each sound source is plotted as a single line against the two axes (the melodic contour is the actual shape of this line); and

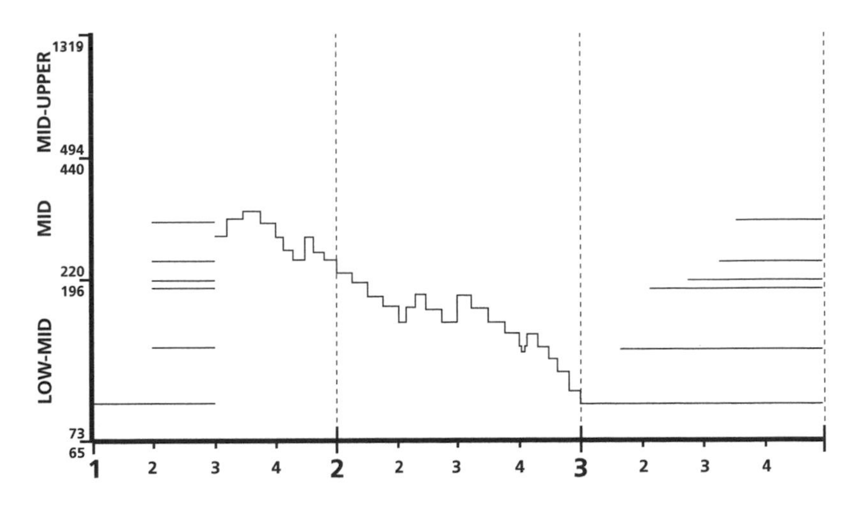

Figure 6-7 Melodic contour graph—The Beatles, "Wild Honey Pie."

4. If more than one sound source is plotted on the same graph, a key should be devised and included with the graph to identify the lines with the sound sources.

Consider the melodic contour of the classical guitar line from *The Beatles* (the "White Album"), that concludes "Wild Honey Pie" and provides transitional material to "The Continuing Story of Bungalow Bill" (Figure 6-7). The melodic line is performed too quickly to be heard as a melody comprised of intervals. Instead, this type of fast melodic gesture fuses into a shape or a contour. As an exercise, check Figure 6-7 for accuracy of shape, detail, and register placement. Determine the tempo of the passage and attempt to identify some of the graphed pitch/frequency levels.

The Melodic Contour Analysis Exercise (Exercise 6-4) will refine this skill. The reader is encouraged to work through several additional recorded examples to become comfortable with this activity.

Melodic gestures of this type are very common in music. They commonly appear from heavy metal guitar solos, to eighteenth- and nineteenth-century keyboard music (especially the works of Chopin, Liszt, and works in the Rococo style). Transcribing fast melodic passages into melodic contour graphs will provide the reader with practice in developing the following skills required of recordists: pitch recognition (estimation), recognizing and calculating pitch changes, and in placing pitch changes against a time line. These skills will be used and further developed in evaluating sound quality and environmental characteristics in later chapters.

Exercises

The following exercises should be practiced until you are comfortable with the material covered.

Exercise 6-1
Pitch Reference Exercise.

Development of a personal pitch reference requires daily attention with a number of 5-, 10-, or 15-minute work sessions. With focused effort this exercise will yield results quickly.

1. First you will need to become aware of your personal pitch reference. You may most likely be able to accomplish this by:

   a. Vocalizing or singing freely to find the pitch-level where your voice causes your chest cavity to resonate. This will require you to pay close attention to the sensations of your body for a fullness to develop in the chest cavity. The voice should then be swept through your comfortable singing range to find the pitch that creates the greatest fullness. That level of greatest fullness would be your pitch.

b. Paying close attention to your natural speaking voice, speak carefully, but deliberately, and try to remove any stress from your voice box. Pay close attention to your inflections and notice when you are causing the pitch of your voice to go up and down. Finally, bring your attention to identify the pitch of your voice where it is operating without resistance, and where it is not being forced up or down in pitch. In this monotone, if properly produced, is the natural pitch of your voice. This can often serve as a reliable pitch reference.

c. People who have played an instrument long enough will often have certain pitches in memory that are related to that instrument (especially when playing it). Guitarists often know the sound of an "E"; trumpet players a "B-flat"; violinists an "A"; and so forth. You can make use of this pitch reference if you have this experience. Think carefully about the tuning pitch of your instrument or some other pitch level you feel drawn to, and clearly identify the pitch. Now play the pitch, listen to it carefully, and work to internalize it.

d. Other pitch references are possible, such as drums or specific pieces of music. These are unusually unique to the individual. At times these can be easily identified or refined.

2. If your pitch reference is based on your voice, make a note of the pitch level you identified. Repeat the process of identifying this level four or five times over two or three days to validate the level. You will eventually identify a specific and consistent pitch level.

3. Now, listen to your reference pitch often throughout several days. A few times per day, stop your normal activity to sing, play, and listen to that pitch level. Use a piano, pitch pipe, tuning fork, or another instrument to play your pitch. Sing it frequently and become accustomed to the placement of that pitch in your voice. Work to bring yourself to hear the pitch in your mind before singing it. With this, you are bringing your sense of pitch into your consciousness.

4. Now, try to consciously carry your reference pitch with you throughout the day. Take five minutes at four or five set times throughout the day to sing your pitch level and check it for accuracy. Before singing, quiet your thoughts and bring your attention to your

voice or to your memory of the pitch. Do not sing a pitch unless you are confident you have a memory of the pitch. If you do not have a memory of the pitch, return to Step 3 for more practice. Sing the pitch in your memory and check it. Don't allow yourself to get frustrated by wrong pitches. Everyone will make mistakes at this stage. By evaluating your mistakes you can learn from them. Keep a record of the pitch you sang—high or low, size of interval off, etc. Look for patterns and try to identify what caused the errors.

5. Remember, this exercise will greatly enhance a skill you will use throughout your career in audio.

Exercise 6-2
Pitch Level Estimation Exercise.

1. Working with a keyboard instrument or a tone generator, practice listening to the boundaries of the registers. Seek to remember the sounds of those boundaries, and remember the pitch names and frequency levels of the several pitches that make up those boundaries.
2. At this point it will be helpful to work with another person. While this other person performs pitches at the boundaries between registers (on a keyboard or other device or instrument), identify those boundaries. Maintain a record of your mistakes so that you can evaluate them and make adjustments.
3. Once confidence has been established in recognizing the general areas encompassed by the registers, begin playing individual pitches against the pitch registers. Identify the pitch register of the sound.
4. Finally, work to identify where pitches are sounding within the registers (i.e., the upper third of low, or the lower quarter of mid). Throughout these steps, you must rely solely on your memory of the pitch/frequency registers in making these judgments.

Exercise 6-3
Pitch Area Analysis Exercise.

Identify an instrument you want to evaluate—most people find drums are easiest in the beginning. A single strike of the instrument being evaluated should be found that is exposed (has no or very few other instruments playing simultaneously). Evaluate that single performance of the sound only, summing all frequency information that occurs over the sound's duration into pitch area(s). You might wish to make a copy of the sound to further isolate it and make repeated playbacks easier.

1. Determine the pitch area of the most prominent pitch area by defining the lower boundary first, then the upper boundary of the area (a steep filter can be helpful in determining these boundaries during beginning studies).
2. Determine any secondary areas of concentrated activity (these will be identified by either width, density, or dynamic prominence of the pitch area), by identifying the lowest then the highest boundary.
3. Repeat the process for all other pitch areas present. The specific frequencies/pitches of the boundaries are often audible, despite the sound not having an audible fundamental frequency. These pitches/frequencies should be identified and noted on the graph.
4. Evaluate the densities of the pitch areas and incorporate that information into the graph (this is the general amount of pitch/frequency activity within the pitch area and is noted on a numbering scale from very dense to very sparse).
5. Finally, identify the dynamic relationships between the pitch areas within the sounds. Describe this information as part of the analysis (this is the general dynamic relationships of the area and is noted on a number scale identifying the relative loudness of the pitch areas).

Exercise 6-4
Melodic Contour Analysis Exercise.

Find a recording with an instrumental melodic line that is performed too quickly to be heard as individual pitches. It is best for the melodic line to be at least two measures, or five seconds, in duration.

1. Determine the time line of the example, including the appropriate time units (clock time or meter) and the length of the time line. Make note of prominent pitch/frequency levels.
2. Begin plotting the melodic contour against the time line by identifying prominent pitch levels and placing them at precise locations on the time line. Check the time line for accuracy of length.
3. Work to establish as many reference pitches as possible, and the highest and lowest pitches of the line (these will identify the upper and lower limits of the Y-axis). Identify the fastest change of pitch level

(this will become the smallest time unit the graph needs to clearly present, and will determine the appropriate division of the X-axis).

4. Draw the melodic contour graph using the X and Y axes determined in Step 3.
5. Locate the reference pitches on the graph at the appropriate locations against the time line.
6. Fill in the remaining pitch information, making certain to check observations regularly. The graph is completed when the last noticeable pitch change is incorporated into the graph.

7 Evaluating Loudness in Audio and Music Recordings

Loudness has traditionally been used in musical contexts to assist in the expressive qualities of musical ideas. This function of dynamics helps shape the direction of a musical idea, helps delineate the separate musical ideas (usually in relation to their importance to the musical message), assist in creating nuance in the expressive qualities of the performance, and/or it may add drama to the musical moment. In all of these cases, dynamics have functioned in supportive roles in the communication of the musical message.

The recording process has given the recordist more precise control over dynamics than exist in live, acoustic performances. This increased control has brought audio recordings to have additional relationships of dynamics and the potential of placing more musical importance on dynamic relationships. An example of a new relationship of dynamics is the occurrence of contradictory cues between the loudness level at which a sound was performed during the initial recording (tracking) and the dynamic level at which the sound is heard in the final musical texture (the mix). The recordist must be aware of all relationships of dynamic levels (both those that are naturally occurring and those caused by the recording/reproduction process), and of the possibility that dynamic levels and relationships may function on any hierarchical level of the musical structure.

The recording process places unique critical and analytical listening requirements on the recordist in the area of the evaluation of loudness. The recordist must be able to focus on changes in dynamics at all levels of perspective and to quickly switch focus between those levels. The recordist must also be able to use the skills of identifying loudness relationships switching between listening to the sound itself, out of context (using critical listening), and listening to the sound within its musical context (analytical listening).

Much confusion often accompanies beginning attempts to perceive, follow, and graph dynamic changes. The listener must remain focused on the act of perceiving and defining changes in loudness levels (which are being applied to music as the artistic element of dynamic levels and relationships), and not allow themselves to be distracted.

The listener must be conscious, and ever mindful, of not confusing other, easily misleading information as changes in dynamic levels. Some aspects of sound that are often confused for dynamics are distance cues, timbral complexity, performance intensity, sound source pitch register, any information that draws the attention (focus) of the listener (such as a text, sudden entrance of an instrument), environmental cues, and speed of musical information.

It is common for sounds most prominent in the listener's focus not to be the loudest sounds in the musical texture. Loudness itself does not create or ensure prominence of the material. It comes as a surprise to many people that the most prominent sound in the listener's attention is often NOT the sound with the highest dynamic level.

This chapter seeks to define actual loudness levels in musical contexts (dynamics), with the exception of the final section. In that section, the actual loudness of the musical parts as musical balance will be compared to performance intensity (the loudness of the sound sources when they were performed in the recording process).

Reference Levels and the Hierarchy of Dynamics

Dynamics have traditionally been described by imprecise terms such as very loud (fortissimo), soft (piano), medium loud (mezzo forte), etc. These terms do not provide adequate information to define the loudness level of the sound source. They merely provide a vocabulary to communicate relative values.

The artistic element of dynamics in a piece of music is judged in relation to context. Dynamic levels are gauged in relation to (1) the overall, conceptual dynamic level of the piece of music, (2) the sounds occurring simultaneously with a sound source, and (3) the sounds that immediately follow and precede a particular sound source. In this way, loudness is perceived as relationships between sound sources and in relation to a reference level. Evaluation is more precise, and it is only possible to communicate meaningful information about dynamic levels, when a reference level is defined.

Reference Dynamic Level

The impression of an overall or global intensity-level of a piece of music will be the primary reference-level for making judgments concerning dynamics. This level is arrived at through our global impression of the intensity level of the performance of the work. It is the *perceived performance intensity* of the work as a whole, conceptualizing the entire work as a single entity out of time. The work's form and essence has a dimension in perceived performance intensity. This is the work's *reference dynamic level.*

Every work can be conceived as having a single overall, reference dynamic level (RDL). This is the dynamic level assigned to the piece when the form of the work is envisioned. The RDL is the single dynamic level that can represent the overall concept (form) of the piece. This overall dynamic level is a realization of the global impression of the performed dynamic level or the intensity level of the performance of a piece of music.

Performers establish this reference dynamic level in their minds before beginning a performance of any piece of music. Often this occurs intuitively. Composers also retain this level in their thoughts (at least subconsciously) throughout the process of writing a piece of music. Recordists must go through a similar process in production work. Recordists establish this level as a reference from which they are able to calculate all other dynamic levels and relationships. In recordings, this level often needs to be consistent for hours, days, weeks or longer, as sessions for a piece of music progress at their own pace.

Performance intensity cues are related to timbral changes of the sound sources. Sound sources will exhibit different timbral characteristics when performed at different dynamic levels and with different amounts of physical exertion. The impressions the listener receives, related to the intensity level of the performance (of all of the musical parts individually and collectively), will be related to actual dynamic relationships of the musical parts.

The perceived performance intensity in the recording and the perceived dynamic relationships of the musical parts will directly shape the listener's impression of the existing RDL of the recording. Dynamic and performance intensity cues (including expressive qualities of the performance) play significant roles in determining RDL as does tempo. The sound sources that present the primary musical materials (or that are at the center of the listener's focus) will often have proportionally more influence than those of sources presenting less significant material will.

Through the perception of these cues and the influence of tempo, a single, conceptual dynamic level will be determined. This is the RDL of the performance (recording/piece of music)—the level at which it is envisioned as existing. This is the reference level that will be used to calculate the dynamic levels of the individual musical parts in relation to the whole, as well as the dynamic contour of the overall program.

Every work will have a specific RDL. When a work is divided into separate major sections, even separate movements (such as a symphony), the work will have an RDL that allows all sections to be related to the overall concept. Even a 90-minute symphony will have only a single RDL.

The reference dynamic level can be perceived as existing anywhere from forte to piano. The RDL will be established as a precisely defined dynamic level that will serve as a reference level throughout the work. This will be discussed further.

The RDL will be used for evaluating/defining:

1. Dynamic relationships of the overall dynamic contour of the program (piece of music),
2. Dynamic levels of the musical ideas (sound sources) of the work, and
3. Dynamic relationships of the individual dynamic contours of the musical balance of the work.

Performance Intensity and Dynamic Markings

Timbre changes of performance intensity are important cues in our perception of dynamic levels of acoustic performances. We apply these same cues to recorded sounds to imagine a live performance, despite the medium. Our reference for performance intensity is our knowledge of the instrument's timbre, as it is played at various levels of physical exertion and with various performance techniques.

The listener's perception of the amount of physical exertion required for the sound quality shapes the perceived performance intensity. Performed sounds that require expending energy are perceived as moderately loud (mezzo forte, m*f*), or above. Performed sounds that appear to be withholding energy are perceived as moderately soft (mezzo piano, m*p*), or below. When the listener imagines a considerable amount of energy (or perhaps an excessive amount) was required to produce a perceived sound quality, the performance intensity will be forte (*f*), or perhaps more. This becomes simpler when remembering to relate dynamic markings and performance intensity cues to energy expended by the performer.

The threshold between mezzo piano (m*p*) and mezzo forte (m*f*) is critical to this understanding. This is the energy level the performer can theoretically sustain indefinitely. Conceptually, at this level the performer is neither putting forth energy, nor holding back—no energy is being exerted. Above this threshold energy is being consumed by pushing forward, becoming more assertive—even if only a very small amount. Below the threshold energy is being held back, or being withdrawn if only in a small amount. Moving further above or below the threshold, the perception becomes a matter of degree, or magnitude, of how much energy is being expended or withheld. The difference then between *ff* and *fff* is the level of intensity and the expectation of the length of time that level of energy can be sustained. Likewise *pp* and *ppp* are distinguished by the level of restraint and the likelihood of the length of time that restraint might be sustained.

The traditional terms for dynamic levels can continue to be of use with a well-defined RDL based on performance intensity information. With a defined RDL, the comparative terms can have more significant meaning. The terms will remain imprecise, but they will be more meaningful. Dynamic levels are more precisely defined when sound sources are compared to one another and placed on the appropriate graph; making the use of the traditional terms a mere starting point for evaluation of loudness levels.

The terms retain their meanings from musical contexts whereas the dynamic terms (such as mezzo forte) describe a quality of performance and an amount of physical exertion and drama on the part of the performer, as well as being a description of the loudness level. When placed on a graph (see Figure 7-1) these general terms are transformed into areas where sounds can be precisely defined against the RDL and in relation to performance intensity.

Dynamic Levels as Ranges

Dynamic levels are not organized into discrete increments in musical contexts. Dynamics are conceived in ranges or areas. Dynamic markings refer to a range of dynamic levels.

The dynamic marking "mezzo forte" does not refer to a precise level. It refers to a range of dynamic levels between "mezzo piano" and "forte." Many gradations of "mezzo forte" may exist in a certain piece of music. Many instruments can be performing at different levels of loudness, yet be accurately described as being in the "mezzo forte" dynamic range. Entire musical works or performances MAY take place within the range of a certain dynamic marking, yet exhibit striking contrasts of levels.

Figure 7-1 presents the vertical axis that will be used for all graphs plotting dynamics. The dynamic markings are centered within the ranges. A number of sound sources are plotted on the graph, and the reference dynamic level is designated on the vertical axis. Each sound source can be compared to the dynamic levels and contours of the other sources, and to the RDL. The unique characteristics of each source are readily apparent from the graph.

If limited to describing the sound sources by traditional dynamic level designations, some sounds would be a "loud mezzo piano," some sounds a "moderate mezzo piano," and others a "soft mezzo piano." The graph, in this instance, circumvents the need for these vague and cumbersome descriptions.

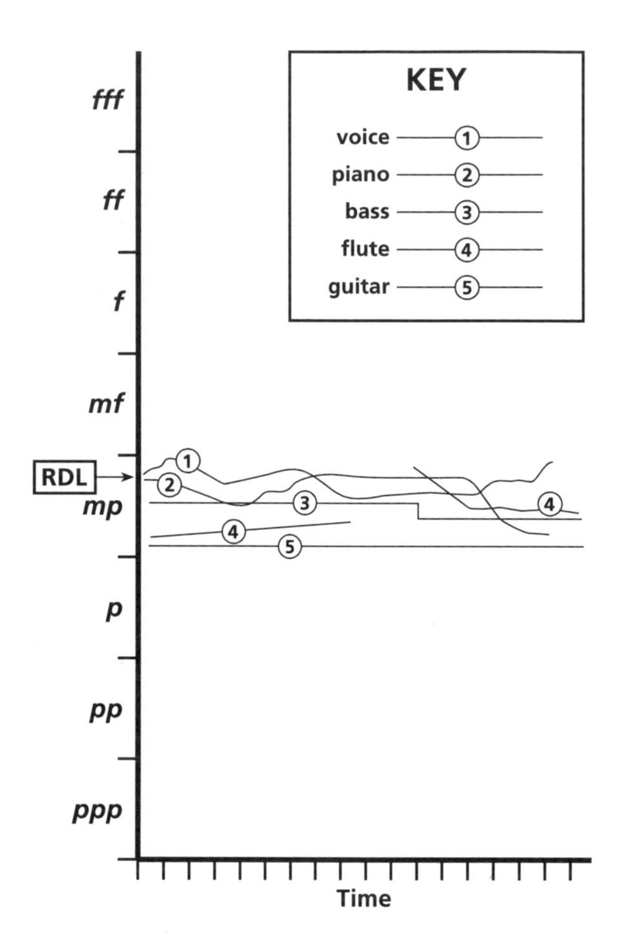

Figure 7-1 Dynamic ranges and dynamic contour.

The most extreme boundaries of dynamic ranges, for nearly all musical contexts, will extend from "*ppp*" to "*fff*" Musical examples that contain material beyond these boundaries are rare. The individual graph should incorporate only those ranges that are needed to accurately present the material, and to leave some vertical area of the graph unused above and below the plotted sounds, for clarity of presentation.

Determining Reference Dynamic Levels

The perceived performance intensity level that serves as our reference for evaluating the dynamic relationships of the work is used at the upper levels of the structural hierarchy. A single RDL will be used (1) at the highest hierarchical level to calculate the dynamic contour of the entire program (the complete musical texture) and (2) at the mid-levels of the structural hierarchy to calculate the dynamic contours of the individual sound sources in musical balance.

At the lower levels of the structural hierarchy, the reference level switches to the global impression of the intensity level of each sound. The intensity level of an appearance of the sound source is used as the reference level to determine the dynamic contours of the individual sound and its component parts. At these levels of perspective, dynamic contours are plotted of (1) a typical appearance of the overall sound source (the dynamic envelope) and (2) the individual components of the spectrum (spectral envelope). This dynamic contour information seeks to define the sound quality or timbre of the sound source, and will be explored in the next chapter, in that context.

The RDL of the piece of music is the reference for determining the dynamic levels of the sound sources and the composite musical texture. In order to make these evaluations, the RDL must first be defined.

The RDL is a precise level, and can be clearly defined. It is not subjective. All listeners putting forth the effort to perceive it will arrive at the same level. The listener will recognize when they have identified the correct level. The level will cause all other dynamic relationships to be understood, to make sense. The listener will perceive the piece of music as existing at the precisely identified dynamic level. The dynamic level is envisioned as a dimension of the essence of the piece of music.

A listener's first attempts to refine RDL are usually difficult. The concept itself alludes many people at first. The reader must remain conscious of trying to understand this large concept. It will require many hearings of the recording/piece of music to learn it well enough to try to define something requiring this depth of understanding. After achieving this level of understanding, the listener can become more comfortable with formulating the impression of a single dynamic level that IS the dynamic/intensity level of the piece. The listener will recognize the RDL of a piece once it has been experienced and understood. It is likely they will then not forget it. Listeners often experience the RDL even in passive listening for entertainment and do not realize it.

It is common for some information to get in the way of formulating this impression of the RDL. Musical materials, lyrics, tempo, and instrument timbres all give cues that the listener will be tempted to factor into this observation. Skilled musicians are even prone to drawing conclusions based on what they would like to be present in the music, rather than listening to what is present. Some instruments may well be performing at intensities that send conflicting cues when considered against the listener's ideas of the potential RDL; this conflicting information enriches art, but makes defining it more difficult. A magic formula does not exist for determining the RDL. This is one of the significant artistic dimensions of a piece of music that defies theoretical analysis and instead

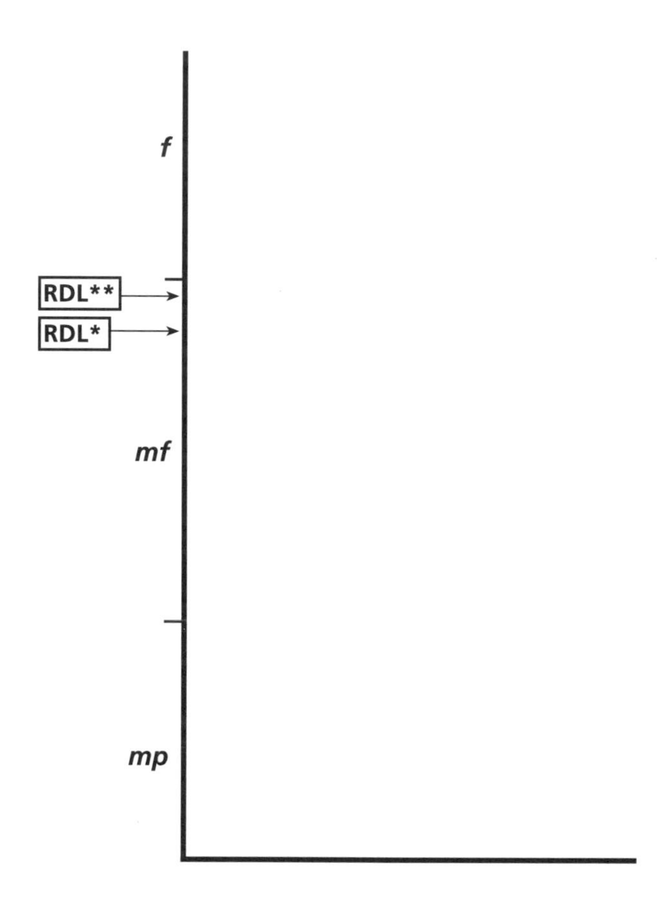

Figure 7-2 Reference dynamic levels of two versions of "Lucy in the Sky with Diamonds;" * original version from *Sgt. Pepper's Lonely Hearts Club Band* and ** 1999 *Yellow Submarine* version.

uses the sensibilities of the listener. This does not make it subjective—only that it cannot be predicted by measured calculation, but is determined through the experience of the music and arrived at through an understanding of the piece. Again, the RDL will be a precise level that all listeners can agree upon (+/- 2 percent) given enough attention to the task.

Figure 7-2 identifies the RDL of "Lucy in the Sky with Diamonds." Two levels are shown: one for the original version from *Sgt. Pepper's Lonely Hearts Club Band* and the other for the 1999 *Yellow Submarine* version. The two versions have slightly different reference dynamic levels. The different levels are caused by the different mixes and presentations of materials, among other factors such as the sound qualities imparted during mastering. The essence of the piece has been slightly altered by the slightly different sound qualities of the sound sources and the recordings. The reference dynamic levels are perceived as clearly within mezzo forte's moderate expending of energy. They are both beyond the level midway between the beginning of **mf** and the threshold of ***f*** (where exertion moves beyond moderate). In fact both exceed the three-quarters level of the area that comprises **mf**. After much listening and contemplation, the levels can be understood as being in the upper 15 percent of the **mf** area, with not more than 10 percent of the area separating the two levels.

RDL is calculated after getting to know the composition, recording, and/or performance well. Two different performances of the same piece by the same performer may each have a different RDL. Each of two different interpretations will nearly certainly have a different RDL.

An exercise in determining the reference dynamic level of a recording/piece of music appears at the end of this chapter, Exercise 7-1.

Program Dynamic Contour

Changes of the dynamic level, over time, comprise dynamic contour. As noted, the dynamic levels and relationships occur at all hierarchical levels. The broadest level of perspective will allow the dynamic contour of the overall program to be plotted. This is the single dynamic level/contour of the composite sound. The dynamic level that results from combining all sounds. This *program dynamic contour* can be envisioned as a mono meter, following the dynamics of the entire program.

Skill in recognizing the dynamic level of the overall program is developed through plotting program dynamic contour. Recordists use this skill in many listening evaluations. This high-level graph, or the associated listening skill alone, will be applied to many analytical and critical listening applications.

Dynamic contour must not be confused with performance intensity, with distance cues, or with spectral complexity. These aspects of recorded sound often present cues that contradict actual dynamic (loudness) level or that alter the perception of the actual dynamic level.

Program dynamic contour information is plotted on a program dynamic contour graph. The graph incorporates:

1. Dynamic area designations for the Y-axis, distributed to complement the characteristics of the musical example;
2. The reference dynamic level is designated as a precise level on the Y-axis;
3. X-axis of the graph is dedicated to a time line is devised to follow an appropriate increment of the metric grid or of real time (depending on the material and context); and
4. A single line is plotted against the two axes (the dynamic contour of the composite program is the shape of this line).

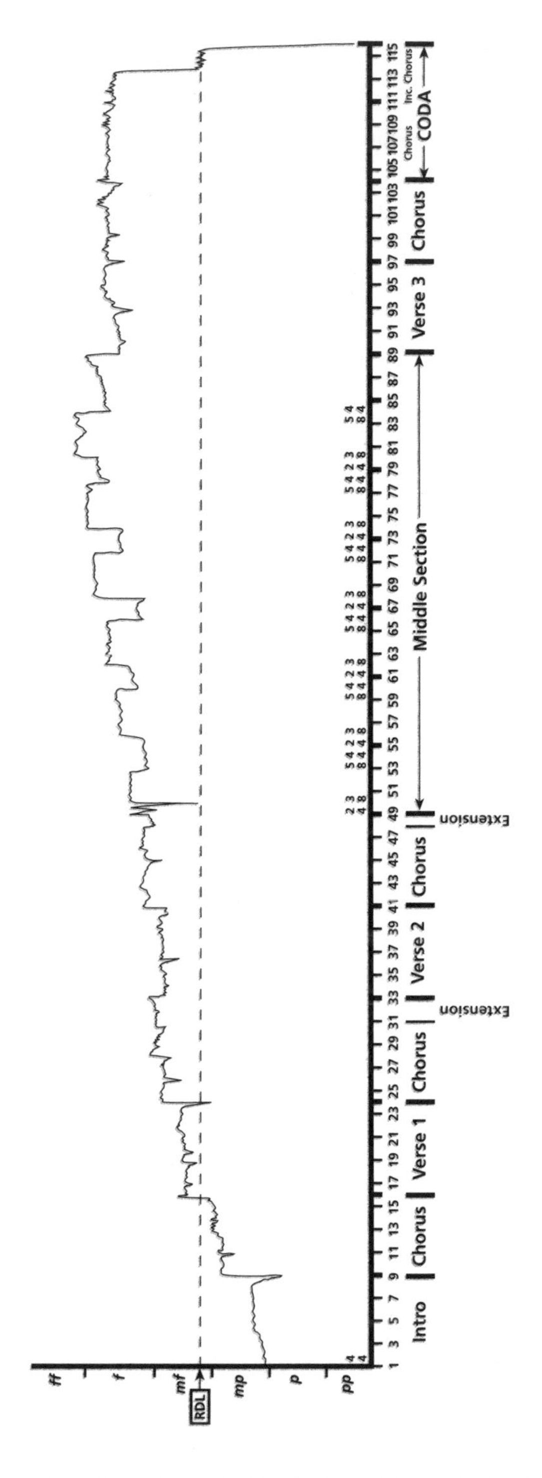

Figure 7-3 Program dynamic contour—The Beatles, "Here Comes the Sun."

The program dynamic contour of The Beatles' "Here Comes the Sun" is plotted in Figure 7-3. The program dynamic contour graph shows the changes in overall dynamic level throughout the work. In listening to the work, the striking changes in the overall dynamic level will be evident. The wide dynamic range goes through many large and many subtle changes of loudness level. Pertinent to reference dynamic level, a sense of arrival happens at the end of the piece. This occurs when the song settles at the RDL for a moment and then the song is over.

It is common for a piece of music to arrive at its RDL as an important occurrence of the work. It might appear in the introduction, the dramatic climax, the final chorus or the final verse. These are all common places an RDL might be reached, but any location is possible. It is also possible that the song's RDL is never sounded—purposefully leaving the listener unfulfilled in this regard. It is possible the song's RDL will be prevalent in a piece or heard only once as in "Here Comes the Sun." Many possibilities exist—including silence, as some songs only make sense when the silence at the end arrives and brings introspection.

Musical Balance

The plotting of all the individual sound sources in a musical texture by their dynamic contours provides the *musical balance graph*. The graph will show the actual dynamic level of the sound sources, in relation to the RDL, as established above. Each sound source will have a separate line on the graph that will allow the dynamic contours of the sources to be mapped against a common time line. This graph will clearly show the loudness levels of all sounds and will represent the mix of the work.

The musical balance graph will not include information on performance intensity, sound quality, or distance. These cues are often confused with perceived dynamic level, and are purposefully avoided here. This graph is solely dedicated to evaluating and understanding the dynamic levels, contours, and relationships of the sound sources. This should be the focus of the listener.

The musical balance graph incorporates:

1. Dynamic area designations for the Y-axis, distributed to complement the characteristics of the musical example;
2. Reference dynamic level is designated as a precise level on the Y-axis;
3. X-axis of the graph is dedicated to a time line that is devised to follow an appropriate increment of the metric grid;
4. A single line plotted against the two axes for each sound source; and
5. A key relating the names of all sound sources to a unique number, color, or line composition to identify all source lines on the graph.

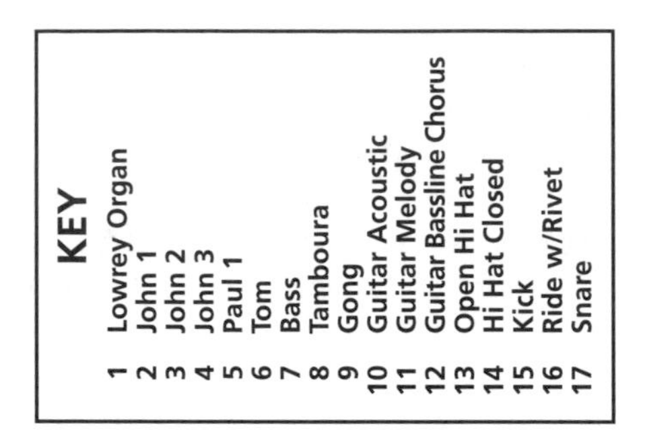

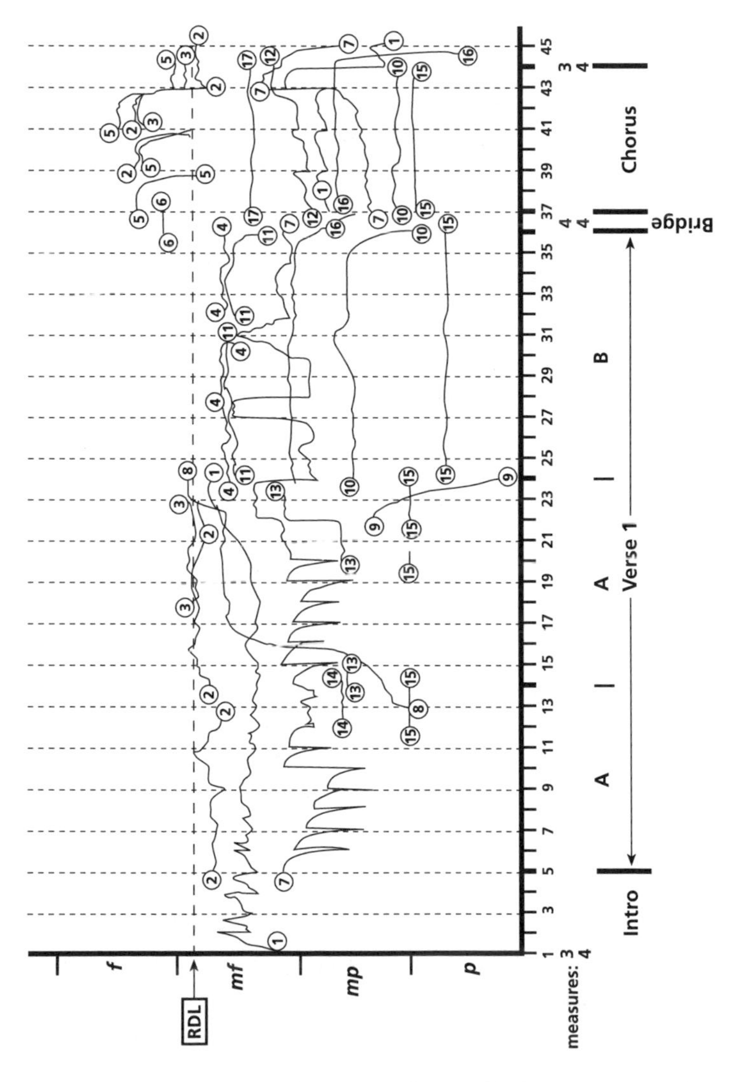

Figure 7-4 Musical balance graph—The Beatles, "Lucy in the Sky with Diamonds."

Figure 7-4 presents a musical balance graph of some of the sound sources in The Beatles' "Lucy in the Sky with Diamonds" *Sgt. Pepper's Lonely Hearts Club Band* version. These are the actual loudness levels of the instruments. The listener will notice that the loudest instrument/voice is not always the most prominent.

Exercise 7-3 at the end of the chapter will lead the listener through the process of creating a musical balance graph. It is suggested the reader become well acquainted with this exercise, as it represents one of the primary skills required of the recordist.

Performance Intensity versus Musical Balance

Performance intensity is the dynamic level at which the sound source was performing when its was recorded. In many music productions, this dynamic level will be altered in the mixing process of the recording. The performance intensity of the sound source and the actual dynamic level of the sound source in the recording will most often not be identical and will send conflicting information to the listener.

The dynamic levels of the various sound sources of a recording will often be at relationships that contradict reality. Sounds of low performance intensity often appear at higher dynamic levels in recordings than sounds that were originally recorded at high performance-intensity levels. This is especially found in vocal lines. This conflicting information may or might not be desirable, and the recordist should be aware of these relationships.

Important information can be determined by plotting performance intensity against musical balance for some or all of the sound sources of a recording. This will often provide significant information on the relationships of sound sources and the overall dynamic and intensity levels of the work, as well as the mixing techniques of the recording.

Performance intensity is plotted as the dynamic levels of the sound sources, as the listener perceives the intensity of the original performance. The listener will judge the intensity of the original performance through timbre cues. The listener will make judgments based on their prior knowledge of the sound qualities of instruments and voices as they exist when performed at various dynamic levels.

The reference for performance intensity is the listener's knowledge of the particular instrument's timbre, as that instrument is played at various levels of physical exertion and with various performance techniques. A reference dynamic level is not applicable to this element.

Musical balance is plotted as the dynamic levels of the sound sources, as the listener perceives their actual loudness levels in the recording itself, as discussed.

The *performance intensity/musical balance graph* incorporates:

1. Dynamic area designations in two tiers for the Y-axis, distributed to complement the characteristics of the musical example (one tier is dedicated to musical balance, one tier is dedicated to performance intensity);
2. Reference dynamic level of the musical balance tier is designated as a precise level on the Y-axis (an RDL is not relevant to the performance intensity tier);
3. X-axis of the graph is dedicated to a time line devised to follow an appropriate increment of the metric grid;
4. A single line is plotted against the two axes for each sound source, on each tier of the graph (each sound source will appear on both tiers; the same number, composition, or color line is used for the source on each tier of the graph); and
5. A key is used to clearly relate the sound sources to their respective source line (the same key applies to both tiers of the graph).

Figure 7-5 will allow the listener to observe some of the differences between the recording's actual loudness levels and the performance intensities (loudness levels) of the sound sources when they were recorded. A few key sound sources are graphed from The Beatles: "Strawberry Fields Forever." Some sound sources are at very different levels in each tier, and others show no significant change. Some sources contain subtle changes of dynamic levels and/or many nuances of performance intensity information, and the Mellotron exhibits few gradations of dynamics and intensity.

An exercise to develop skills in recognizing and evaluating performance intensity versus musical balance information appears as Exercise 7-4 at the end of the chapter. As an additional exercise, the reader might determine the musical balance and performance intensities of the other sound sources in "Strawberry Fields Forever" during the measures of Figure 7-5.

The musical balance and the performance intensity graphs will function at the same level of perspective as the pitch density graph. When evaluated jointly, these three artistic elements will allow the listener to extract much pertinent information about the mixing and recording processes, and the creative concepts of the music.

Exercises

The following exercises should be performed with care. You should become comfortable with the material covered.

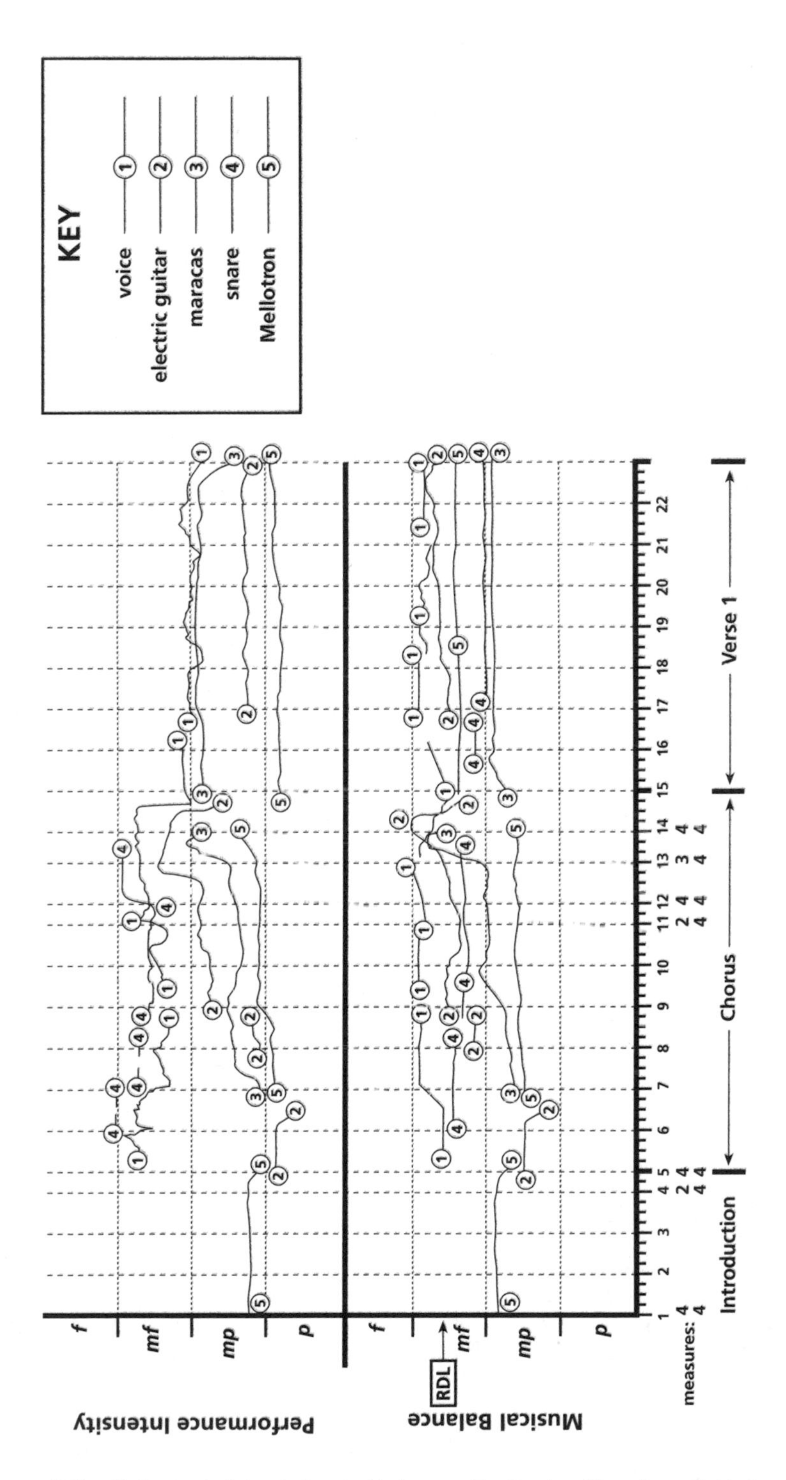

Figure 7-5 Performance intensity/musical balance—The Beatles, "Strawberry Fields Forever."

Exercise 7-1
Reference Dynamic Level Exercise.

Select a recording you know well for initial attempts at determining the reference dynamic level of a piece of music. It would be best for the work to be less than four minutes duration.

1. Before listening to the piece, spend some time thinking about the overall character of the piece; consider the overall energy level, performance intensity, concept or message, and other important aspects of the song.
2. Listen to the song several times to confirm that the observations in your memory are reflected in the actual music and recording.
3. Reconsider your observations with each new hearing of the recording.
4. Attempt to determine a precise dynamic level for the RDL. Begin this process by working from the extreme levels—*ppp* and *fff*—asking if the level exists in those areas.
5. Once the dynamic area has been determined, work to define a precise level by asking if the RDL is below 50 percent in the level, or above. Continue to work toward a specific level by narrowing the area further.
6. Leave the example and your answer for a period of time (several hours or several days). Listen to the song again. Reconsider the RDL previously defined.

If you do not know a piece of music, many hearings will be required before initial observations can be made.

Exercise 7-2
Program Dynamic Contour Exercise.

Select a short song for initial attempts at creating a program dynamic contour graph. The entire song should be graphed for overall dynamic contour. The dynamic level of the entire recording, or the composite dynamic level of all sounds, will be the focus of this exercise.

1. During the first hearing(s), listen to the example to establish the length of the time line. At the same time, notice prominent instrumentation and activity that will provide cues as to the overall intensity level of the performance.

2. Check the time line for accuracy and make any alterations. Establish the RDL of the work by working through the previous exercise.
3. Notice the activity of the program dynamic contour for boundaries of levels of activity and speed of activity. The boundary of speed will establish the smallest time unit required to accurately plot the smallest significant change of the element. The boundary of levels of activity will establish the smallest increment of the Y-axis required to plot the smallest change of the dynamic contour.
4. Begin plotting the dynamic contour on the graph, continually relating the perceived dynamic level to the RDL. First, establish prominent points within the contour. These reference points will be the highest or lowest levels, the beginning and ending levels, points immediately after silences, and other points that stand out from the remainder of the activity. Use the points of reference to judge the activity of the preceding and following material. Focus on the contour, speed, and amounts of level changes to complete the plotting of the contour.
5. The evaluation is complete when the smallest significant detail has been perceived, understood, and added to the graph.

Exercise 7-3
Musical Balance Exercise.

Select a popular song with at least three instruments and voice. The first 32 bars will be evaluated for musical balance. This exercise will graph the dynamic contours—actual loudness levels—of all sound sources. Musical balance is the relationships of sound sources to one another. Initial attempts should use pieces of music with only a few sound sources.

The exercise will follow the sequence:

1. During the first hearing(s), listen to the example to establish the length of the time line. At the same time, notice prominent instrumentation and activity of their dynamic levels against the time line.
2. Check the time line for accuracy and make any alterations. Establish a complete list of sound sources (instruments and voices), and sketch the presence of the sound sources against the completed time line. A key should be created, assigning each sound source with its own number, color, or line format.
3. Determine the reference dynamic level of the sound using the process previously presented.

4. Notice the activity of the dynamic levels of the sound sources (instruments and voices) for boundaries of levels of activity and speed of activity. The boundary of speed will establish the smallest time unit required to accurately plot the smallest significant change of dynamic level. The boundary of levels of activity will establish the smallest increment of the Y-axis required to plot the smallest change of dynamics.
5. Begin plotting the dynamic contours of each instrument or voice on the graph. Keeping the RDL clearly in mind, establish the beginning dynamic levels of each sound source. Next, determine other prominent points of reference. Use the points of reference to judge the activity of the preceding and following material. Focus on the contour, speed, and amounts of level changes to complete the plotting of the dynamic contours.
6. You should periodically shift your focus to compare the dynamic levels of the sound sources to one another. This will aid in developing the dynamic contours and will keep you focused on the relationships of dynamic levels of the various sources. The evaluation is complete when the smallest significant detail has been incorporated into the graph.

It is important to remain focused on the actual loudness of instruments, making certain your attention is not drawn to other aspects of sound.

As you gain experience and confidence in making these evaluations, songs with more instruments should be examined and longer sections of the works should be evaluated.

Exercise 7-4

Performance Intensity versus Musical Balance Exercise.

Select a multitrack recording of a popular song with at least five sound sources. The first 16 bars will be evaluated for performance intensity and musical balance. Select five sound sources to graph for this exercise. The graph will have two tiers: one will graph musical balance (the actual loudness levels in the recording), the other performance intensity (the loudness levels of the instruments when they were recorded).

1. The musical balance exercise should first be completed as in the previous section. This will generate the graph's time line, as well as provide all information and contours for the musical balance tier.

2. Performance intensity will now be determined for each sound source, for the performance intensity tier. Sound sources will have the same number, color, or line format as on the musical balance tier.
3. Begin plotting the performance intensity of each sound source on the graph. Start by establishing the beginning performance intensity levels of each sound source. Next, determine other prominent points of reference. Use the points of reference to judge the activity of the preceding and following material. Focus on the contour, speed, and amounts of level changes to complete the plotting of these performance intensity contours. The evaluation is complete when the smallest significant detail has been incorporated into the graph.

You can now compare the two tiers, and learn significant information on how the instruments were tracked, and how the instruments were altered by the recording and mixing processes.

As you gain experience in making these evaluations, all of the sound sources of songs with many instruments should be examined for longer sections of songs that have significant changes in the mix.

8 Evaluating Sound Quality

This chapter will present a process for evaluating sound quality that can also be used to define the characteristics of timbre. This process is projected towards establishing a vocabulary for audio professionals to communicate meaningful information about sound.

By directly describing the physical dimensions of sound, the process can be easily adapted to any sound evaluation. The reader will learn to describe sound quality (accounting for the various contexts within which sounds exist), and to evaluate sound out of context (as an abstract, sound object).

The critical listening process and the technical needs of the audio industry are often juxtaposed with creative applications and analytical listening processes. These differences will be articulated, in addition to the perception of sound quality at various hierarchical levels.

The perception of sound quality for its global form or shape exists at a number of levels in perspective and types of listening. These concepts are central to the evaluation process, as they allow for our realization of sound as an object (available for evaluation out of time) at all levels of detail in our perception.

Communicating information about sound quality is central to all facets of music production. Nearly all positions in the entire audio industry need to communicate about the content or quality of sound. Yet a vocabulary for describing sound quality, or a process for objectively evaluating the components of sound quality, do not exist. Meaningful communication about sound quality can be accomplished through describing the values and activities of the physical states of the components of timbre. Sounds will be described by the characteristics that make them unique. These characteristics are the activities and states that occur in the component parts of the sound source's timbre. The characteristics have been reduced to the definition of fundamental frequency, dynamic envelope, spectrum (spectral content), and spectral envelope. Meaningful information about sound quality can be communicated verbally through the physical states of its characteristics. This can be done in great detail or in a general way. Information is communicated in a more detailed and precise manner through graphing sound quality. The following sections will lead the reader to develop skill and language to objectively describe sound.

Sound Quality in Critical Listening and Analytical Listening Contexts

Describing sound quality in and out of musical contexts are skills that are important for audio professionals. In both contexts, sound quality evaluations occur at all levels of our perceptual hierarchies and at all perspectives.

Outside musical contexts, the audio professional is concerned with understanding the characteristics of sound quality for its own sake and as the integrity of the audio signal. This first approach to examining sound quality looks at the unique character of a particular sound. This character of the sound may be what separates one microphone's imprint from another, one person's snare drum sound from another's, or even one monitor speaker from others. Through critical listening skills, sound quality of a particular sound is evaluated for its unique qualities. The evaluation is in relation to the activities and states of the components of sound that occur within that sound only. This use of sound quality evaluation will seek to define the individual sound so it can be understood, and the resulting information made available for other uses (i.e., comparisons with other sources). The examination of the crash cymbal sounds from Chapter 6 is an example of this type of comparison. Critical listening evaluation is also often concerned with the technical quality, or integrity, of the recording.

The technical quality of recordings reflects the degree of signal degradation or the presence and amount of unwanted sound. This is often referred to as the integrity of the signal within a recording. It is usually the goal of recordings to be of the highest technical quality, and to be void of all degradations of signal quality/integrity and all unwanted sound (although Internet compression seems to be modifying this goal by accepting degradations of signal in exchange for speed of transmission). The technical quality of recordings is defined by unwanted sounds and sound events, and of numerous types of ways the dimensions of frequency/pitch and amplitude/loudness are undesirably altered by the recording and reproduction equipment and processes (often malfunctions or miscalibrations).

The focus of the listener may be at any level of listening perspective, with the listener able to quickly and accurately shift perspectives. The listener may be evaluating the technical quality of the sound for information related to frequency response (or spectral content, or spectral envelope), or the listener may be listening for transient response (or dynamic envelope, or dynamic contour of a specific frequency area). As examples of extremes of perspective, the listener may focus on the sound quality of the overall program or may be listening at the close perspective of focusing on the sound quality of a particular characteristic of a single, isolated sound source.

Critical listening involves the evaluation of sound quality to define what is physically present, to identify the characteristic qualities of the sound being evaluated, or to identify any undesirable sounds or characteristics that influence the integrity of the audio signal. This process and conceptualization is performed without consideration of the function and/or meaning of the sound and without taking into account the context of the sound.

Analytical listening involves the evaluation of sound quality to identify its characteristic qualities in relation to the context of the sound. Analytical listening will seek to define the sound quality in terms of what is physically present, but will then relate that information to the musical context in which the sound material is presented and perceived. It involves defining sounds then comparing the sound to others.

Pitch density is the evaluation of pitch-related characteristics of sound quality, at a high-level in the structural hierarchy. This evaluation may or may not occur in a musical context. Pitch area analysis (such as the ones performed on percussion sounds in Chapter 6) is another evaluation of pitch-related characteristics of sound quality, only at a lower hierarchical level. It also does not necessarily take place within the musical context.

The pitch density analysis that appears in Chapter 10 is in a musical context. The pitch-area analyses of the percussion sounds in Chapter 6 are out of musical context.

Both of these studies are simple (or more general) approaches to timbre and sound quality evaluation. They both define the pitch-component information that leads to defining timbre, or evaluating sound quality. This process is related to evaluating the spectral content of a sound. In the same way, dynamic contour analyses are related to sound quality analysis, at various structural levels.

Sound Quality and Perspective

Sound quality is our perception of sound as a single concept or entity. The listener conceives sound in its global form or shape at a number of levels of their perceptual hierarchies. Sound quality is recognized at all levels of perspective (perceived detail). This allows for the understanding of sound quality as an object (available for evaluation out of time) or in relation to its musical context, at all levels of perceptual detail.

People are able to recognize sound quality as the global qualities of (1) overall program (or the entire musical texture), (2) groupings of similar or similarly acting sounds within the overall program (such as a brass section within an orchestra, or the rhythm section of a jazz ensemble), (3) overall impressions of individual sound sources (instruments, voices, synthesizer patches, special effects), or (4) as specific sounds generated by individual sound sources (individual expressive vocal sounds, a specified voicing of a guitar chord, a single sound's timbre characteristics). Listeners even recognize the sound quality of acoustic spaces, as environmental characteristics (explored in the next chapter).

At these very different levels of the perceptual hierarchy, we recognize sound quality as the concept that makes a sound a single, unique entity, as an overall form of the sound. This global quality is evaluated to determine specific information, to identify the aspects that make all sounds unique.

This evaluation may take place out of musical contexts. In these critical listening applications, evaluation is performed out of the time line of musical context. Instead, clock time is used to evaluate the changes that have occurred over time. While we are conceiving sound quality as the shape of the sound in an instant, or out of time, sound only exists in time. Sound can only be accurately evaluated as changes in states or values of the component parts, which occur over time. Sound quality in critical listening applications approaches the sound as an isolated, abstract object.

The evaluation may also take place within musical contexts. In these instances, the time of the metric grid will be used, if it is present. Evaluation of sound quality will be focused on the musical relationships of the material. The pitch density analysis of Chapter 10 is a suitable example of the evaluation of the pitch aspects of sound quality in relation to the musical material. When the pitch density graph was coupled with a musical balance graph, much information on the sound quality of the entire program (in a musical context) is available to the listener. Sound quality in analytical listening applications approaches the sound for its relationship to other sounds, to the musical texture as a whole, and to the musical message of the work.

The sound quality of the entire program, or of any individual sound source, may be supplying the most significant musical information in certain pieces of music. This concept of music composition (that can be explained through equivalence) is quite prominent in many very different styles of music. Throughout the twentieth century, a type of writing, *sound mass composition*, evolved through the work of composers Edgar Varèse, George Antheil, Krzysztof Penderecki, Karlheinz Stockhausen (whose photo appears on the album cover of *Sgt. Pepper's Lonely Hearts*

Club Band), and Luciano Berio (to name only a few). Their music of this type places an emphasis on the dimensions of the overall musical texture (or sound mass) and/or on the sound quality relationships within the overall musical texture.

The concept of giving musical significance to the sound quality of the entire program, the relationships of sound qualities, and to pitch density can be found in a wide variety of popular works from the past 40-plus years. Many examples of these ideas exist, although this concept is used in isolated areas in most works.

The Beatles' "A Day in the Life" is one such work. The concept of sound quality and pitch density is what shapes the music and the dramatic motion of the song's transition section and its conclusion. The sound mass concept of pitch density is the primary musical element, causing timbre/sound quality to be the dominant artistic element during those sections.

Evaluating the Characteristics of Sound Quality and Timbre

Sound quality will be evaluated in relation to spectral content, spectral envelope, and dynamic contour, in both musical contexts and in critical listening evaluations. Sound quality evaluation will be approached in this way at all levels of the perceptual hierarchy. At the highest levels, individual sound sources making up the texture of the whole program can be conceived as individual spectral components. At the lowest level, individual spectral components are evaluated, and individual evaluations may be performed for each occurrence of a sound source.

Individual sound sources may be analyzed for their contributions to the sound quality of the overall program. In musical contexts, this evaluation will compare the sound sources to the overall sound quality through their individual dynamic contours (creating musical balance), their pitch area (creating pitch density evaluations), and their spatial characteristics (Chapter 9). In critical listening, the contributions of the individual sound sources to the overall program will be approached in relation to the same dimensions, but without relation to musical time or context.

Most often, the audio professional will be concerned with evaluating the individual sound source. Individual sound sources are evaluated for their unique sound quality as sound objects. The sounds are evaluated to define their unique characteristics, through an evaluation of the states and activities of their component parts, out of the musical context.

This is the most widely applied use of the evaluation of sound quality. It is used for many activities from setting signal processors to evaluating the performance of audio devices, and from defining the general characteristics of a sound source (such as the sound quality of a guitar part in a recording) to a detailed evaluation of a particular guitar sound. Many other possibilities exist.

Often, a sound quality evaluation will be performed on a single, isolated presentation of the sound source. An isolated presentation will have its unique pitch-level, performed dynamic level, method and intensity of articulation, etc. Among many uses, examining a particular isolated presentation of a sound source allows for meaningful comparison between different performances of the same source, or of a different sound source performing similar material.

Evaluations of sound quality will seek to define the states and activities of the sound source's (1) dynamic envelope, (2) spectral content, and (3) spectral envelope. It will also make use of the listener's carefully evaluated perception of (4) pitch definition.

Defining the Four Components of Sound Quality Evaluations

Sound quality is defined by the physical dimensions of the sound source (1, 2, and 3), and their unique states and levels. The listener's perception of the definition of fundamental frequency (4) aids in defining information of those values, especially the loudness-level of the fundamental frequency in relation to the remainder of the sound's spectrum, and the dominance of harmonic partials. The four components of sound quality evaluations are examined throughout the duration of the sound material and are plotted against a single time line. It is important to note that pitch definition and dynamic envelope exist at the perspective of the overall sound, and that spectrum and spectral envelope are internal components of the sound and are recognized at a lower level of perspective.

The reader will first define the time line of the sound. A clock, counter, or stopwatch might be used as a reference. Next increments within the time line and suitable reference points within the time line will be identified. Skill needs to be acquired in performing these tasks. The Time Judgment Exercise of Chapter 5 will greatly assist this development. While determining a time line will require a number of hearings, this number will be reduced with experience and increased ability.

Dynamic Contour

The dynamic contour of the sound, or the sound's overall dynamic level as it changes throughout its duration, is reasonably apparent at first hearings. Difficulties may arise with confusing loudness changes and spectral complexity changes. The listener must remain focused on actual loudness and not be pulled to other perceived parameters of sound.

A reference dynamic level (RDL) will be required for mapping the dynamic contour. The RDL will be determined by the intensity-level at which the source was performed. The intensity-level itself is the RDL; it will be transferred to a precise dynamic level (a specific point in a dynamic area such as "mezzo forte" or "forte"). The same reference dynamic level will be used in the spectral envelope tier, explained below. In this way, the same reference level functions on two levels of perspective, just as one reference dynamic level functions for both program dynamic contour and for musical balance (in Chapter 7).

The dynamic contour is easily described. This is accomplished by discussing the shape and speed of the dynamic envelope, and its dynamic levels at defined points in time. By discussing how loudness changes and by defining the levels and speed of those changes, the listener is describing the physical elements of sound that are experienced by others. These comprise important components of the unique objective character of the sound.

Spectral Content

The reader may hear few or no spectral components at the beginning of their studies. Harmonics and overtones fuse to the fundamental frequency, and we have been conditioned to perceive all of this information as part of a whole (the global sound quality). To a great extent, the evaluation of sound quality works against all of our learned listening techniques and our previous listening experiences. Much patience will be required. Practice and repetitive listening must be undertaken to acquire the skills of accurately recognizing spectral components and of accurately tracking dynamic contours of the components that make up spectral content.

The harmonic series can be used to assist in identifying spectral components. The listener can envision the harmonic series as a chord above the fundamental frequency. Once the listener has learned the sound of this chord it will be possible to imagine the pitches of the harmonic series while listening to a sound. The listener will then be in a better position to identify pitches/frequencies of the harmonic series that are present. Frequencies/pitches other than harmonics will also be noticed; the listener will ultimately be able to quickly calculate where the overtones fall in relation to the envisioned harmonic series. In this way the harmonic series is used as a template, to which the frequencies/pitches present can be compared and identified. This will make evaluating spectral content much more approachable.

The reader is encouraged to return to the discussion of the harmonic series in Chapter 1 to study its content and to spend time learning the sound of the series. The reader is also encouraged to review the conversion of pitch levels into frequency levels and the reverse.

A tone generator or keyboard may be used to assist in identifying frequency-levels or pitch-levels of prominent harmonics and overtones. This will especially prove helpful in initial studies. A steep, tunable filter may also be of use in determining the pitch aspects of the spectrum. Whatever assists the reader in bringing their focus to the perspective of the spectrum should be used.

The reader will be able to describe much about a sound by addressing the spectrum. Defining which harmonics are present and those that are prominent, as well as indicating overtone content, provides a significant amount of objective information about a sound.

Spectral Envelope

Describing the entrances and exits of the partials (harmonics and overtones) as well as the individual dynamic contours of these partials (the spectral envelope) will provide additional important information on sound quality. The spectral envelope will be calculated against the reference dynamic level that was identified for the overall dynamic contour and against a common time line. All of the spectral components identified, as spectral content, will be present in the spectral envelope tier—including the fundamental frequency.

The spectral envelope includes the fundamental frequency, subtones, and subharmonics, overtones and harmonics that comprise spectral content. In mapping the dynamic levels and contours of these partials, significant information on pitch definition and the character of the sound is obtained. By describing this activity, the listener is communicating very pertinent and meaningful information about the sound that is completely objective, and can be experienced and understood by others.

Pitch Definition

The definition of the fundamental frequency is useful in making preliminary and general observations of a sound. Definition of fundamental frequency is often somewhat stable during the sustain portion of a given sound. Changes in pitch-quality are most commonly found between the onset and the body of the sound. Pitch-quality is placed on a continuum between the two boundaries of well-defined in pitch or as precisely pitched (as a sine wave) through completely void of pitch or nonpitched (as white noise). A dominance of harmonics will bring the sound to have a more defined pitch quality. With increased presence of overtones (in either number or loudness level) comes a more nonpitched character to the sound.

Often the definition of fundamental frequency can be verbally described as having a certain pitch quality for a certain portion of its duration, then another certain quality for the remainder of its duration. In effect, a contour of pitch definition is present. The pitch definition tier of the sound quality characteristics graph is not always required, but this aspect of the sound should always be addressed, if only during the early stages of the evaluation. Examining pitch definition supplies many clues of the content of the spectrum and the spectral envelope.

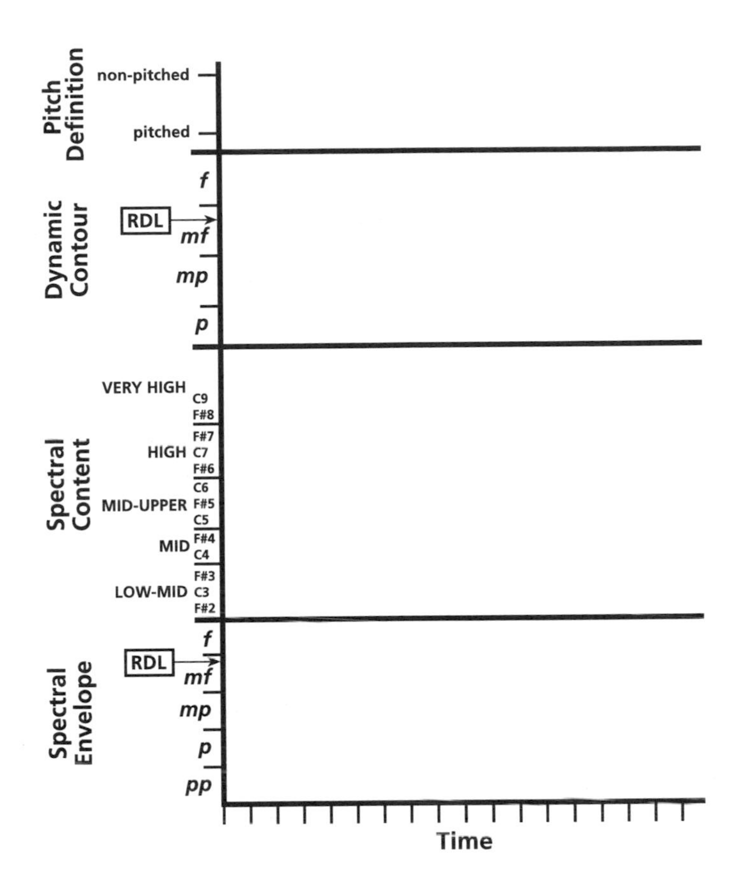

Figure 8-1 Sound quality characteristics graph.

Process of Evaluating Sound Quality

The *sound quality characteristics graph* will be used to assist in making sound quality evaluations. The graph will greatly assist a detailed evaluation of the sound and will allow the reader to record observations of any sound, whether or not a detailed evaluation is undertaken. Pertinent information can be written down that can provide a resource for objective descriptions formulated at a later time.

The process of evaluating sound quality should follow the sequence of events:

1. During the first hearing(s), listen to the example to establish the length of the time line. At the same time, notice prominent states and activity of the components (especially dynamic contour and spectral content) against the time line.
2. Check the time line for accuracy and make any alterations.

3. Notice the activity of the component parts of sound quality for their boundaries of levels of activity and speed of activity. The speed boundaries will establish the smallest time units required in the graph to accurately present the smallest significant change of the element. The boundaries of levels will establish the smallest increment of the Y-axis required to plot the smallest change of each component (dynamic contour, spectral content, spectral envelope).
4. Pitch definition calculations are made as an initial step in sound quality evaluations of sound sources. Begin by making general observations about the pitch-quality of the sound. Identify precise points in time when the sound is at its most pitched and least pitched states, and assign a relative value. Use these levels as references to complete this contour of pitch definition. The prominence of the fundamental frequency and the number and relative loudness levels of harmonics greatly influence the pitch quality of the sound source. Knowing when a sound is more pitched than other times provides information that can be used in determining the spectral content and spectral envelope of the sound.
5. Begin plotting the activity of the dynamic envelope on the graph. First, determine the RDL of the sound by defining its performance intensity. Second, establish the beginning and ending dynamic levels of the sound. The highest or lowest dynamic levels are the next to be determined then place them against the time line. Use these levels as points of reference to judge the activity of the preceding and following material. Alternate your focus on the contour, speed, and amounts of level changes to complete the plotting of the dynamic contour. The evaluation is complete when the smallest significant detail has been perceived, understood, and added to the graph. The smallest time increment of the time line may need to be altered at this stage to allow the dynamic contour to be clearly presented on the graph.
6. Plot the spectral content on the graph. Often, spectral components remain at the same pitch/frequency level throughout the sound's duration. Certain instruments have prominent overtones that change in pitch-level over the duration of the sound. First, identify the frequency/pitch levels of the prominent spectral components and the fundamental frequency. Knowledge of the sound of the harmonic series will prove valuable here. Map the presence of these frequencies against the time line, clearly showing their entrances and exits from the spectrum. Finally, map any changes in the pitch/frequency levels of these partials against the time line. Certain spectral components may not be present throughout the duration of the sound. It is not unusual for harmonics and overtones to enter and exit the spectrum.

The evaluation is complete when all of the spectral components that can be perceived by the listener are added to the graph. Accuracy and detail will increase markedly with experience and practice on the part of the listener. With time and acquired skill, this process will yield much significant information on the sound. Initial attempts may not yield enough information to accurately define the sound source, but will improve substantially over time.

7. Plot the dynamic activity of the partials on the spectral envelope tier of the graph. Use the same RDL as the dynamic envelope tier. First, establish the beginning and ending levels of each of the spectral components that were identified in Step 6. For each of the spectral components, determine the highest or lowest dynamic levels and any other prominent points of reference within the dynamic contours. Use these points of reference to evaluate the preceding and following material. Alternate focus on the contour, speed, and amounts of level changes to complete the plotting of the activity. The dynamic envelopes of all of the spectral components are plotted on this tier. The evaluation is complete when the smallest significant dynamic level change has been incorporated into the graph.

Many hearings will be involved for each of the above steps. Each listening should seek specific new information and should confirm what has already been noticed about the material. Before listening to the material, the listener must be prepared to extract certain information, to confirm their previous observations, and to be receptive to new discoveries about the sound quality. The listener should check their previous observations often, although their listening attention may be seeking new information.

The sound quality characteristics graph incorporates:

1. A multitiered Y-axis, distributed to complement the characteristics of the musical example: one tier with dynamic areas for dynamic envelope (with notated RDL), one tier with pitch area register designations for spectral content, one tier with dynamic areas for spectral envelope (with notated RDL), and a fourth tier designating a pitched to nonpitched continuum;
2. The X-axis of the graph is dedicated to a time line that is devised to follow an appropriate increment of clock time (the increment will vary depending on the material);
3. Each spectral component is plotted as a single line, against the two axes; its pitch characteristics on the spectral content tier, and its dynamic contour on the spectral envelope tier;
4. Each spectral component's line should have a different color or composition, to allow the viewer of the graph to compare the two tiers.

Sample Evaluations

Several sound quality evaluations follow. Two are synthesized sounds and the other is a highly modified (feedback, etc.) electric guitar sound. These invented timbres are evaluated to determine their unique characteristics for critical listening use and to better understand their relationships to other sounds in the music.

Appearances of the sound sources where their characteristics are not being masked by other sound sources or by performance technique were selected. Each evaluation is of one identified sounding of the instrument.

Figure 8-2 shows the great complexity of the pitch definition of the opening guitar sound from "Its All Too Much" (*Yellow Submarine,* 1999). The pitch definition changes are reflected in changes to the spectral envelope. They are closely associated.

Figures 8-3 and 8-4 are sound quality evaluations of Moog synthesizer sounds from *Abbey Road*. The different waveforms used for the two sounds make for differences in spectral content. The simplicity of the early instrument is reflected in the basic contours of the dynamic envelope and spectral envelope, and the inclusion of only harmonics in the spectrum of each sound.

As an exercise, add detail to these three sound quality evaluations, while checking the existing information for accuracy.

The reader should perform the Sound Quality Evaluation Exercise at the end of this chapter. A very important skill will be gradually acquired with some focused effort. As skill develops over a period of time, the recordist will gain considerable awareness to the content of sound qualities. The recordist will begin to hear things they previously could not imagine. A new world of sound will present itself—as will a new way of understanding that world.

Summary

The ability to evaluate and communicate about sound quality and timbre is extremely important for the synthesist, sound designer, recording engineer, and producer. It is also required of nearly all positions in the audio industry.

The sound quality evaluations performed by following the method presented in this chapter will allow the reader to readily recognize the unique characters of sounds. Learning to recognize and describe the activities of the component parts of sound quality will allow the reader to talk about sound articulately and to share information that is pertinent and readily understood by others.

The skills gained through the previous chapters are brought together in the many steps of sound quality (and timbre) evaluation. Pitch and

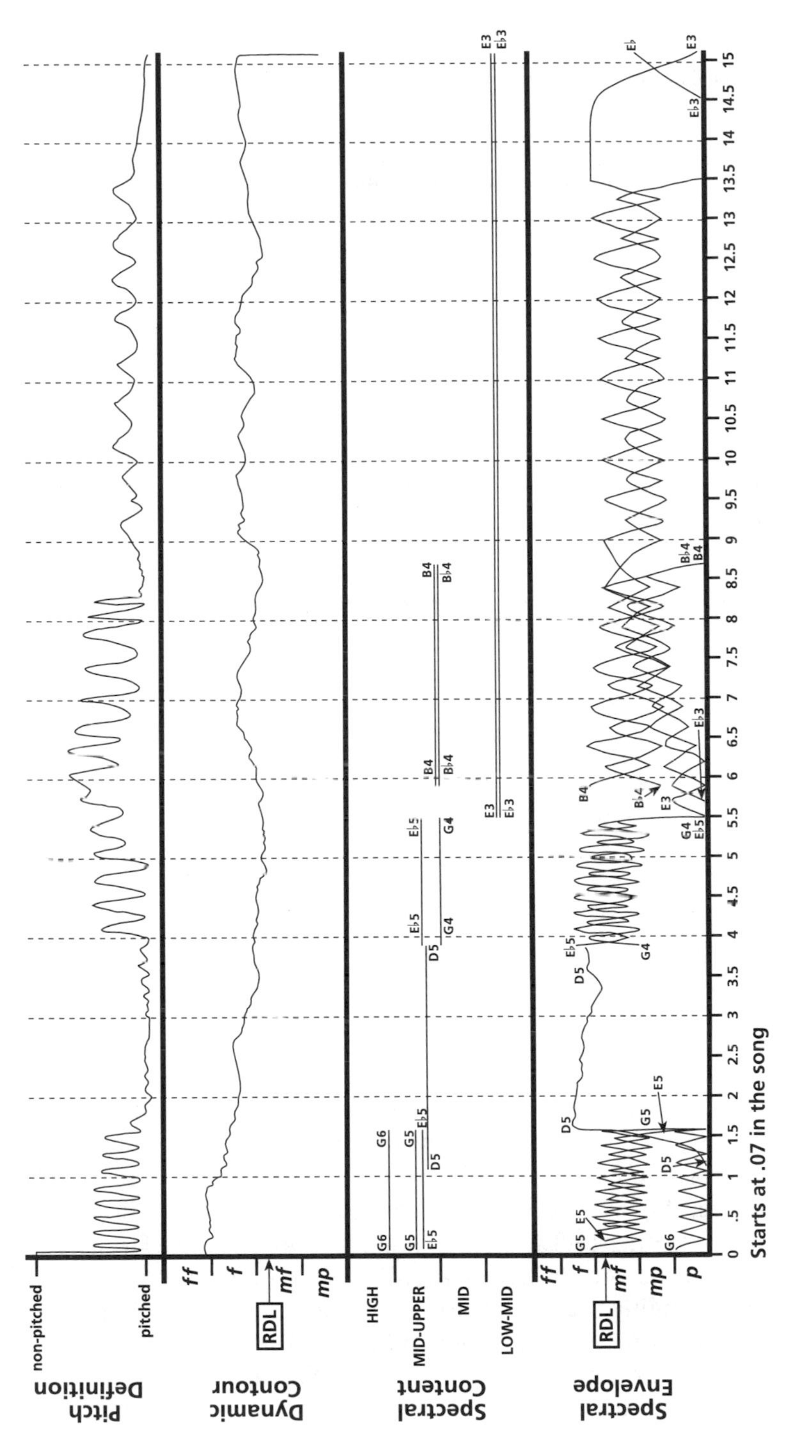

Figure 8-2 Sound quality evaluation of the opening guitar sound from The Beatles, "It's All Too Much" (*Yellow Submarine*, 1999).

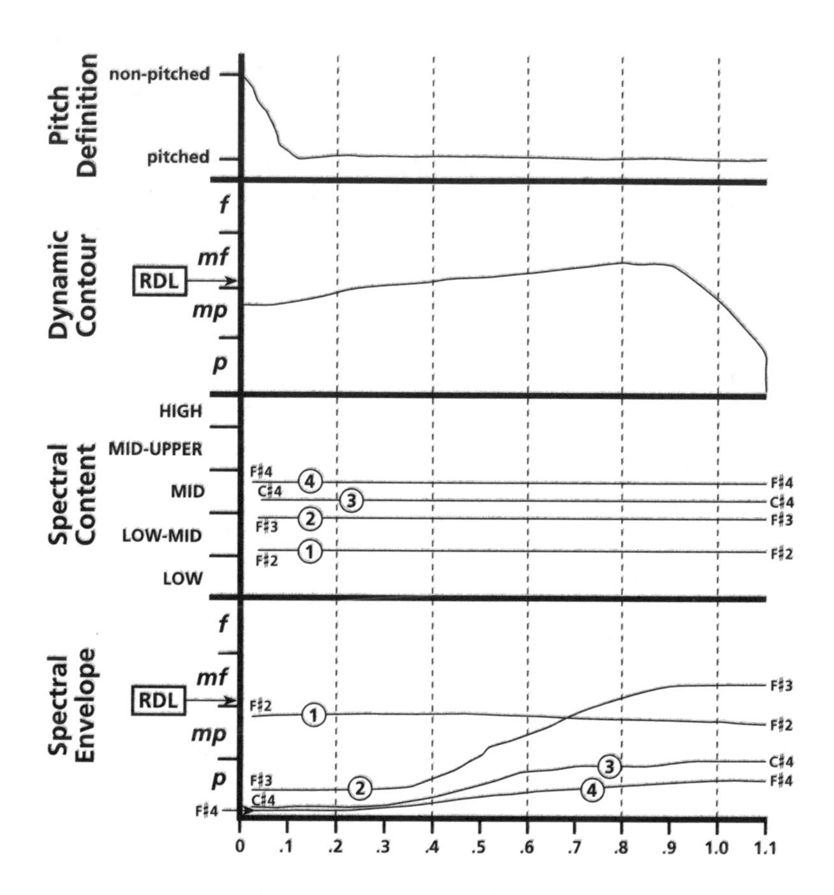

Figure 8-3 Sound quality evaluation of the Moog synthesizer sound from The Beatles, "Maxwell's Silver Hammer," at 51.1 seconds (*Abbey Road*).

pitch-area estimation, melodic and dynamic contour mapping, and judging time increments are all now used for a more demanding task, and a very important one. These skills should become highly refined, and continually developed through carefully considering how the recording process is altering, capturing, or creating sound quality. The audio professional will be continually engaged in evaluating sound. Recognizing and understanding the characteristics of sound quality are the first steps towards communicating accurate and relevant information about sound. Learning to hear and recognize the components of sound quality/timbre will allow the recordist to control their craft in shaping recordings.

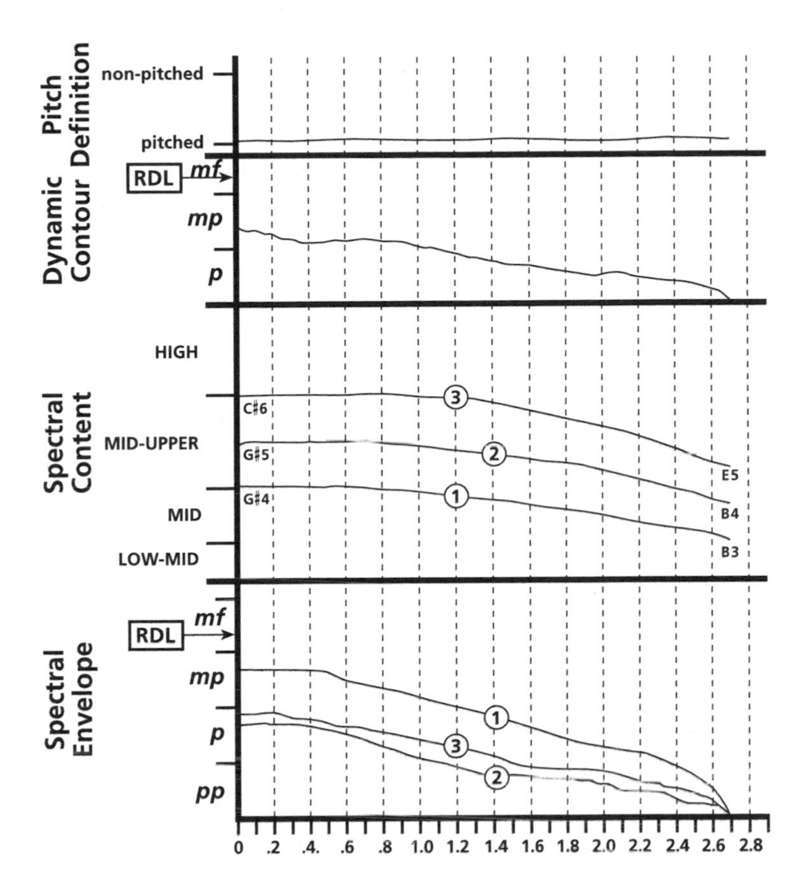

Figure 8-4 Sound quality evaluation of the first Moog synthesizer sound from The Beatles, "Here Comes the Sun," at 0:12 (*Abbey Road*).

Exercises

Exercise 8-1
Sound Quality Evaluation Exercise.

Find a sound that has some components of sound quality that change over time and some that do not. The sound should also be a complex waveform (containing overtones as well as harmonics) and have a duration of at least five seconds. A single occurrence of the sound should be identified and evaluated. A presentation of the sound that does not have other sounds occurring simultaneously will be easiest to evaluate.

Follow the sequence of events listed for the process of evaluating sound quality in the chapter to evaluate the sound quality of the sound selected. Making a recording of the single presentation of the sound that will be evaluated will allow the listener to more easily repeat hearings of the sound.

Use the Sound Quality Evaluation Graph for the evaluation, and summarize and describe findings using an objective vocabulary based on the physical dimensions of the sound.

---

9 Evaluating the Spatial Elements of Reproduced Sound

The spatial characteristics and relationships of sound sources are an integral part of music and audio productions. Spatial elements are precisely controllable in audio recording, and sophisticated ways of using these elements have developed in audio and music productions. Spatial relationships and characteristics contribute significant information in current music productions, and are important in critical listening applications.

The evaluation of the spatial characteristics of stereo recordings covers three primary areas: (1) localization on a single horizontal plane in front of the listener, (2) localization in distance from the listener, and (3) the recognition of environmental characteristics. Surround sound recording replaces localization in front with localization 360° around the listener. The elements of environmental characteristics and distance illusions further interact and create other sound characteristics that must be evaluated in both formats.

The recordist must be able to evaluate these characteristics to properly evaluate recorded/reproduced sound. The skills required to evaluate the spatial characteristics of a recording have been gradually developed throughout the previous four chapters. These skills of sound quality evaluation, time judgments, pitch estimation, and dynamic contour mapping will be used again (from a new perspective) to recognize and evaluate the spatial elements of reproduced sound. The further development of these skills will again require patience and practice.

An accurate evaluation of the spatial elements is only possible under certain conditions. The listener must be located correctly with respect to the loudspeakers of the playback system. This is critical to accurately hear directional cues. The sound system must interact correctly with the listening environment to complement the reproduced sound. Reproduced sound can be radically altered by the characteristics of the playback room and the placement of loudspeakers within the room. Further, the sound system itself must be capable of reproducing frequency, amplitude, and spatial cues accurately.

Many of the concepts of the spatial elements have not previously been well defined. The length of this chapter is the result of the number of important spatial elements of sounds in recordings, the methods one must use to perform meaningful evaluations of these elements, and the explanations required of new concepts.

Understanding Space as an Artistic Element

Spatial characteristics and relationships are used as artistic elements in music productions. They are used as primary and secondary elements that help to shape the unique character of musical ideas. Space has the potential of being the most important artistic element in a musical idea, but most often serves to support other elements. It may support other elements by delineating musical materials, by adding new dimensions to the unique character of the sound source or musical idea, and/or by adding to the motion or direction of a musical idea.

Perceived Performance Environment

The listener perceives the spatial relationships of reproduced sounds through an awareness of sound stage and imaging. They will imagine a performance space wherein the reproduced sound can exist during the re-performance of listening to the recording. The listener will perceive individual sound sources to be at specific locations within this *perceived performance environment*.

The recording represents an illusion of a live performance. The listener will conceive the performance as existing in a real, physical space, because the human mind will interpret any human activity in relationship to the known, physical experiences of the individual. The recording will appear to be contained within a single, perceived physical space (the perceived performance environment), because in human experience we can only be in one place at one time.

The perceived performance environment will have an audible, characteristic sound quality that is established in one of two ways. The qualities of the perceived performance environment may be established by applying a set of environmental characteristics to the overall program (i.e., putting the final mix through a reverb). Most often, the perceived performance environment is a composite of many perceived environments and environmental cues. In these instances, the listener formulates an impression of a perceived performance environment through interrelationships of many environmental characteristics cues. These cues may be (1) common or complementary between the environments of the individual sound sources, (2) prominent characteristics of the environments of prominent sound sources (source that presents the most important musical materials, or the loudest, the nearest, or the furthest sound sources, as examples), and/or (3) the result of environmental characteristics found in both of these areas.

Further, the listener will perceive themselves to be at a specific location within the perceived performance environment. The listener might be aware of their relationship to the sidewalls and any objects (balconies, seating, etc.) in the performance environment, to the wall behind and the ceiling above their location, and of their relationship to the front wall of the performance environment.

The listener will also calculate their location with respect to the front of the sound stage.

Sound Stage and Imaging

Within this perceived performance environment is a two-dimensional area (horizontal plane and distance) where the performance is occurring—the *sound stage*. The sound stage is the location within the perceived performance environment, where the sound sources are perceived

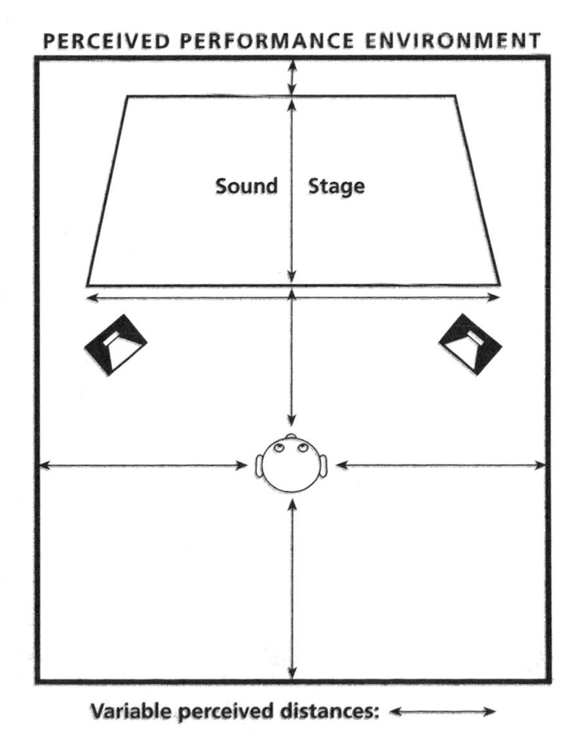

Figure 9-1 The sound stage within the perceived performance environment.

to be collectively located, as a single ensemble. The listener will unconsciously group all sources into a single performance, emanating from a single location.

The area of the sound stage may be any size. The size of the sound stage may appear to be anything from a small well-defined point (an infinitesimally small world), to a space occupying an area extending from immediately in front of the listener to a location (spanning a distance) well beyond our reality or imagination (perhaps conceived as being an area beyond the size of anything known within human experience) and, within the horizontal plane, filling an area beyond the stereo array.

The sound stage may be located at any distance from the listener. The placement of the front edge of the sound stage may be immediately in front of the listener, or at any conceivable distance from the listener.

The perceived distances of the sound sources from the listener determine the depth of the sound stage. The sound source that is perceived as nearest to the listener will mark the front edge of the sound stage. The sound source that is perceived as being furthest from the listener will define the back of the sound stage and will also help to establish the rear wall of the perceived performance environment. The perceived location of the rear boundary (wall) will be determined by the relationship of the furthest sound source to its own host environment. The rear wall of the perceived performance environment may be located immediately behind the furthest sound source, or some space may exist between the furthest sound source and the rear wall of the sound stage/perceived performance environment.

All sound sources will occupy their own location in the sound stage. Two sound sources cannot be conceived as occupying the same physical location. Our sensibilities will not allow this to occur. It is possible for different sound sources to occupy significantly different locations within the sound stage, anywhere between the two boundaries. *Imaging* is the perceived location of the individual sound sources within the two perceived dimensions of the sound stage (see Figure 9-2). Sources are located within the sound stage by their angle (on the horizontal plane) and distance from the listener.

Environments of Individual Sources

Imaging will be influenced by the characteristics of the unique performance environments of each individual sound source, and the placement of each source within its own environment. In current music productions, it is common for each instrument (sound source) to be placed in its own host environment. This host environment of the individual sound source (a perceived physical space) is further imagined to exist within the per-

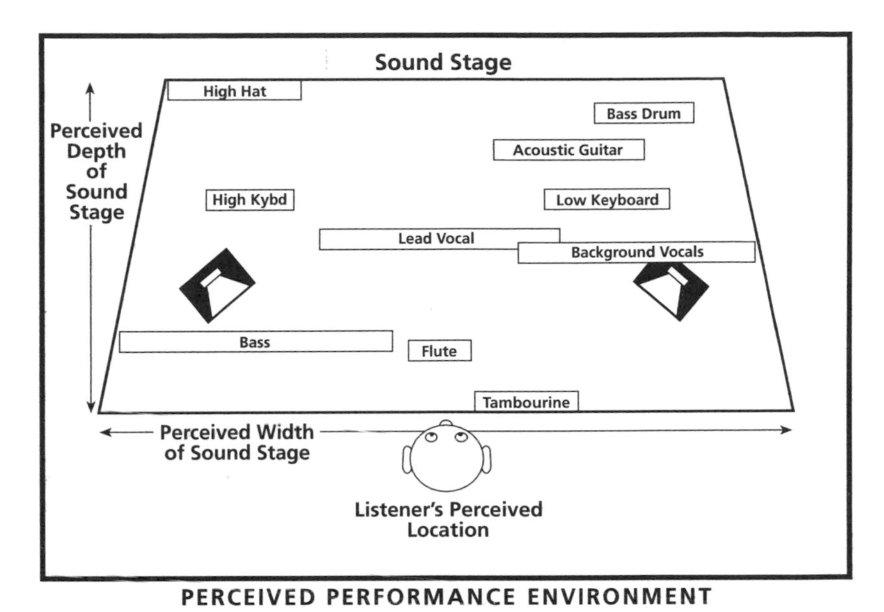

Figure 9-2 Imaging of sound sources within the sound stage.

ceived performance environment of the recording (space). This creates an illusion of a *space* existing *within* another *space*.

The environments of the sound sources and the overall program may be of any size. The acoustical characteristics of any space may be simulated by modern technology. The sound sources may be processed so that the cues of any acoustical environment may be added to the individual sound source, to any group of sound sources, or to the entire program. Not only is it possible to simulate the acoustical characteristics of known, physical spaces, it is possible to devise environment programs that simulate open air environments (under any variety of conditions) and programs that provide cues that are acoustically impossible within our known world of physical realities.

Figure 9-3 presents an easily accomplished set of environmental relationships, with individual sound sources appearing in very different and unique environments:

- Timpani placed in an open-air environment
- A stringed instrument placed in a large concert hall
- A vocalist performing in a small performance hall
- A piano sounding in small room
- A cymbal appearing to exist in a very unnatural (perhaps otherworldly or outer space), remarkably large environment

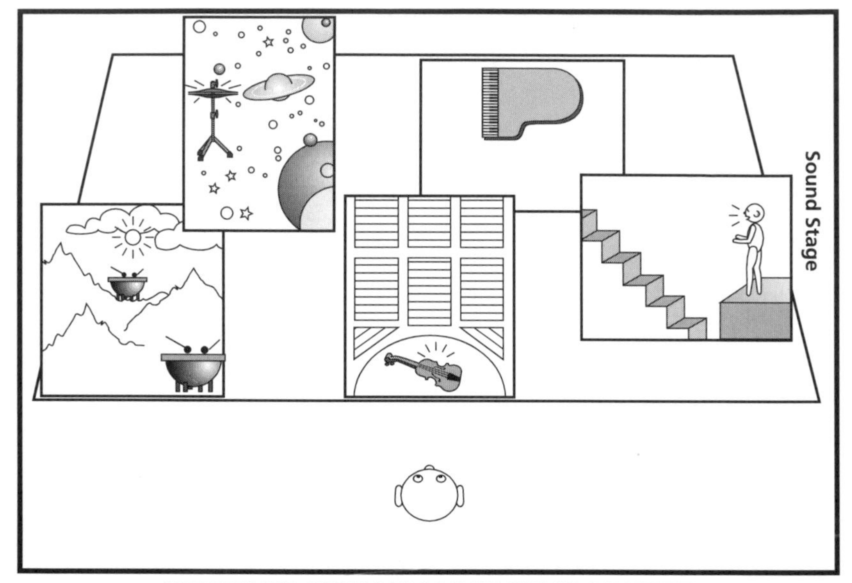

Figure 9-3 Space within space.

These many simulated acoustical environments are perceived as existing within the overall space of the perceived performance environment. The spaces of the individual sources are within the space of the perceived performance environment. It is possible for more than one source to be placed within an environment. Sources contained within the same environment may have considerably different distance locations, or they may be similar.

The environments of the sound sources and the overall program may be in any size relationship to one another. The environment of a sound source may have the characteristics of a physically large space, and the perceived performance environment may have the characteristics of a much smaller physical environment. This is a common relationship, and the reverse is also possible (though more difficult to achieve). The spaces of the individual sound sources are understood (by the listener) to exist within the all-encompassing perceived performance environment, no matter the perceived physical dimensions of the spaces involved.

The spaces of the individual sound sources are subordinate spaces and exist within the overall space of the recording. A further possibility (not commonly used, at present) exists for subordinate spaces to appear within other subordinate spaces, within the perceived performance environment. *Space within space* is a hierarchy of environments existing within other environments. Its creative applications have not been fully exploited in current music production practices.

The characteristics of the perceived performance environment function as a reference for determining the characteristics of the individual environments of the individual sound sources. All of the environments of a recording will have common characteristics that are created by the perceived environmental characteristics of the perceived performance environment (as discussed above). These characteristics provide a reference for determining the unique characteristics of the individual performance environments of the individual sound sources.

The characteristics of the perceived performance environment also function as a frame of reference for the listener in determining the distance locations of the individual sound sources within the sound stage.

Distance in Recordings

Distance is perceived as a definition of timbral detail. It is further calculated in relation to the characteristics of the environment in which the sound exists, as well as the perceived location of the sound source and the listener within that environment. The listener will perceive the distance of the sound source as it is sounding within its unique environment. The listener will then unconsciously transfer that distance to the perceived performance environment, combining any perceived distance of the source's environment from the listener's location in the perceived performance environment.

The actual distance location placement of the sound source within the sound stage is determined by (1) the distance between the sound source and the perceived location of the listener within the individual source's environment combined with (2) the perceived distance of that environment from the listener's location in the perceived performance environment. All this information blends into a single impression of distance. Through this process, sound sources (with and within their environments) are conceived at specific distances from the listener. This is all accomplished subliminally.

The placement of sounds at a distance and at an angle from the listener (imaging) takes place at the perspective level of the perceived performance environment.

Directional Location

Sound sources (with their individual environments and conceived distance locations) will be located at an angle from the listener. Directional location is used differently in stereo and surround formats.

The *stereo location* of the sound sources will place them on the sound stage, within the stereo loudspeaker array, at an angle of direction from the listener. The size of the sound source will be a narrow and precisely

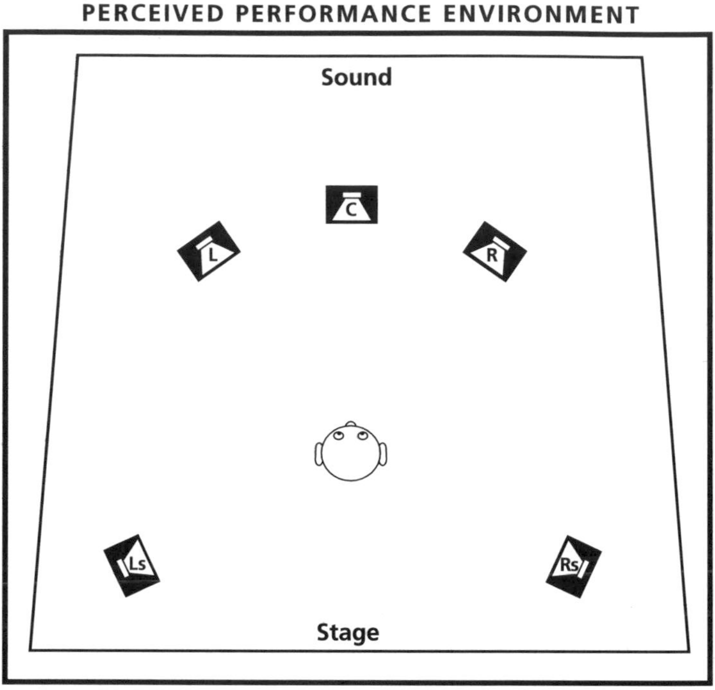

Figure 9-4 Listener within the sound stage in surround sound.

defined point in space, or it will be an area between two boundaries. Sources that occupy an area may be of any reproducible width and may be located at any reproducible location within the stereo array. Further, under certain production practices, it is possible for sound sources to appear to occupy two separate locations or areas within the stereo array.

Surround location of sound sources is more complex. Location of sound sources may be at any angle from the listener. The listener may actually be placed within the sound stage (Figure 9-4), or the surround speakers may be used solely for environment information, allowing the sound stage to remain in front of the listener (Figure 9-5). Sound source size (width) remains variable, but now may be anything from a single point in space to completely enveloping the listener. Further, sources may readily occupy several locations simultaneously.

Production practice for surround is currently being defined, and the potentials of the medium are being explored. How the complex potential of sound location around the listener ultimately translates into our music listening experiences will be defined over the upcoming years. The listener and audio professional must only remain receptive to possibilities, as this new technology shapes music, music listening and the creation of recordings.

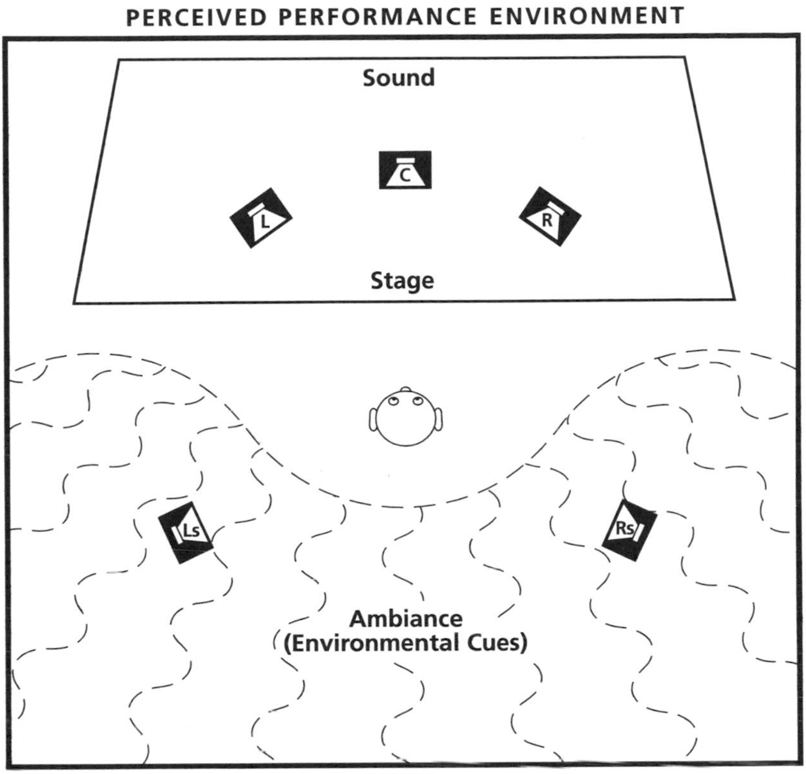

Figure 9-5 Listener enveloped by environment cues, sound stage in front.

Stereo Sound Location

Sound location is evaluated within the stereo array to determine the location and size of the images of the sound sources. These cues will hold significant information for understanding the mix of the piece and may contribute significantly to shaping the musical ideas themselves. Phantom images may also change locations or size during a piece of music. These changes may be sudden or gradual. The changes may be prominent or subtle.

The *stereo sound-location graph* will plot the locations of all sound sources against the time line of the work. The graph portrays the direction of sources from the listener and the size of the phantom images.

Left and right loudspeaker locations, and the center position are identified on the graph. The actual boundaries of the vertical axis extend slightly beyond the loudspeaker locations (up to 15°). Source angle from the listener is represented by placing the sound source's location within the L-R speaker-location boundaries. Precise degree-increments of angle are not incorporated into the graph.

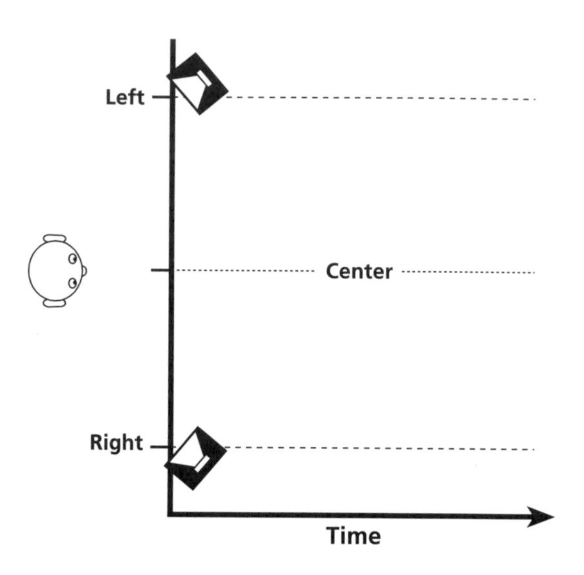

Figure 9-6 Continuum for stereo sound location.

The *stereo sound location graph* incorporates:

1. The left and right loudspeaker locations, a designation for the center of the stereo array, and space slightly beyond the two loudspeaker locations as the Y-axis;
2. The graph will become unclear if too many sound sources (especially many spread images) are placed on the same tier; the Y-axis may be broken up into any number of similar tiers (each the same as listed in Step 1) to clearly present the material on a single graph;
3. The X-axis of the graph is dedicated to a time line that is devised to follow an appropriate increment of the metric grid, or is representative of a major section of the piece (or the entire piece) for sound sources that do not change locations;
4. A single line is plotted against the two axes for each sound source, the line will occupy a large, colored/shaded area in the case of the spread image; and
5. A key will be required to clearly relate the sound sources to the graph. It should be consistent with keys used for other similar analyses (such as musical balance and performance intensity) to allow different analyses to be easily compared.

A sound source may occupy a specific point in the horizontal plane of the sound stage, or it may occupy an area within the sound stage. The graph will dedicate a source line to each sound source and will plot the stereo locations of each source against a common time line.

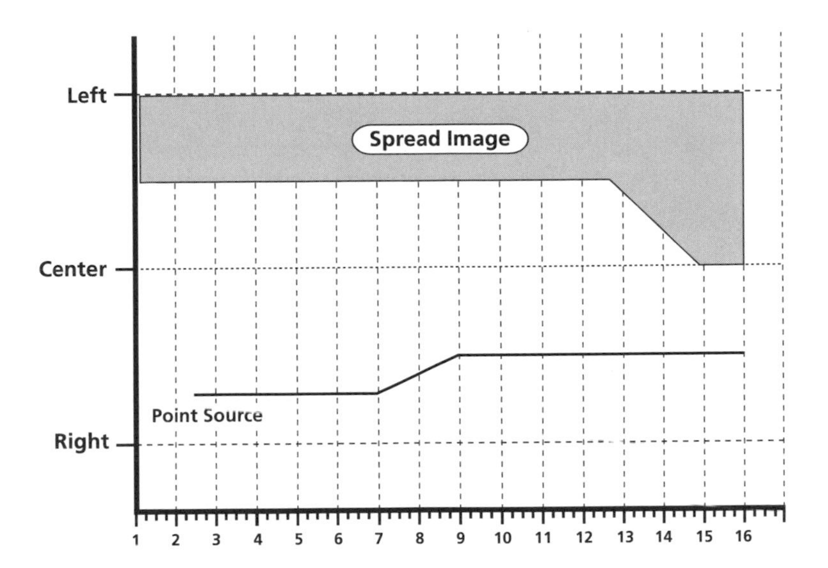

Figure 9-7 Stereo sound-location graph.

The source line for point source images will be a clearly defined line at the location of the sound source. The source line for the spread image will occupy an area of the sound stage and will extend between the boundaries of the image itself.

It is common for stereo sound-location graphs to be multitier, placing spread images on separate tiers from point sources or providing a number of separate tiers for spread images. Figure 9-7 presents a single-tier stereo sound-location graph, containing two sound sources: one spread image and one point source.

Figure 9-8 presents the stereo sound location of a number of the sound sources from The Beatles' work, "A Day in the Life." The location, size, and movements of the sound source images directly contribute to the character and expression of the related musical materials. As an exercise, listen to the recording and notice the placement of the percussion sounds. The stereo locations of the percussion sounds complement the placements of the voice, bass, piano, guitar, and maracas to balance the sound stage.

Locations and image size do not often change within sections of a work. Changes are most likely to occur between sections of a piece, or at repetitions of ideas or sections (where changes in the mix often occur). The listener should, however, never assume changes will not occur. Gradual changes in source locations and size are present in many pieces and sometimes in pieces where such events are not expected. The changes in source size and location from *Abbey Road* works discussed in Chapter 2

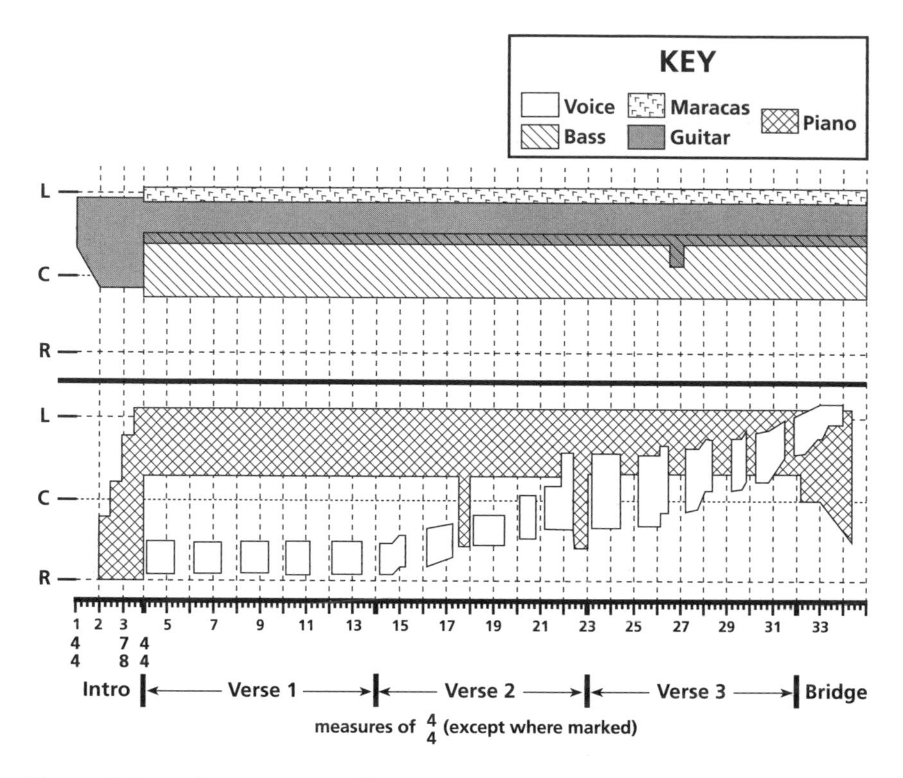

Figure 9-8 Multitier stereo sound-location graph—The Beatles, "A Day in the Life."

will not be recognized unless the listener is willing to focus on this artistic element and is prepared to hear these changes.

The reader should work through the Stereo Location Exercise (9-1) at the end of this chapter to refine this skill.

Distance Localization

Distance localization and stereo localization combine to provide the imaging of the sound source. Figure 9-9 is an empty stereo sound stage, onto which sound sources are imagined to be located. Placing sounds on this empty sound stage will allow the listener to make quick, initial observations regarding imaging. These observations can then lead to the more detailed evaluations of stereo location and of distance. It is important to note, these location observations relate to specific moments in time or to sections of a work (specific periods of time). While location changes cannot be written on this figure, it is very useful for initial observations, for graphing sources that do not change, and for planning mixes.

Distance is the perceived location of the sound source from the listener. It is a location where the listener envisions the sound to be placed

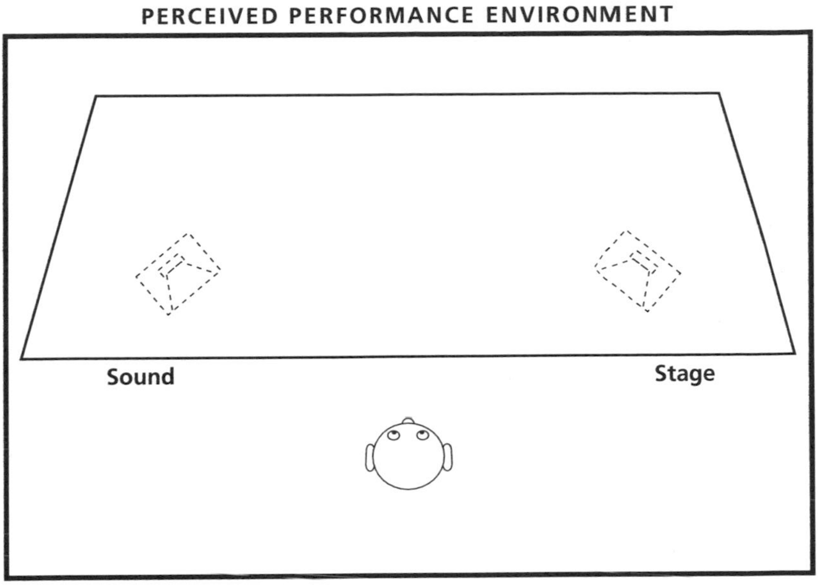

PERCEIVED PERFORMANCE ENVIRONMENT

Sound Stage

Figure 9-9 Empty sound stage.

along the depth of the sound stage. Humans perceive sounds to occupy a precise distance location. Sounds do not occupy distance areas. A source environment may provide an area of depth to the image, but the source will be heard as located at a precise point within that environment. The environment and source fuse into a single sound impression that will occupy an area of distance, with the source localized in front.

The perceived source locations nearest to and furthest sources from the listener establish the front and rear boundaries of the sound stage. The front edge of the sound stage may be immediately in front of the listener, or at any distance. The depth of the sound stage will be conceived as the area encompassing all sound sources.

Understanding Distance Location

The reader must approach distance location carefully. Distance cues are often not accurately perceived. Many activities of other artistic elements are confused with distance. Further, humans mostly rely on sight to calculate distance and are not normally called upon to focus on aural distance cues.

Distance is NOT loudness. In nature, distant sounds are often softer than near sounds. This is not necessarily the case in recording production. Loudness does not directly contribute to distance localization in audio recordings. At times loudness and distance cues are associated in recordings, but this is often not the case—especially in multitrack and synthesized productions. A "fade out" can be accomplished without causing source distances to increase. Conversely, a fade out may cause sound sources to be perceived as increasing in distance. The distance increase will be the result of a diminishing level of timbral detail, it will not be the result of decreasing dynamic level.

Very often people will describe a sound as being "out in front," implying a closer distance. The sound may actually be louder than other sounds or may stand out of the musical texture because of the prominence of some other aspect of its sound quality. Much potential exists for confusing distance with dynamic levels, and other aspects of the sound or the musical context that might draw the listener's attention.

Distance is NOT determined by or the result of the amount of reverberation placed on a sound source. In nature, distant sounds are often comprised of a high proportion of reverberant energy in relation to direct sound. Reverberant energy does play a role in distance localization, but not so prominent a role that it can be used as a primary reference. Reverberant energy is most important as an attribute of environmental characteristics and in placing a sound source at a distance within the individual source environment. Humans perceive distance within environments, through time and amplitude information extracted from processing the many reflections of the direct sound. The ratio of direct to reflected sound influences distance location. Thus, reverberation contributes to our localization of distance, but it is NOT the primary determinant of distance location, in and of itself.

Distance is NOT the perceived distance of the microphone to the sound source that was present during the recording process. The only exception to this statement occurs when the initial recording is performed with a single stereo-pair of microphones, and no signal processing is performed on the overall program. Microphone to sound source distance is a contributor to the timbral characteristics of the sound source. Microphone to sound source distance will determine the amount of definition of the sound source's timbre (how much timbral detail is present in the sound) captured by the recording process. It will also determine the amount of the sound of the initial recording environment that has become part of the sound source's timbre. Generally, the closer the microphone to the sound source, the greater the definition of timbral components captured during the recording process. This will provide distance localization information, in such a way that very close microphone placement will cause the image to be perceived very close to the listener (if no timbral modifications or signal processing is performed in the mixing process). The sound quality may be significantly altered in the mix, significantly altering microphone to sound source distance cues. Microphone to sound source distance contributes to the overall sound quality of the source's timbre. It contributes to our localization of

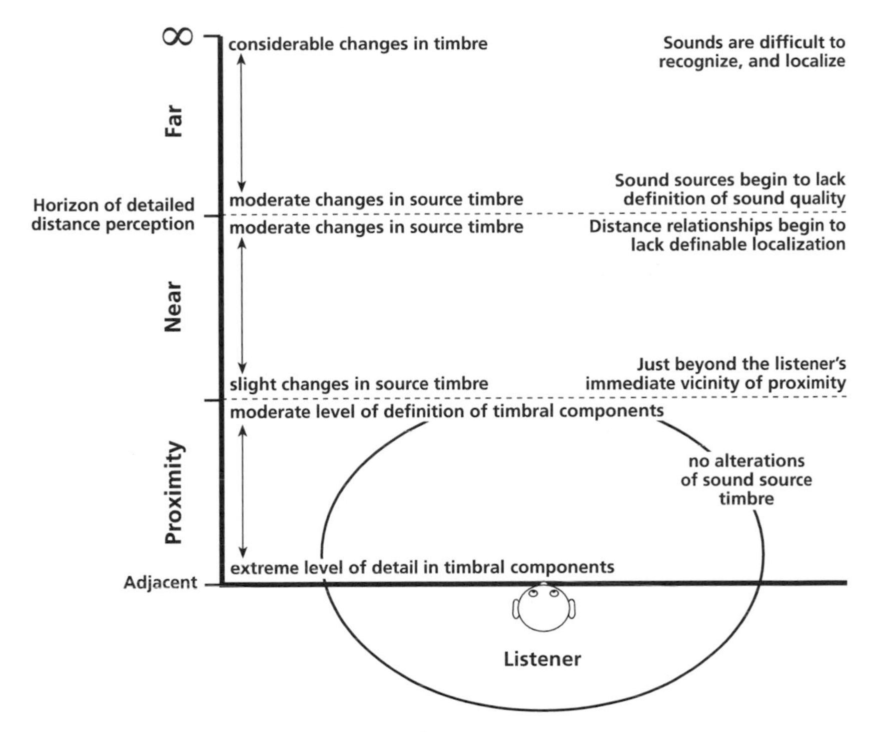

Figure 9-10 Continuum for designating distance location.

distance through definition of timbral detail, but it is NOT a primary determinant of distance localization.

Distance location IS primarily the result of timbral information and detail. Timbre differences between the sound source as it is known in an unaltered state and the sound as it exists in the host environment of the recording primary determine distance localization. The listener is aware of how timbres are altered over various distances. It is through the perception of these changes that we formulate the distance of a source from our listening location. Humans are unable to estimate the physical distance (meters, feet, etc.) of a sound source from their location. We perceive distances in relative terms and compare locations to one another.

We rely on timbral definition for most of our distance judgments and use the ratio of direct-to-reverberant sound to a lesser degree. The extent to which we rely on either factor depends on the particular context. Distance localization is a complex process, relying on many variables that are inconsistent between environments.

The listener knows the sound qualities of sound sources within the area immediately around them. The listener has a sense of occupying an area, encompassing a space immediately around them. This area serves as a reference from which we judge "near" and "far." Within this area, sounds have no changes in timbre. All characteristics of timbral content are present, and sounds will have more detailed definition the closer they are to the listener. The overall sound quality of the sound source may be somewhat altered by the characteristics of the host environment, but the level of detail present in the timbre causes the listener to perceive the source as being within their immediate area of *proximity*. This space that immediately surrounds the listener is called proximity and is the listener's own personal space. Sounds in proximity are perceived as being close, as occurring within the area the listener occupies. The actual size or area of proximity may be perceived as being rather large or very small, depending on the context of the material and the perspective of the listener.

The listener knows the sound qualities of sound sources at "near" distances. Humans conceive "near" as being immediately outside of the area that they perceive themselves as occupying. Throughout this "near" area, the listener is able to localize the sound's distance with detail and accuracy. Timbres are very slightly altered in the closest of sounds considered near and are moderately altered in the furthest of sounds considered near. An area will exist between these two boundaries where sounds are readily compared as being closer or farther than other similar sounds. Sounds cease to be considered near when the listener begins to have difficulty localizing distances in detail.

"Far" sound sources lack definition of sound quality. The closest of far sounds will have moderate alterations to sound quality, with little or no definition. Few low amplitude partials will be present, and amplitude and frequency attack transients will be difficult to detect. The furthest of far sounds will have considerable alterations to sound quality; the sounds will lack all definition. The furthest far sounds may even be difficult to recognize. An area will exist between these two boundaries of "far." It could conceivably be quite large, perhaps stretching to infinity.

Evaluating Distance Location

The listener will initially focus on distance cues of the sound source at the perspective of the source within its own host environment. The listener will intuitively transfer that information to the perspective of the sound stage. There, the sound source's degree of timbre definition, the perceived distance of the sound source within its host environment, and the perceived distance of the source's host environment from the perceived location of the listener blend instinctively into a single perception of distance location. This process will determine the actual perceived distance of the sound source, at the perspective of imaging. Fortunately, this all happens quite instinctively.

Again, the definition of, or the amount of timbral detail present will play the central role in determining perceived distance.

The continuum for distance location extends from "adjacent" to "infinity." Adjacent is that point in space that is immediately next to the space the listener is occupying. It should be conceived literally as being the next molecule available beside the listener, as a sound may be localized at that location.

The continuum for distance localization consists of three areas. The areas represent conceptual distance, not physically measurable distance increments. Distance is judged as a concept of space between the sound source and the listener. An area of proximity surrounds the listener. This area serves as a reference for judging near and far distances. Human experience of the nature of sound is used as a reference to conceptualize the amount of space (distance) between the source and the listener.

The three areas of the continuum are:

1. An area of "proximity," the space in which the listener perceives as their own area, is the area immediately surrounding the listener that may be extended in size to be conceived as the size of a small to moderately-sized room. The listener will perceive the proximity area as being their own immediate space;
2. A "near" area is the area immediately outside of the space that the listener perceives themselves as occupying, extending to a horizon where the listener begins to have difficulty localizing distances in detail; and
3. A "far" area, beginning where perception dictates that space ceases to be "near;" where detailed examination of the sound is difficult, and extending to where sounds are almost impossible to recognize. Extreme far sound sources contain very little definition of sound quality.

These three areas are not of equal physical size. The amount of physical distance contained in the conceptual area of proximity will be considerably different than the physical distance encompassed by the conceptual area of far. All three areas of the continuum occupy a similar amount of conceptual space, but represent significantly different amounts of physical area. The vertical axis of the distance location graph must clearly divide the three areas.

The size of the three areas may be adjusted between appearances of the distance location graph. The amount of vertical space occupied by the areas may be adjusted to best suit the material being graphed, with certain areas being widened in certain contexts and narrowed in others. The far area may even be omitted in certain contexts. The area of proximity should always be included (although if necessary it may be narrowed to occupy less vertical space), to clearly present the conceptual distance between the perceived location of the listener and the front edge of the sound stage.

Sound sources will be placed on the graph (1) by evaluating the definition of the sound quality of each sound source (the amount of detail present in the timbre of the sound source), and incorporating information on the ratio of direct-to-reverberant sound and the quality of the reverberant sound as appropriate, and (2) by directly comparing the sound source to the perceived distance locations of the other sound sources present in the musical context (using proportions of different locations between three or more sound source distances, to make more meaningful comparisons).

The individual listener's knowledge of timbre and environmental characteristics, and their ability to recognize the sound source are variables that may cause the listener to inaccurately estimate distance. For example, a very close tamboura may sound like a far sound to a person who does not know the sound of a tamboura. As the life experience of listeners varies, so does an individual's ability to conceptualize the distance relationships of sounds.

During initial studies, distance judgments may be difficult to conceive and perceive. Distance is, however, a central concern of sound source imaging, and thus of music production. Skill in this area can be refined and should become highly developed.

The *distance location graph* incorporates:

1. Continuum from adjacent through infinity (divided into three areas) as the Y-axis;
2. The X-axis of the graph is dedicated to a time line that is devised to follow an appropriate increment of the metric grid;
3. A single line is plotted against the two axes for each sound source; and
4. A key will be required to clearly relate the sound sources to the graph. The key should be consistent with keys used for other similar evaluations (such as musical balance or stereo location) to allow different elements to be easily compared.

The locations of all sources are plotted as single lines. Sources are precisely located at a specific distance from the listener. No two sources can be at the same distance level unless they are at clearly different lateral locations (placement of phantom image in stereo or surround formats).

Sound sources do not often change distance locations in real-time or within sections of a work. Changes are most likely to occur between sections of a piece, at entrances or exists of individual sound sources, or at repetitions of ideas or sections (where changes in the mix often occur). The listener should, however, never assume changes will not occur. Gradual changes in distance are present in many pieces. The reader is encouraged to work through the Distance Location Exercise at the end of the chapter.

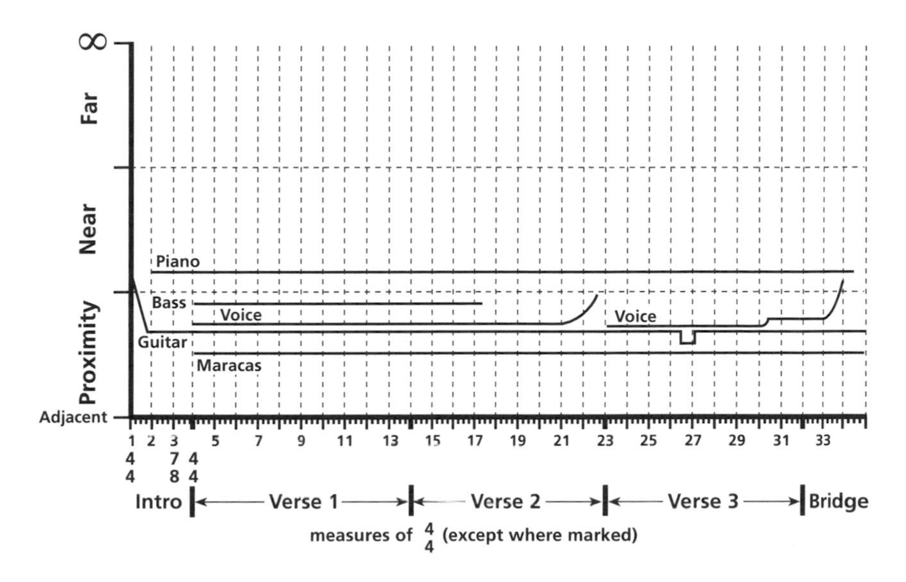

Figure 9-11 Distance location graph—The Beatles, "A Day in the Life."

Figure 9-11 is a distance location graph from The Beatles: "A Day in the Life." While the vocal line has a significant percentage of reverberant sound, its timbral detail brings it to be located in the rear-third of the proximity area. The other sound sources have widely varied distance locations, giving the sound stage great depth. The guitar, maracas, and bass are also in the proximity area, and the piano is at the front of the near area. As an additional exercise, the reader should place the percussion sounds in an appropriate distance location. Notice the great conceptual distances between the various instruments of the drum set, as some sounds are located toward the rear of the far area.

The fade of "Lucy in the Sky with Diamonds" is graphed in Figure 9-12. The vocal lines and bass do not change distance locations with diminishing loudness. The snare drum and organ are quite different, and do change distance locations as their loudness levels decrease. It is interesting to note that the speed and amounts of perceived distance location changes are different for the two sounds, as they decrease in loudness almost in parallel.

Environmental Characteristics

The characteristics of the sound source's host environment are important in shaping four qualities of the recording: (1) the overall quality of the sound source, (2) the perceived performance environment, (3) space within space, and (4) the imaging of the sound stage.

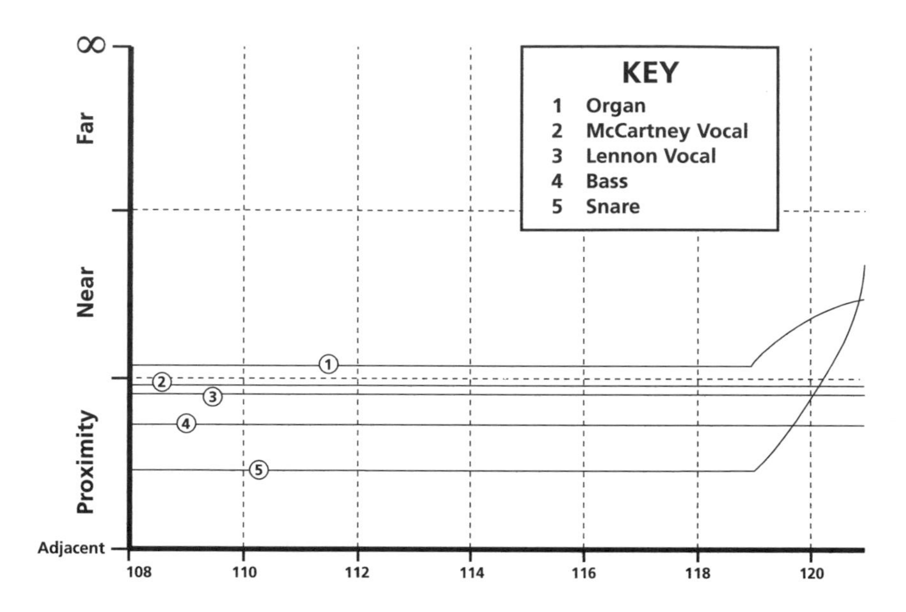

Figure 9-12 Distance location graph of the fade out from The Beatles, "Lucy in the Sky with Diamonds" (*Yellow Submarine*, 1999).

The environmental characteristics of the entire program (the perceived performance environment) shape the conceived space in which a performance is occurring. The characteristics of the conceived space of the envisioned performance environment will greatly influence the conceptual setting for the artistic message of the work.

Environmental characteristics of both the host environments of the individual sound sources and the perceived performance environment play significant roles in music production. These artistic elements have the potential to provide significant information for enhancing and communicating the musical message of the piece of music. Currently, they are most often used in supportive roles. They are coupled with sound quality in defining the unique characters of individual sounds (enhancing their sound quality and providing each sound with a sense of depth). Environmental characteristics are used as a separate element in creating depth of sound stage, in providing a resource for space within space, in creating the illusion of the perceived performance environment, and in giving breadth and depth to phantom images.

The recordist needs to be able to recognize the characteristics of the environments within which sound sources exist, and the characteristics of the perceived performance environment. This will lead to an understanding of the influence of the environment on the overall sound qualities and on the effectiveness of the sound sources and its musical materials.

Evaluating Environmental Characteristics

A composite sound of environmental characteristics occurs with the sounding of a sound source within the environment. The sound source and the environment interact and fuse to create a composite sound—a new overall sound quality. To understand the influence of the host environment on the sound source in making this composite sound, the environment and the sound source must be evaluated separately.

We perceive environmental characteristics as an overall sound quality that is comprised of a number of component parts. As global sound qualities, environmental characteristics are conceptually similar to timbre. The evaluation of timbre (sound quality) and environmental characteristics will be similar, in that both will seek to describe the states and activities of the physical components of sound. While people might recognize large halls, small halls, and other spaces as having common environmental characteristics, each environment is unique. To meaningfully communicate information about environmental characteristics, the audio professional must define the environment by its unique sound characteristics. These characteristics can only be objectively described through discussing the levels and activities of the component parts of the environment's sound.

Environmental characteristics appear as alterations to the sound source's timbre, created by the interaction of the sound source and the environment. The evaluation of the characteristics of the environment is therefore a definition of the changes that have occurred in the sound source's timbre after being placed within the environment. Evaluating environmental characteristics will engage activities that are contrary to our natural tendency to fuse the environment's sound with the source's sound. Care in focusing on the correct perspective and aspects of sound will be required.

Environmental characteristics are determined by the listener through comparing their memory of the sound source's timbre outside of the host environment to the sound source's timbre within the host environment. The listener must go through this comparison process carefully, scanning the composite sound for information and then comparing that information with their previous experiences with the timbre of the sound source (at times considering how the source appeared within other environments). Differences in the spectrum and spectral envelope of the sound source as remembered by the listener, and as heard in the host environment, form the basis for determining most environmental characteristics.

If the listener does not recognize the sound source (timbre) or has no prior knowledge of the sound source, they will be at a disadvantage in calculating the characteristics of the host environment. The listener will have no point of reference in determining how the environment has altered the timbre of the original sound source. They must rely on their knowledge of what they presume to be similar sounds to calculate estimations of the characteristics of the environment. This may or may not turn out to be accurate.

In evaluating environmental characteristics, the listener is seeking to define the characteristics of the environment itself. The listener must make certain they are NOT identifying characteristics of the sound source and must make certain they are NOT identifying characteristics of the sound source within the environment. The characteristics of the environment have specific component parts that contribute to its own unique sound quality. These characteristics are what must be determined by identifying the differences between the sound quality of the sound source itself and the sound quality of the sound source within the environment.

The component parts of environmental characteristics are (1) the reflection envelope, (2) the spectrum, and (3) the spectral envelope. The environmental characteristics graph (Figure 9-13) allows for the detailed evaluation of these three components.

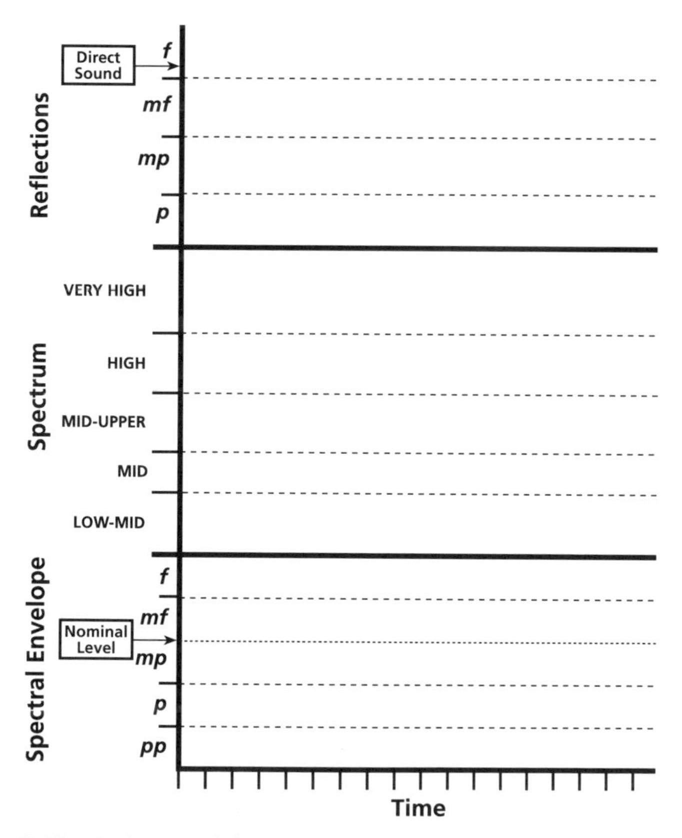

Figure 9-13 Environmental characteristics graph.

Reflection Envelope

The *reflection envelope* is comprised of the amplitudes of the initial reflections and the reverberant energy of the environment throughout the duration of the environment. This envelope is comprised of many reiterations of the sound source. The reiterations vary in dynamic level and in spacing between one another (time density).

The spacing of the reiterations of the sound source will be dramatically different over the duration of the sound of the environment (for example, the spacing of the early reflections will be considerably different from the spacing of reflections near the end of the reverberant energy). This portion of the graph can clearly show the time of arrival of the reflected sounds to the listener location, the density of the arrival times of the reflected sounds, and the amplitude of those arrivals in relation to the amplitude of the direct sound.

The amplitude of the direct sound is included to serve as a reference level for the calculation of the dynamic levels of the reflected sound. The recordist should use their skills at pattern recognition to extract time information from the sound of the environment. The reflections portion of the graph will show the following information. The listener should organize their listening to the time elements of the environmental characteristics to recognize this information:

- Patterns of reflections created by dynamics
- Patterns of reflections created by spacings in time
- Spacing of reflections in the early time field
- Dynamic contour of the entire reflections portion
- Density (number and spacings of reflections) of reverberant sound
- Dynamic relationships between the direct sound, individual reflections (of the early time field), and the reverberant sound
- Dynamic contour shapes within the reverberant sound

An isolated appearance of the sound source in the host environment must be found for all of the time and reflection-amplitude information to be accurately evaluated. This is especially true for the decay of the reverberant energy and the spacings of the early reflections (which play key roles in the characterization of the environment). A short (staccato) sound, from the sound source, will allow the reflection information to be most audible, since it will not have to compete with the sound of the source itself.

An exercise to develop the reader's skill in this area appears at the end of this chapter. The Reflections and Reverberation Exercise helps the reader understand and recognize this characteristic of environmental sounds even if they never wish to perform an evaluation as detailed as the reflection envelope of the environmental characteristics graph. The exercise will lead the reader to important observations that can serve as the basis for their understanding and use.

When listening to sounds in actual music recordings, hearing these characteristics is a much greater challenge—and may at times be impossible. Without the opportunity to hear the environment's complete presentation, information related to the reverberant energy might never be completely audible. The listener should try to find several appearances of a sound source where it can be heard alone, without other sound sources, and where it is playing short durations.

Many hearings to the sound in a wide variety of presentations will be necessary to compile an accurate evaluation of all of the time characteristics of the environment.

Environment Spectrum and Spectral Envelope

The spectrum of the reverberant sound and the initial reflections are a composite of the frequencies or bandwidths of pitch areas that are emphasized and de-emphasized by the environment itself. This spectrum will be only those frequencies that are altered by the environment. The environment may emphasize or de-emphasize bandwidths of frequencies or specific frequencies. Often the spectrum of the environment will only contain a small number (three to seven) of prominent frequencies or pitch areas that are either emphasized or de-emphasized.

These frequencies are determined through a careful evaluation of many appearances of the sound source in the environment, by listening to the way the sound source's timbre is changed by the environment over a wide range of pitch levels. Some appearances of the sound source will not have frequency information in certain frequency areas that are emphasized or de-emphasized by the environment. The listener must scan many pitch-levels of the sound source to determine the spectral content and the spectral envelope of its environment.

The spectral envelope of the environment is how the frequencies that are emphasized and de-emphasized by the environment (spectrum) vary in loudness level over the duration of the sound of the environment. The spectral envelope and spectrum portions of the graph are coordinated to present different activity of the same sound components (as with sound quality evaluation).

A *nominal level* is used as a reference for plotting the dynamic contours of the spectral components. The nominal level will vary in loudness/amplitude over the sound's duration. The nominal level *is* the dynamic envelope of the environment, where the sound source's frequency components are unaltered. The dynamic envelope of the environment changes over time. It is the dynamic contour that is outlined by the reflections envelope. This dynamic envelope is represented as a fixed, steady state level on the spectral envelope portion of the environmental characteristics graph.

The nominal level is placed at the dynamic level precisely between mezzo forte and mezzo piano. Frequencies or pitch-areas that are emphasized by the environment will be plotted as activity above the nominal level. Frequencies or pitch-areas that are de-emphasized by the environment will be plotted as activity below the nominal level.

The Environmental Characteristics Spectrum and Spectral Envelope Exercise appears at the end of this chapter. In a similar way to the reflections and reverberation exercise, this exercise will increase the listener's ability to recognize these important aspects of environmental characteristics. Undertaking the detailed task of creating an environmental characteristics graph is not necessary to develop the skills needed to describe the spectrum and spectral envelope of an environment. This exercise will, however, aid the listener in developing the skills to make such observations accurately and with as much detail as the audio professional's position requires.

Environmental Characteristics Graph

The *environmental characteristics graph* allows for a detailed evaluation of the reflection envelope, the spectrum and the spectral envelope (see Figure 9-13). Creating environmental characteristics graphs will greatly assist in understanding the nuance of any environment. When created with much detail, this graph requires great skill that will be acquired over an extended period of practice and patience. Using this graph for general observations and beginning studies will also prove very helpful to the beginner and audio professional alike. Observations can be recorded for future reference and to assist in learning, understanding, and recognizing this artistic element.

The environmental characteristics graph incorporates:

1. Three tiers as the Y-axis: reflections (a continuum of dynamic level), spectrum (a continuum of pitch level), and spectral envelope (a continuum of dynamic level);
2. The reflections portion of the graph is comprised of a vertical line at each point in time that a reflection occurs. The height of the vertical line corresponds to the amplitude of the reflection. The dynamic level of the direct sound is indicated on the vertical axis and serves as a reference for calculating the dynamic levels of the reflections. This portion of the graph presents information on the dynamic contour of the reflections of the environment and the spacings, in time, of the reflections throughout the sound of the environment;

3. The spectrum portion of the graph is comprised of the registers established in Chapter 6. Spectral components are placed against the Y-axis by pitch/frequency level. A single line is plotted against the two axes for each spectral component, and it will occupy a large, colored/shaded area in the case of pitch area and a narrow line in the case of a specific frequency;
4. The spectral envelope portion of the graph depicts the dynamic contours of the spectral components, using dynamic areas as the Y-axis;
5. The X-axis of the graph is dedicated to a time line that is devised to incorporate an appropriate time increment (usually needing to allow millisecond increments to be observed) to clearly display the smallest change of a duration, dynamics, or pitch present in the characteristics of the environment; and
6. A key will be required to clearly relate the components of the spectrum and spectral envelope tiers of the graph.

The perspective of the environmental characteristics graph will always be of either the individual sound source or of the perceived performance environment.

It is not always possible to compile a detailed evaluation of environmental characteristics. The information of the environment is often concealed by the other sounds in the musical texture and is not easily separated from the sound quality of the sound source itself. The ability to recognize environmental characteristics involves much practice. It relies on a knowledge of many sound sources, on an ability to evaluate sound quality of the sound source within the host environment, and on an acquired skill for comparing and contrasting a previous knowledge of the sound source, with the appearance of the sound source within the environment that is to be defined.

Using the graph for general evaluations of environmental characteristics is often the most feasible approach to evaluating environmental characteristics. This approach will not require as advanced a skill level as a detailed graph and will provide a good amount of significant information. These general evaluations are acceptable for most applications. They provide pertinent information quickly and without detail that is difficult and time intensive to identify.

General evaluations of environmental characteristics will include (1) the contour and beginning level of the reverb, (2) the level of the direct sound, and (3) approximately two frequencies or frequency bands that are emphasized or attenuated. If possible, and after practice, they should include a general description of the spectral envelope and an indication of the content of the early time field.

The complexity of environmental characteristics can vary widely. Certain environments will have very few frequency differences from the original sound source. Some environments will not have reflections present between the early time field and the reverberant energy, and will have the reverberant energy increase in density through a simple, additive process. Other environments may be quite sophisticated in the way they were created—with time increments of the early time field precisely calculated at different time intervals, with spectral components at precisely tuned in patterns of frequencies (designed to complement the sound source), and with spectral envelope characteristics reacting accordingly.

It is possible for all perceived environmental characteristics to be changed in real time, with our current technology. It is also possible for the perceived environment of a sound source to be generated solely by a delay unit, by a simple reverberation unit, or by any similar process that would provide easily calculated cues. Although these environments would not be perceived as natural spaces, the listener would proceed to imagine an environment created by the impressions of those simple characteristics. The listener simply will not perceive a sound as being void of environmental characteristics. If no environment is present, it will be imagined from whatever information is present.

Figure 9-14 presents an environmental characteristics evaluation of McCartney's vocal from the opening of "Hey Jude." The graph shows several alterations of spectrum and spectral envelope, and the environment's subtle time elements. The sparse texture during the beginning of the song

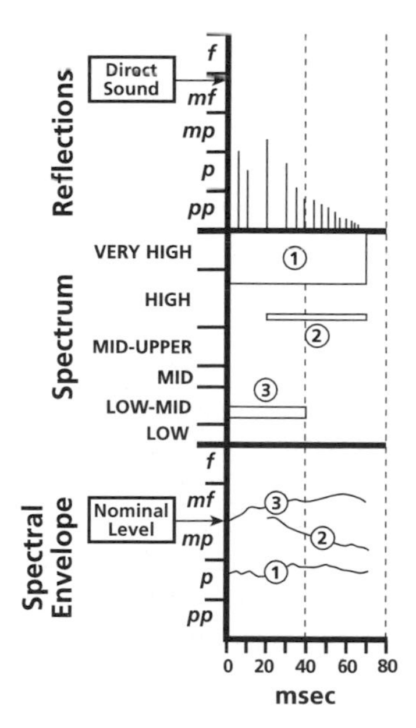

Figure 9-14 Environmental characteristics graph of Paul McCartney's vocal from The Beatles' "Hey Jude."

allows the characteristics of all environments to be perceived quite clearly.

Exercise 9-5 at the end of this chapter provides guidance in learning to evaluate environmental characteristics. The reader will gain much from attempting this exercise on several different sounds from several different recordings.

Space Within Space

The overall environment of the program provides a setting within which the subordinate environments of the individual sound sources will appear to exist. This overall environment (or perceived performance environment) is a constant that equally influences the individual environments of all sound sources. The perceived performance environment becomes part of the overall character of the recording/piece of music.

The overall environment is either (1) perceived by the listener as being a composite of the dominant, predominant, and/or common characteristics of the individual environments of the sound sources, or (2) is a set of environmental characteristics that is superimposed on the entire program.

Works will be perceived to have a single overall environment that is present throughout the piece. The listener will imagine a single space (the perceived performance environment) in which the performance (recording) occurs. When this overall environment is created with the addition of environmental characteristics to the entire musical texture, it is possible for the overall environment to change during the course of a work. In such instances, abrupt changes (usually at a major division of the form of a piece, such as between verse and chorus) are most common. The various environments will be perceived as having occurred within a single overall environment, even if a single environment is not present.

The perceived performance environment may be a composite of many perceived environments and environmental cues. The listener will perceive the overall performance environment in this way, if an overall environment has not been applied. In these instances, the perceived performance environment is imagined by the listener through a perception of environmental characteristics (1) that are common or complementary between the environments of the individual sound sources, (2) that are prominent characteristics of the environments of prominent sound sources (a source that presents the most important musical materials, or the loudest, nearest, or furthest sound sources, as examples), and/or (3) that are the result of environmental characteristics found in both of these areas.

Within this overall environment, the individual environments of the individual sound sources are perceived to exist. This is the illusion of space within space. If reverberation has been applied to the overall

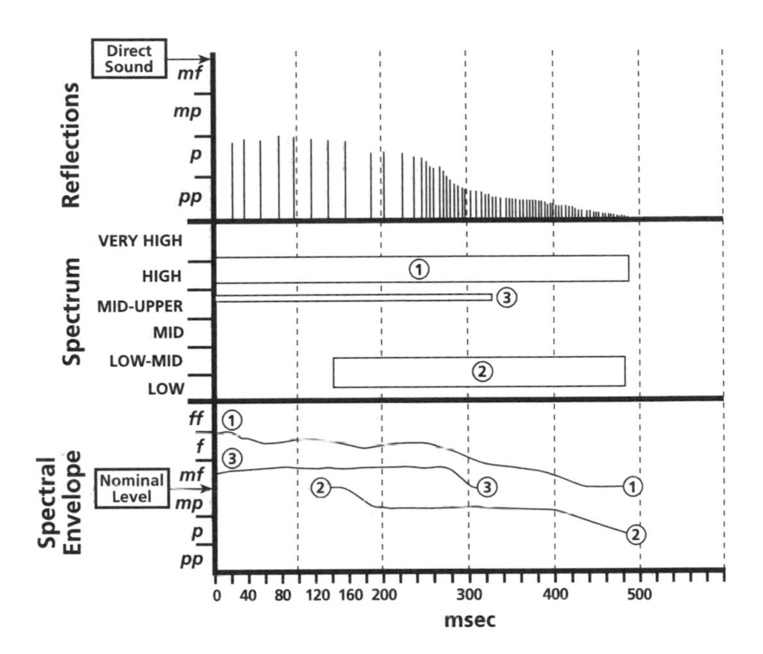

Figure 9-15 Perceived performance environment of The Beatles' "Hey Jude."

program to create a perceived performance environment, all of the recording's sound sources and their host environments will be altered by those sound characteristics.

Figure 9-15 is the perceived performance environment of The Beatles' "Hey Jude." The lead vocal, with its fused environmental characteristics of Figure 9-14, appears contained within this overall environment of the recording/performance. This space within space illusion is convincing in bringing the listener to accept the small space of the lead vocal contained within a mid-sized, natural sounding performance space (perceived performance environment).

A complete space within space evaluation will include the environmental characteristics of all sound sources and the perceived performance environment. Relationships between the environments and the perceived performance environment can then be understood and evaluated, among other possible observations. This complete evaluation might not be undertaken often, but this type of attention is often a level of focus in the mastering process and the final mixdown. The skill to compare environments also leads to an understanding of how environments can be used to enhance sound sources and the recording in complementary ways.

Distance location and environmental characteristics are related. The interrelationships of distance location and space within space should always be considered when evaluating a music production. They work in a complementary way to give depth to the sound stage. The two are

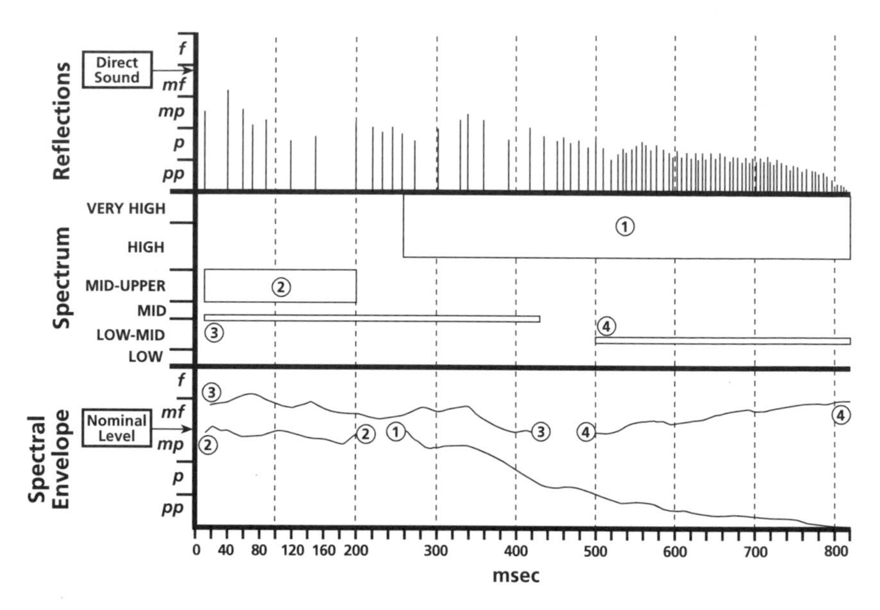

Figure 9-16 Environmental characteristics evaluation of the first appearance of the piano from The Beatles' "Hey Jude."

closely interactive, and comparing the two will offer the listener many insights into the creative ideas of the recording.

Additional environments from The Beatles' recording "Hey Jude" are presented in Figures 9-16, 9-17, and 9-18. The graphs allow us to recognize the very different decay times of the three environments. When comparing the graphs of Figures 9-14 and 9-15 we can also recognize the uniqueness of all four host environments and their perceived performance environment. It is interesting to note that the high and very high areas are attenuated (in different bands) in the guitar and piano environments, and the spectral alterations of the tambourine environment are very subtle. The reflections component of the sounds are sparse, to varying degrees, with fairly regular reflections occurring in all three environments.

Surround Sound

As noted before, surround recording is in its infancy. While we have some well-conceived recordings on the market, we will surely see profound developments in how surround sound is used to deliver and enhance music. Recording practice is still being defined, ways that surround locations can be used to mix music are still being discovered and undergoing experimentation, and reproduction formats remain in flux.

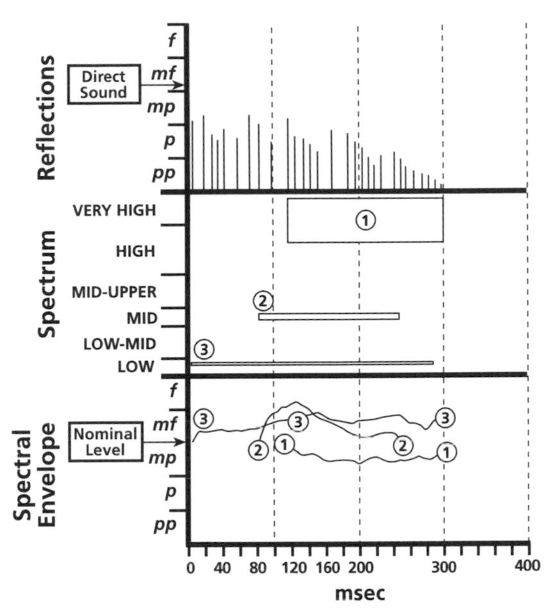

Figure 9-17 Environmental characteristics evaluation of the first appearance of the guitar in The Beatles' "Hey Jude."

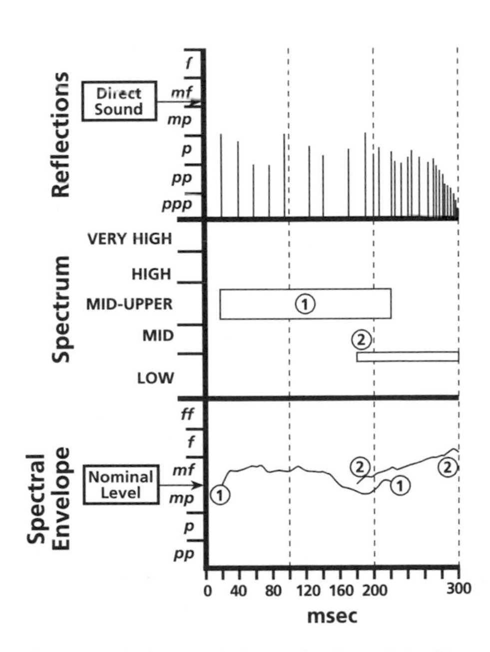

Figure 9-18 Environmental characteristics evaluation of the first appearance of the tambourine in The Beatles' "Hey Jude."

This section will present a way to document and evaluate the directional location information of surround recordings. It is expected that the ways the audio professional will need to evaluate surround recordings will change to reflect developments in production practice and whatever becomes the accepted consumer format(s). How we document and evaluate surround will need to be adapted to reflect future developments.

Format Considerations

Much debate and deliberation is occurring regarding surround formats. Many different channel and loudspeaker combinations and placements have been proposed for surround; too many to accurately count let alone cover here. These include formats from four channels to seven or eight; most with subwoofers, and some with bipolar surround speakers. Some formats have all speakers at ear level, others have the surround speakers higher, and one format uses a wonderfully effective sixth overhead channel (providing subtle but convincing ambience and some impressive vertical cues). While time will tell which format will become the standard, the five-channel system with a subwoofer for low frequency effects has seemingly emerged from this fray of formats.

Our need for a common format has likely already been determined. Widespread consumer adoption of the 5.1 cinema format is likely, as it is already the most commonly purchased system. Using the specifications of the International Telecommunications Union (ITU) Recommendation 775 (see Figure 9-19), the format has proven stable, and was created after a great deal of thought and experimentation. Further, it can accomplish almost all of what is reasonable to expect of surround (albeit with the sad loss of the overhead channel). While the wide spread angle of the surround speakers make rear phantom imaging unstable and can quickly pull images forward, the equidistant placement of all five speakers have advantages for dynamic balance and time-based considerations, and provide convincing ambience.

Another positive aspect of this format is that it is compatible with current two-channel playback concerns. The 60° angle between the left and right speakers provides for accurate listening to stereo recordings (see Figure 9-20). It is the recommended listening relationship for accurate stereo reproduction and can therefore also be used for evaluating two-channel recordings.

This format in the ITU specification (also defined by the Audio Engineering Society) was used in making the evaluations of surround recordings that appear in this book. It appears likely that further refinements will be made in surround sound reproduction systems and formats in coming years. Format incompatibilities, complexities of system installation and setup, and other issues will need to be addressed.

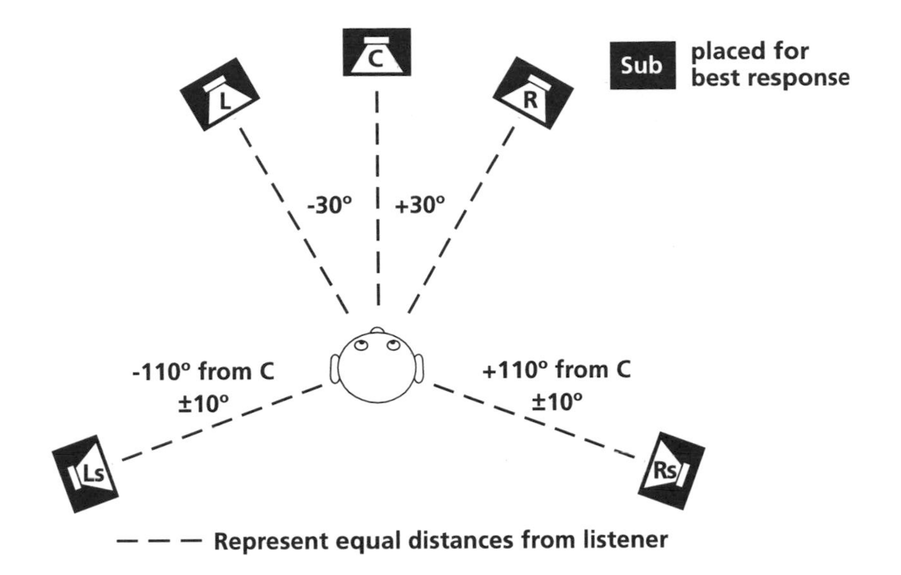

Figure 9-19 ITU recommended speaker layout for surround sound.

The audio industry will surely see profound developments in how surround sound is used to deliver and enhance music. Use of surround sound in music production is in its infancy. Here we will look at the sound stage dimensions of location and distance, and at environmental characteristics in terms of their potential and current use. We have no certain way to predict how artistic expression will bring surround production into maturity and can only examine what is currently before us.

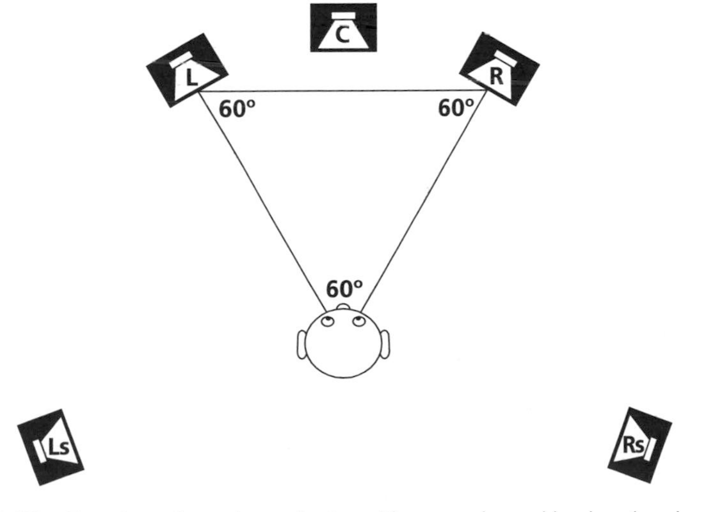

Figure 9-20 Two-channel sound reproduction with surround sound loudspeaker placement.

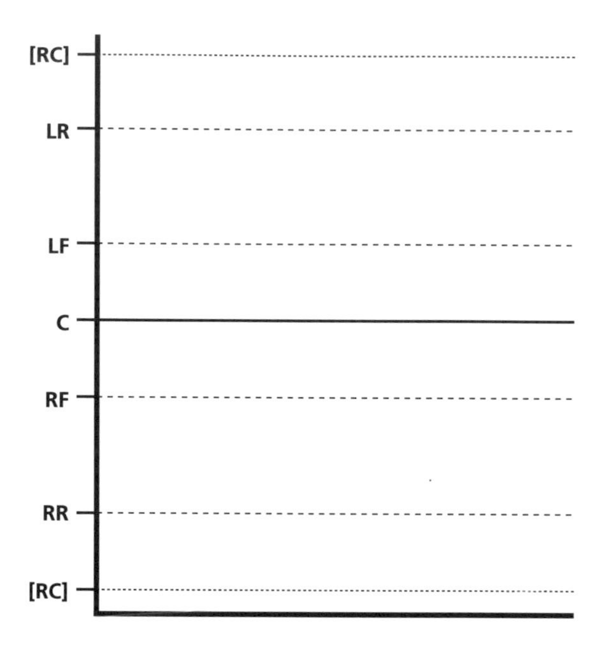

Figure 9-21 Surround sound location graph.

Evaluating Location in Surround Sound

Sound location is evaluated in surround to identify the location and size of the phantom images of sound sources. These cues hold significant information for many surround productions, and may contribute significantly to shaping the musical ideas themselves. These are the cues that separate surround recordings from two-channel (stereo) recordings. Phantom images may change locations or size at anytime during a piece of music. These changes may be sudden or they can be gradual, and may be prominent or subtle.

The *surround sound location graph* will allow the reader to plot the locations of all sound sources against the time line of the work. The graph can portray the direction of sources from the listener and the size of the phantom images.

"Left," "right," "center," "left rear," "right rear," and "rear center" locations are identified on the graph. The sound source's location is represented by placing a mark on the graph at the location of the phantom image. The angle of the sound source from the listener can be determined from the centerline out 180° up or down, but precise degree-increments of angle are not incorporated into the graph. The rear center location is placed at the very top and at the very bottom of the graph. This allows sound movement and spread images across the rear sound field to be graphed—although not as clearly as we might wish. As Figure 9-22

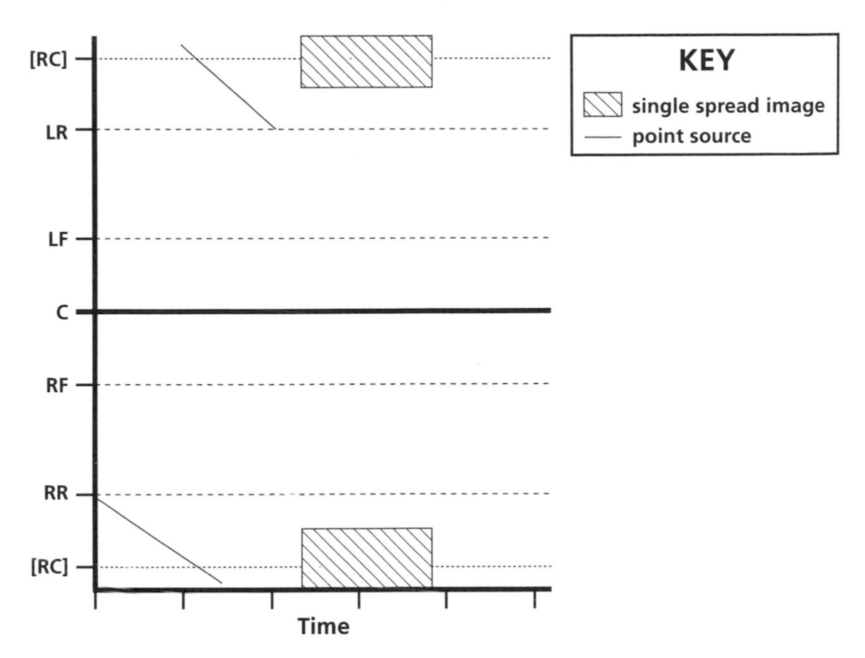

Figure 9-22 Rear center sound source movement and spread images on surround sound location graph.

shows, the movement of sound sources across the rear and the locations of spread images would move off the top and bottom of the graph to wrap the source to the other rear center location.

The surround sound location graph incorporates:

1. "Left," "right," "center," "left rear," "right rear," and "rear center" as the Y-axis;
2. The X-axis of the graph is dedicated to a time line that is devised to follow an appropriate increment of the metric grid or is representative of a major section of the piece (or the entire piece) for sound sources that do not change locations;
3. A single line is plotted against the two axes for each sound source and will occupy a large, colored/shaded area in the case of the spread image; and
4. A key will be required to clearly relate the sound sources to the graph. The key should be consistent with keys used for other similar analyses (such as musical balance and performance intensity) to allow different analyses to be easily compared.

A sound source may occupy a specific point in the horizontal plane of the sound stage, or it may occupy an area within the sound stage, as was

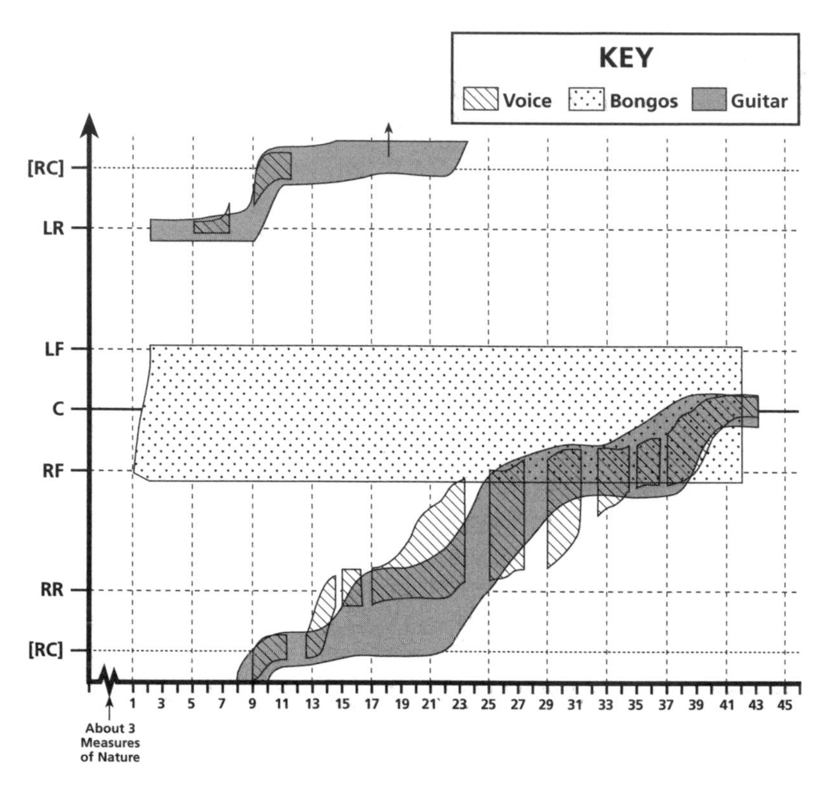

Figure 9-23 Surround sound location graph—Alan Parsons' "Blue Blue Sky."

earlier found in stereo sound location. Similarly, the surround location graph will dedicate a source line to each sound source and will plot the phantom image locations of each source against a common time line. The source line for point source images will be a clearly defined line at the location of the sound source, and the source line for the spread image will occupy an area that will extend between the boundaries of the image itself.

Figure 9-23 displays the surround sound location of three sound sources from Alan Parsons' "Blue Blue Sky." The location, size, and movements of the sound source images directly contribute to the character and expression of the musical materials. The guitar accompaniment leads the vocal line in moving from the rear left, across the rear sound field to the right rear, then circling the right side of the listener until it reaches the center position at measure 43. A number of drums are spread across the front sound field at defined locations and present a movement of point sources within the designated area. The reader is encouraged to listen to the recording, and notice the placement and movement of these sounds, as well as the stunning change of the performance ensemble relationship and musical texture that occurs at measure 43.

Locations and image size do not often change within sections of a song/piece of music. Changes are most likely to occur between sections of a piece, or at repetitions of ideas or sections. These are places where changes in the mix make musical sense and often occur. The listener should, however, never assume changes will not occur. Gradual changes in source locations and size are present in many pieces, and sometimes in pieces where such events are not expected.

When sound source changes of location do not occur within a section, a stationary graph without a time line can appropriately be used. Figure 9-24 presents a surround location image that can be used to notate the locations of sound sources in such instances. Sizes of images and placements are clearly shown. It is also possible for the listener to sketch the sound stage—the area where the performance appears to emanate. This allows the listener to recognize when the surround speakers are being used for ambience only.

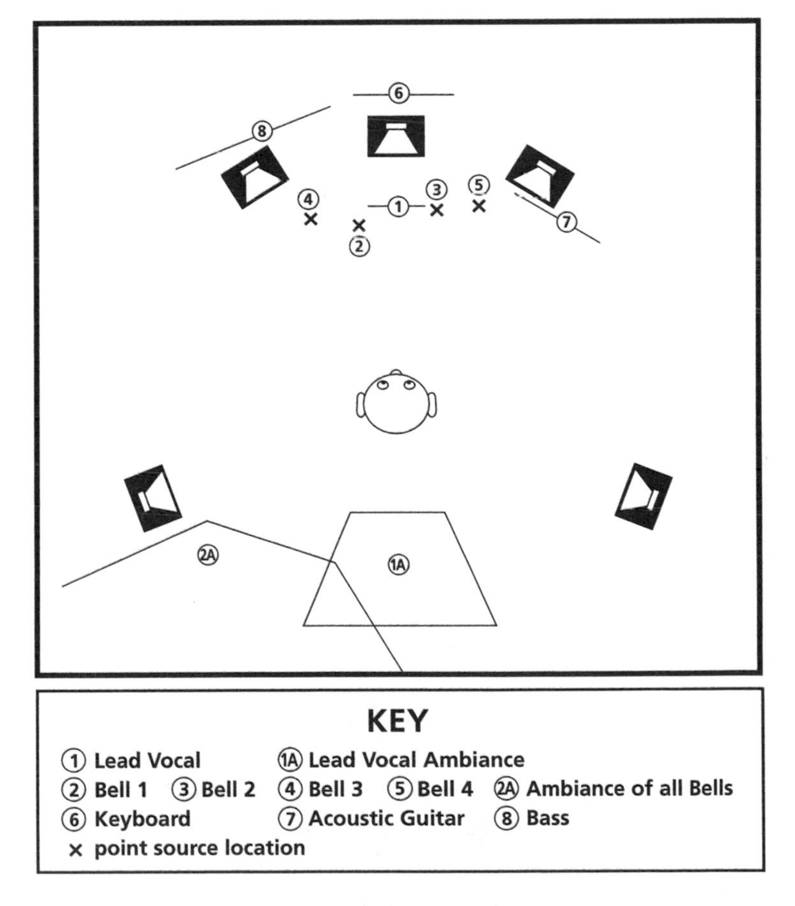

Figure 9-24 Surround sound imaging of a hypothetical mix.

This figure also has the advantage of allowing the listener to notate sound source phantom images created by nonadjacent pairs of loudspeakers and by groups of speakers. These images cannot be incorporated into the surround sound location graph. The graph can also show separations between a sound source phantom image and its ambiance (environmental characteristics) found in many surround productions. It can also show image location areas, containing breadth and depth. These can envelop the listener or surround the listener.

Figure 9-24 depicts the surround imaging of a hypothetical mix containing a number of sources separated from their environmental cues.

The reader should work through the Surround Sound Location Exercise at the end of this chapter to refine their skill in localizing sources in surround playback. The exercise should be performed on different pieces of music, and should use both the Surround Sound Location Graph and the Surround Sound Imaging figure.

Exercises

Exercise 9-1
Stereo Location Exercise.

Find a work that displays significant changes in stereo location of sound sources. Plot the location of the sound source that displays the greatest amount of change, and at least one spread image and one point source. Do this throughout the first two major sections of the piece.

The process of determining stereo sound location will follow this sequence:

1. During the initial hearing(s), listen to the example to establish the length of the time line. At the same time, notice the presence of prominent instrumentation, with placements and activity of their stereo location, against the time line.
2. Check the time line for accuracy and make any alterations. Establish the sound sources (instruments and voices) that will be evaluated and sketch the presence of the sound sources against the completed time line.
3. Notice the locations and size of the sound sources for boundaries of size, location, and any speed of changing locations or size of image. The boundaries of sources' locations will establish the smallest increment of the Y-axis required. The perspective of the graph will always be of either the individual sound source or of the overall sound stage.

4. Begin plotting the stereo location of the selected sound sources on the graph. The locations of spread images are placed within boundaries; the boundaries may be difficult to locate during initial hearings, but they can be defined with precision; the listener should continue to focus on the source until it is defined. The locations of point source images are plotted as single lines. These sources are often easiest to precisely locate and are the sources most likely to change locations in real time.
5. Continually compare the locations and sizes of the sound sources to one another. This will aid in defining the source locations and will keep the listener focused on the spatial relationships of the various sound sources. The evaluation is complete when the smallest significant detail has been incorporated into the graph.

As you gain experience in making these evaluations, songs with more instruments should be examined and longer sections of the works should be evaluated.

Exercise 9-2
Distance Location Exercise.

Select a recording with at least five sound sources that exhibit significantly different distance cues. Plot the distance locations of those sources throughout the first three major sections of the work.

The process of determining distance location will follow this sequence:

1. During the initial hearing(s), establish the length of the time line. Notice the selected sound sources and any prominent placements and activity of distance location, especially as they relate to the time line.
2. Check the time line for accuracy and make any alterations. Clearly identify the sound sources (instruments and voices), and sketch the presence of the sound sources against the completed time line.
3. Make initial evaluations of distance locations. Notice the locations of the sound sources to establish boundaries of the sound stage (the location of the front and rear of the sound stage). Notice any changing distance locations and calculate any speed of changing locations. The placement of instruments against the time line, more than the boundary of speed of changing location (which are quite rarely used), will most often establish the smallest time unit required in the graph to accurately show the smallest significant change of location. The amount of activity in each area will establish the amount of Y-axis

space required. The perspective of the graph will always be of either the individual sound source or the overall sound stage.

4. Begin plotting the distance of the selected sound sources. Sound sources will be placed on the graph by (1) evaluating the timbre definition of each sound source by focusing on the amount of detail present, while being aware of the amount and characteristics of the reverberant sound; (2) transferring this evaluation into a distance of the source from the listening location, and penciling in the sound on the distance location continuum for reference; (3) reconsidering the definition of the timbre (is the source in the listener's own space, or proximity? Is it near or far?), then placing the sound in relation to the sound stage; (4) precisely locating the distance location of the sound source by comparing the sound source's location to the locations of other sound sources.
5. Once several sounds are accurately placed on the distance continuum, identifying additional distance locations is most readily accomplished by directly comparing the sound source to the perceived distance locations of the other sound sources present in the music. Use proportions of differences between the locations of three or more sound source distances to make for more meaningful comparisons. Is sound "c" twice or one-half the distance from sound "a," as "a" is from sound "b?" How does this compare to the relationship of sounds "d" and "c?" Sounds "c" and "b?" Continually compare the distance locations of the sound sources to one another. The evaluation is complete when the smallest significant detail has been incorporated into the graph.

Remain focused on the distance location of the sound sources, making certain your attention is not drawn to other aspects of sound.

As you gain experience in making these evaluations, you should examine songs with more instruments and evaluate longer sections of the works.

Exercise 9-3
Reflections and Reverberation Exercise.

Find a snare drum sound that was recorded without environmental cues. A recording of a drum might be made for the purpose of this assignment, with the sound repeated many times, over a period of 10 or more minutes. Route the sound through an appropriate reverb unit.

1. Make the reverb unit emphasize the items listed below one at a time, and make radical (perhaps unmusical) settings of these parameters to learn their characteristic sound qualities. Listen carefully to individual snare drum hits while adjusting the device.
2. Seek to create a pronounced early time field from the unit. Establish a clear set of two reflections and create a setting that will repeat this pattern. A recurring pattern of reflections is the result. Listen carefully, and alter the speed of the reflections and spacings of the reflections and patterns.
3. Repeat this sequence with a clear set of 3, 4, and then 5 reflections, establishing recurring patterns while gradually increasing the number of reflections and the complexity of the pattern. Listen carefully to create and recognize:
   a. Patterns of reflections created by dynamics
   b. Patterns of reflections created by spacings in time
   c. Spacing of reflections in the early time field
   d. Dynamic contour of the entire reflections portion
   e. Density (number and spacings of reflections) of reverberant sound
   f. Dynamic relationships between the direct sound, individual reflections (of the early time field), and the reverberant sound
   g. Dynamic contour shapes within the reverberant sound

You will begin to notice and recognize that certain spacings in time have a certain, consistent and unique sound quality. A "sound of time" can be understood and recognized for delay times and reverberation rates. With patience and practice, this skill can become highly refined—as many room designers will attest.

Exercise 9-4
Environmental Characteristics Spectrum and Spectral Envelope Exercise.

Find a high quality sampled acoustic instrument sound that can be controlled by a keyboard. Route the sound through an appropriate reverb unit.

1. Establish a reverb setting with three or more seconds of decay and with a high proportion of reverb signal (or only reverb signal).

2. Alter the frequency response, equalization, or any similar frequency processing control on the reverb to emphasize and de-emphasize (attenuate) several specific frequencies or frequency bands.
3. Play single pitches with short durations on the keyboard. Listen carefully to how changes of settings alter the sound quality of the instrument. Keep track of the settings played. Repeat this process while moving through the entire frequency range(s) the unit will alter and listening (and learning) carefully.
4. In a separate process, listen carefully to pitches played throughout the keyboard range, played through an unchanging reverb setting. Notice how the qualities of some pitches are altered differently than others. Changes in the environment's spectrum and spectral envelope will occur only if the particular pitches performed have spectral energy at the frequencies being altered.

Repeat this process again several hours, then several days later. During these sessions try to anticipate what the modification will sound like before you listen to it. Check your memory and your recognition of many different spectrum changes. Keep returning to this exercise to become comfortable with the material.

Exercise 9-5

Environmental Characteristics Exercise.

Return to the work or works evaluated in the Distance Location Exercise. Carefully select three of the five sound sources previously evaluated for distance and perform environmental characteristics evaluations on those sounds as outlined below.

As an alternative, look for a suitable surround sound recording with few sound sources. Identify three to five sources that have separate locations for their direct sound and environmental characteristics. This will greatly assist you in comparing the direct sound and the environment, and in isolating environmental characteristics.

While these evaluations are most easily accomplished for short duration percussive sounds, environmental characteristics evaluation is possible for any sound source as long as the reverberant energy of the environment is exposed (not accompanied by or masked by other sound sources) after the sound source has ceased sounding.

The process of determining environmental characteristics will follow this sequence:

1. During initial hearings of the entire work, listen to each sound source to identify a location where the sound is isolated throughout the duration of the environment. Nearly always the graph's time increments on the time line will need to show milliseconds. Estimate the length of the time line for that presentation of each sound source.
2. Check the time line for accuracy and make any alterations. Work in a detailed manner to establish a complete evaluation of the reflections of the sound. First, sketch the presence of the most prominent reflections against the completed time line; then, establish the precise time placement and the dynamic levels of these prominent reflections against the time line. Use the prominent reflections as references to fill in the remaining reflections in the early time field. After the early time field is plotted, complete the reflections portion of the graph by plotting the dynamic envelope and spacing of reflections (density) of the reverberant energy.
3. Notice the locations and size of any emphasized or de-emphasized pitch areas or frequencies. Scan the entire piece of music, listening to how the sound source is altered by the environmental characteristics by listening to many different pitch levels. Throughout these hearings, keep track of pitch areas or specific frequencies that appear to be emphasized or de-emphasized. With a running list of observations, regularly identified pitch areas/frequencies will begin to emerge. Further hearings will allow you to more accurately identify these frequencies and pitch areas (that make up the spectrum of the environmental characteristics), and to place the presence of these frequencies or pitch areas against the time line.
4. You will now plot the dynamic contours of the components of the spectrum against the time line. This process is the same as the process of plotting the spectral envelope of sound quality evaluations. Each component of the spectrum is plotted as a single line, and these components are listed in a key, so their dynamic contours may be related to the spectral envelope tier of the graph.
5. Continually compare the dynamic levels and contours of the spectral components to one another. This will aid in remembering the nominal dynamic level (where the amplitude of the spectral components of the sound source are unaltered by the environment), will aid in keeping the dynamic levels and contours consistent between spectral components, and will keep you focused on the relationships of the sound source and its host environment. The evaluation is complete when the smallest significant detail has been incorporated into each tier of the graph.

This evaluation can be detailed and time intensive. It is not proposed that these detailed evaluations be undertaken in normal, daily activities of audio professionals. As a learning tool, this study will be very successful at bringing you to hear, understand, recognize, and remember these important aspects of sound. You are encouraged to return to this exercise. Once speed and accuracy improve, you should undertake evaluations of more complex environments and sounds that are partially masked.

Exercise 9-6
Exercise in Determining the Environmental Characteristics of the Perceived Performance Environment.

This exercise will seek to define the environmental characteristics of a recording's perceived performance environment. A multitrack recording should be selected that contains no more than three or four sound sources, a sound stage that clearly separates the images, and an overall sound that appears to envelop the sound stage.

1. Identify the sound sources and the different environments of the piece.
2. Perform general environmental characteristics evaluations of the environments. These initial evaluations should be general in nature, seeking prominent characteristics rather than detail.
3. Compare the environments for similarities of time, amplitude, and frequency information to identify common traits between the individual environments. (1) When traits are common to all sounds, an applied, overall environment is present. The traits will be present in all environments equally. If the common traits are not applied to all sources equally, a single environment has not been applied to the entire program. (2) Then you must look at other factors as well. Next, identify the predominant traits of the environments of musically significant sound sources. They also directly contribute to the characteristics of the overall environment.
4. Listen to the work again to identify an overall environment of the program. An applied overall environment will be most easily detected by its detail in spectral changes of the reverberant sound, and in the clarity of the initial reflections of the early time field. The characteristics of these environments will be perceived by listening for detail at a close perspective of slight changes to the predominant characteristics of the environment. Overall environments that are an illusion (created by the composite and predominant characteristics of the individual sound sources) will have characteristics that are not

readily apparent. The characteristics of these environments will be perceived by listening at the more distant and general perspective of the dominant characteristics of the environment.

5. Compile a detailed environmental characteristics evaluation of the perceived performance environment. The evaluation is complete when the smallest significant detail has been incorporated into each tier of the graph.

Repeat this exercise on other recordings until you have evaluated a recording with an applied overall environment and a recording with a perceived performance environment that is the perceived result of the environmental characteristics of the individual sound sources.

Once skill and confidence are improving, repeat this exercise on recordings that have more activity and with less pronounced characteristics in the perceived performance environment.

Exercise 9-7
Space Within Space Exercise.

Select a multitrack recording containing a small number (five or six) sound sources. A recording with a sparse texture (few instruments sounding simultaneously) and pronounced environments on the individual sound sources will be easiest to evaluate during initial studies.

The process for determining space within space follows this sequence:

1. Identify the various environments of the piece. Some sound sources may share environments with other sound sources (at the same or different distances), and some sources may change environments several times in the piece.
2. Perform general environmental characteristics evaluations of the environments. These initial evaluations should be general in nature, seeking prominent characteristics rather than detail.
3. Compare the environments for similarities of time, amplitude, and frequency information. This observation will determine common traits between the individual environments of the sources. These common traits will signal a possible applied, overall environment if they are present in all environments equally. If the common traits are not applied to all sources equally, other factors are in play as well. Identify the predominant traits of the environments of musically significant sound sources. They also directly contribute to the characteristics of the overall environment.

4. Listen to the work again to identify the characteristics of the overall environment of the program (the perceived performance environment). Compile a detailed environmental characteristics evaluation of the perceived performance environment.
5. Begin the master listing of environments with this environmental characteristics evaluation of the perceived performance environment.
6. Perform detailed environmental characteristics evaluations of the individual host environments of each sound source. The characteristics of the overall environment may or may not be present in these evaluations, depending on the nature of the overall environment and the nature of the individual sound sources' environments. The evaluation of each source is complete when the smallest significant detail has been incorporated into each tier of the graph.
7. Number each environment and enter the evaluation to the master listing of environments. Note on the master listing the sound source or sources that are present within the environment.

Once skill and confidence are improving, repeat this exercise on more sound sources in recordings that have more activity and with less pronounced environmental characteristics.

Exercise 9-8
Surround Sound Location Exercise.

Two approaches can be used for surround sound location (A and B). You should work through both approaches, as one will be more suitable to any sound material than the other. Determining which approach is most suitable will be a valuable undertaking in itself.

A. Find a surround recording with sources located around the array, but listening at the audience perspective. Perform an evaluation of the locations of four or five sources for the first two major sections of the work.

   The process of determining surround sound location will follow this sequence:

1. During the initial hearing(s), listen to the example to establish the length of the time line. At the same time, notice the presence of prominent instrumentation, with placements and activity of their surround location, against the time line.
2. Check the time line for accuracy and make any alterations. Establish a complete list of sound sources (instruments and voices), and sketch the presence of the sound sources against the completed time line.

3. Notice the locations and size of the sound sources (instruments and voices) for boundaries of size, location, and any speed of changing locations or size of image. The placement of instruments against the time line will most often establish the smallest time unit required in the graph to accurately exhibit the smallest significant change of location. The boundaries of the sound sources' locations will establish the smallest increment of the Y-axis required. The perspective of the graph will always be of either the individual sound source or of the complete array.
4. Begin plotting the surround location of each source on the graph. The locations of spread images are placed within boundaries. The boundaries may be difficult to locate during initial hearings, but they can be defined with precision. Continue to focus on the source until it is defined. The locations of point source images are plotted as single lines. These sources are easiest to precisely locate and are most likely to change locations in real time.
5. Continually compare the locations and sizes of the sound sources to one another. This will aid in defining the source locations and will keep you focused on the spatial relationships of the various sound sources. The evaluation is complete when the smallest significant detail has been incorporated into the graph.

Locating sound sources originating from behind normally causes a listener to move their head. You should consciously keep your head still and focus on the direction and size of the image.

B. This exercise should be repeated on other recordings. Find a recording with the listener located within the ensemble or the performance.

   When sound sources do not change locations or when sources appear to envelop you, a stationary sound location figure will be substituted for the surround location graph. Figure 14-2 will be used for these evaluations and will also allow you to localize phantom images generated by nonadjacent pairs of loudspeakers (as described in the chapter), and represents a fixed period of time. Identify and graph several point sources and spread image sound sources in the song. A separate graph will be used whenever source size or locations change. Graph several different sections of the work on separate graphs to note and understand the characteristics and changes of imaging that occur in the song.

10 Complete Evaluations and Understanding Observations

Evaluations of artistic elements from previous chapters will be drawn together and compared here. Observations will be made from examining evaluations and comparing the artistic elements to the musical materials.

The evaluations made have been on three different levels of perspective: (1) the characteristics of an individual sound, (2) the relationships of individual sound sources, and (3) the overall musical texture, or overall program.

This chapter will examine relationships of the various dimensions of the overall texture and explore pitch density. The information offered by comparing the various evaluations of individual sources is explored next. An examination of how the artistic elements shape a recording will lead to a summary of the system for evaluating recorded/reproduced sound.

The Overall Texture

The highest level of perspective brings the listener to focus on the composite sound of a recording. At this level, all sounds are summed into a single impression. Recreational listening usually is mostly focused on this level of perspective, with shifts of focus moving to text and melody—and other aspects attractive to the listener, such as beat or pulse—in a random manner and at undirected times.

This single impression of the overall texture has a variety of dimensions that provide the recording and music with its unique character. These dimensions are the piece of music/recording's form, perceived performance environment, reference dynamic level and program dynamic contour, and its pitch density. All but one of these dimensions has been explored in previous chapters.

Pitch Density

The process of evaluating pitch density is directly related to pitch area analysis from Chapter 6. *Pitch density* is the amount and placement of pitch-related information within the overall pitch area of the musical texture. It is comprised of the pitch areas of the musical materials fused with the sound qualities of the sound sources (voices, instruments, or groups of instruments and voices) that present the material.

The concept of pitch density allows each musical idea/sound source to be perceived as having its own pitch area in the musical texture. The range of pitch that spans our hearing (and the musical texture) can be conceived as a space. Within this space, sound sources appear to be layered according to the frequency/pitch area they occupy. The range of pitch is divided into areas occupied by the sound sources, or left empty. The size and placement of these areas are unique to each piece of music, and may remain stable or change at any time.

With this approach, the concept of pitch density is often applied to the processes of mixing musical ideas and sounds in recording production. This is similar to the traditional concept of orchestration, where instruments are selected and combined based on their sound qualities, and the pitch area the sounds occupy. The recording medium and various formats provide new twists to this traditional approach to combining sounds.

Pitch density will be evaluated to determine the pitch areas occupied by musical materials, fused with the sound quality of the sound source(s) presenting the material.

Musical materials are a single concept or pattern. These materials will fuse in our memory and perception as a group of pitches, which occupy a pitch area. The pitch area of the musical material is defined by its boundaries—its highest and lowest pitches. Within the boundaries, the number of different pitch levels that comprise the musical material and their spacing creates a density in the pitch area.

The pitch area of the sound source will also influence the pitch area of the musical idea. The primary pitch area of the sound source will often be the fundamental frequency of an instrument or voice, perhaps with the addition of a few prominent lower partials. A primary pitch area may also contain environmental cue information (delayed and reverberated sound). These cues may add density to the sound without adding more pitch information. Pitch-shifted information, and other processing effects, may also provide additional spectral information and added density.

Often, the primary pitch area of the sound source is rather narrow, often slightly more than the fundamental frequency alone. This is especially true when an instrument or voice is being performed at a moderate to low dynamic level. As the dynamic level of a sound source increases, lower partials will often become more prominent, and the width of the pitch area will tend to widen.

The pitch area of each musical idea must be determined by (1) defining the length of the idea. It is then possible to define (2) the lowest boundary of the pitch area, then (3) the highest boundary of the area. This pitch information is determined by the listener calculating the melodic and harmonic activity of the musical idea (to determine the highest and lowest pitch-levels), then adding detail pertaining to the sound qualities of the sources performing the idea. Most listeners can easily accomplish this sequence by simply asking "when does this idea end, and when does the next idea begin?"

The lowest boundaries of pitch-areas are most often the fundamental frequencies of the sound sources performing the musical material (roughly following the melodic contour or the lowest notes of a chord progression). The upper boundary is either the highest pitch-level of the musical material, or is the top of the predominant pitch-area of the highest sound source (and pitch-level) of the musical idea. The boundaries of pitch areas of musical ideas may or may not change over time.

The pitch area of a musical idea is a composite. It contains all of the appearances of the sound source performing all of the pitch-material within the time period of the musical idea. It is the sum of all of the pitch levels and the primary pitch areas of its sound source(s).

The individual sound source is conceived as a single pitch area and is identified by boundaries that conform to the composite pitch material. The relative density of the idea is determined by the amount of pitch information generated by the musical idea and the spectral information of the sound source(s).

The process of determining pitch area is repeated for each individual musical idea. All sounds will appear on the graph.

Pitch areas may be mapped against a time line. If a time line is used, changes in pitch area as the musical idea unfolds in time can be presented (as appropriate to the musical context).

All sound sources are plotted on the *pitch density graph*. The graph may take two forms, with or without a time line. The graph may simply plot each sound source's pitch area against one another (as the pitch area graph, in Chapter 6). Most helpful is when the sound sources are plotted against the work's time line, allowing the graph to visually represent changes in pitch density as the work unfolds.

In either form, the pitch density graph contains:

- Each sound source is represented by an individual box denoting the pitch area
- The Y-axis is divided into the appropriate register designations
- The density of the pitch areas should be denoted on the graph through shadings of the boxes or verbal descriptions

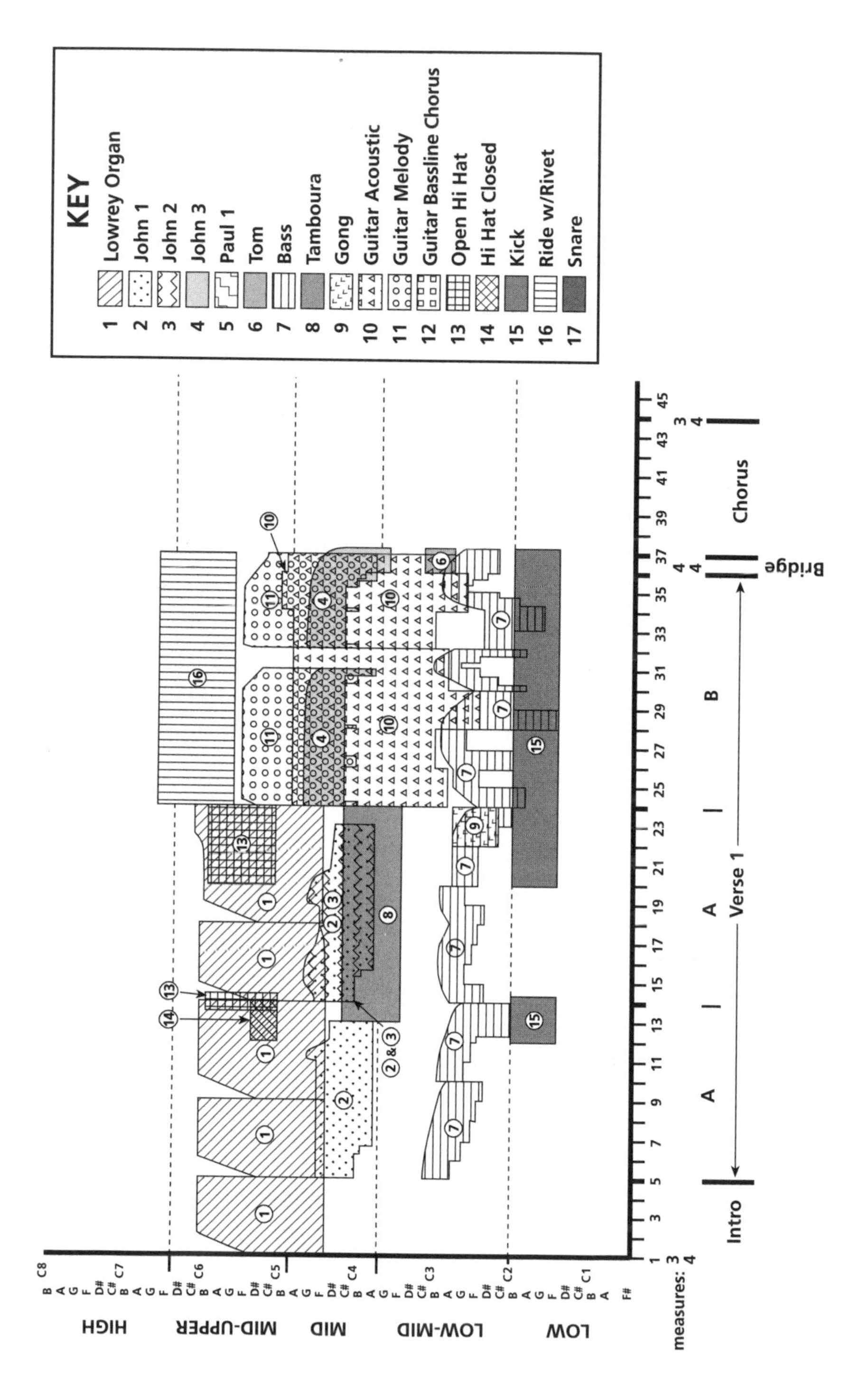

Figure 10-1 Pitch density graph—The Beatles' "Lucy in the Sky with Diamonds."

The pitch areas of all of the sound sources may be compared to one another and to the overall pitch-range of the musical texture. Pitch density allows the pitch area of all sound sources to be compared. Thus, the recordist is better able to understand and control the contribution of the individual sound source's pitch material to the overall musical texture and of the mix.

The pitch density of the beginning sections of The Beatles' "Lucy in the Sky with Diamonds" (1999 *Yellow Submarine* version) appears in Figure 10-1. The work uses pitch density and the register placement of sound sources and musical ideas to add definition to the musical materials and sections of the music. Pitch density itself helps create directed motion in the music. The musical ideas are precisely placed in the texture, allowing for clarity of the musical ideas. The expansion and contraction of bandwidth of the overall pitch range and textural density of the musical ideas (and sound sources) add an extra dimension to the work and support it for its musical ideas.

A pitch density exercise appears at the end of this chapter. The reader is encouraged to spend enough time with this exercise to feel comfortable with this concept, which is so important to the mixing process.

Examining Characteristics of the Overall Texture

The characteristics of the overall texture provide many fundamental qualities of the music and recording. These greatly shape the music and its sound qualities, and communicate most immediately to the listener. The framework for the music and the context of the message of the recording are crafted at this level of perspective.

The characteristics of the overall texture are:

- Perceived performance environment
- Reference dynamic level
- Program dynamic contour
- Pitch density
- Form

The perceived performance environment creates a world within which the recording exists. This adds a dimension to the music recording that can substantially add to the interpretation of the music. The level of intimacy of the recording is related to the level of intimacy of the message of the music. This element will largely be defined by the sound stage as placed in the perceived performance environment.

The reference dynamic level represents the intensity and expressive character of the music and recording. What the music is trying to say is translated into emotion, expression, and a sense of purpose. These are reflected in recordings and performances as a reference dynamic level. Understanding this underlying characteristic of the music allows the recordist to calculate the relationships of materials to the inherent spirit of the music.

The program dynamic contour allows us to understand the overall dynamic motion of the entire piece of music. Actual dynamic level will impact the recording process in many ways, and it also shapes the listener's experience. While program dynamic contour depicts how the work unfolds dynamically over time, this contour is often closely matched to the drama of the music. The tension and relaxation, the points of climax and repose, movement from one major idea to another, and more are contained in the contour of this sum of all dynamic information.

Pitch density provides information on the spectrum and spectral envelope of the entire musical texture. The movement of musical ideas through the "vertical space" of pitch provides a sense of place for musical materials and adds an important dimension to the character of the overall texture. The actual sound quality of the overall texture is largely shaped by pitch density. Pitch density can be envisioned as providing spectral information—materials that are harmonically related, and those that are not all add their unique qualities to the production as they change over time or provide a continual presence that is part of the overall sound of the recording/music.

Finally, the form of music is created by all of these characteristics, plus the text and musical materials. This essence of the song is what reaches deeply into the listener. It resonates within the listener when the song is understood. Form is the overall concept of the piece, as understood as a multidimensional, but single idea.

Figure 10-2 presents a time line and the structure of "Lucy in the Sky with Diamonds." This is an outline of the materials that contribute to the form of the piece. The shape of the music and interrelationships of parts, as well as elements of the text, can be incorporated into this graph and made available for evaluation.

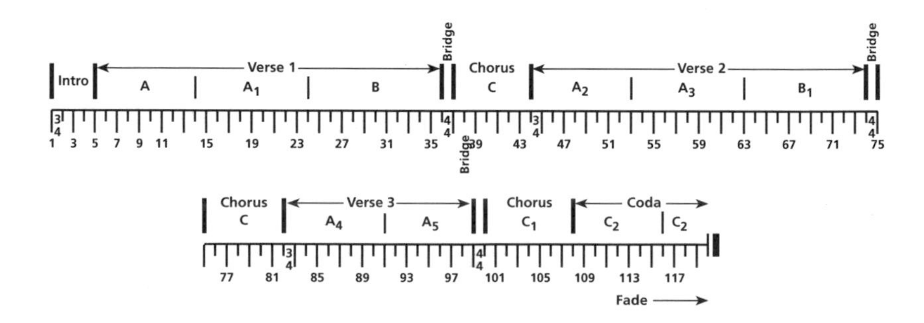

Figure 10-2 Time line and structure—The Beatles' "Lucy in the Sky with Diamonds."

Matching the text against the structure of the piece will allow the reader to notice recurring sections of text/music combinations—verses and choruses. The musical materials enhance nuances of the meaning of the text as they were captured and enhanced by the recording process.

Dynamic relationships between the various sections of "Lucy in the Sky with Diamonds" are present. These are clearly observed in the program dynamic contour graph (Figure 10-3). The graph also shows the overall shape and dynamic motion of the song. The reader can also explore how that motion relates to the song's text and sense of drama.

The reference dynamic level of the song is also identified in Figure 10-3, as a high mezzo forte.

The pitch density graph of the song in Figure 10-1 allows the reader to identify and understand how the song emphasizes one pitch area for a time, then more evenly distributes pitch area information—moving from one texture to another between sections. An evaluation of the perceived performance environment will provide the last of the characteristics of the overall texture, allowing all to be compared and considered. Through this process important and fundamental characteristics of the song can then be more readily understood and communicated to others.

Relationships of the Individual Sound Sources and the Overall Texture

The mix of a piece of music/recording defines the relationships of individual sound sources to the overall texture. In the mixing process, the sound stage is crafted by giving all sound sources a distance location and an image size and stereo/surround location. Musical balance relationships are made during the mix, and relationships of musical balance with performance intensity are established. The sound quality of all of the sound sources is finalized at this stage also, as instruments receive any final signal processing to alter amplitude and frequency elements to their timbre and environmental characteristics are added.

These elements crafted in the mix exist at the perspective of the individual sound source. Many important relationships exist at this level. This focus is common and important for the recordist, but is not common in recreational listening. The many ways sound is shaped at this level brings the recordist to often focus on this level. Learning to evaluate these elements, and to hear and recognize how these elements interact to craft the mix is very important for the recordist.

Figures 7-4 and 10-4 present two musical balance graphs of the beginning sections of "Lucy in the Sky with Diamonds." These are evaluations of two separate versions of the song. Mixing decisions brought certain sounds to be at different dynamic levels in each version. Listening

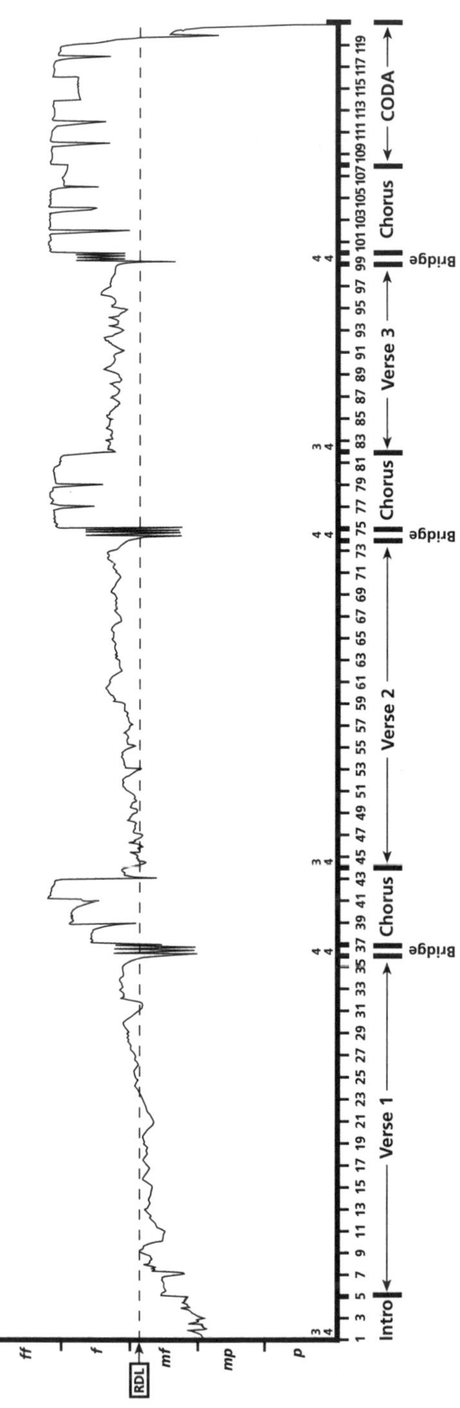

Figure 10-3 Program dynamic contour—The Beatles' "Lucy in the Sky with Diamonds" (1999 *Yellow Submarine* version).

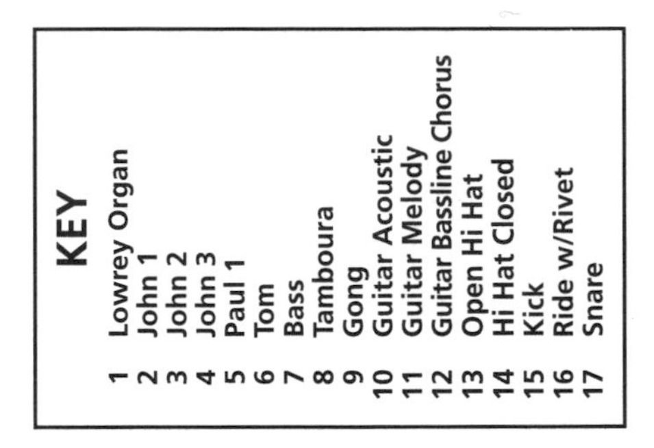

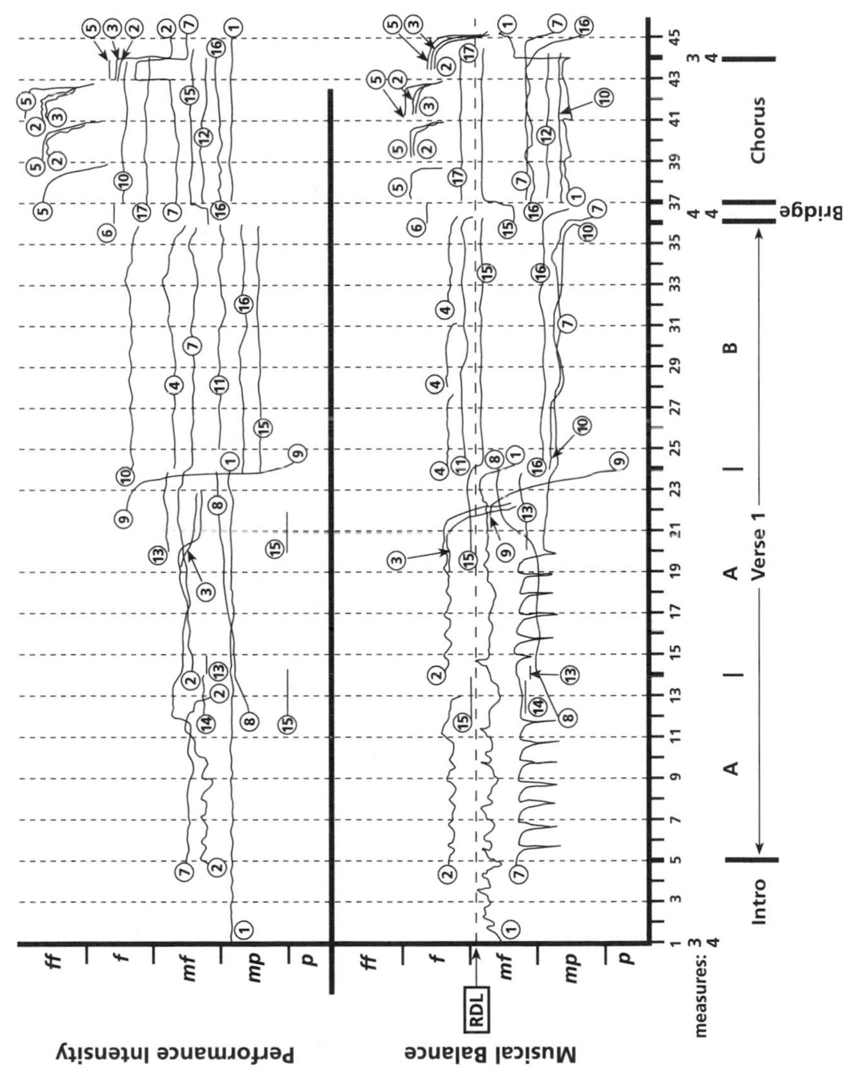

Figure 10-4 Musical balance versus performance intensity graph—The Beatles' "Lucy in the Sky with Diamonds" (1999 *Yellow Submarine* version).

with a focus on several different sound sources, while comparing the two graphs to what is heard, will provide the listener with insight into these very different mixes. The performance intensity tier of Figure 10-4 should also be examined and then compared to the final dynamic levels. Here the reader is able to identify how the dynamics of the sounds were transformed by the mixing process. This gives much insight into the performance of the tracks and their sound qualities, and the musical balance decisions that followed.

The sound stage provides each recording with many of its unique qualities. In examining the structure of sound stage the listener will learn many things about a recording. Among the variables are:

- Distribution of sources in stereo or surround location
- Size of images (lateral and depth)
- Clearly defined sound source locations, or a highly blended texture of locations (wall of sound)
- Depth of sound stage
- Distribution of distance locations
- Location of the nearest sound source
- Sound stage dimensions that draw the listener's attention or are absorbed into the concept of the piece
- Changes in sound stage dimensions or source locations

Figure 10-5 presents the stereo location of sound sources in the 1999 *Yellow Submarine* version of "Lucy in the Sky with Diamonds." The phantom image locations and sizes add definition to the sound sources and musical materials, and the width of several images changes between sections.

Comparing this stereo location graph to surround placements in Figures 10-6 and 10-7 allows some interesting observations. The lead vocal, Lowrey organ, and tamboura have been graphed from the first verse in the surround version, and both graphs need to be used to clearly define their dimensions. The image sizes and locations in a surround mix are quite different from the two-channel versions, and offer a very different experience of the song. Comparing these location graphs will allow the reader insight into what makes each mix different.

The reader is encouraged to also evaluate the stereo location of the sources in the original *Sgt. Pepper's Lonely Hearts Club Band* version of the song. Then compare the three location evaluations for similarities and differences of image location and sizes, and note when and how images change sizes or locations. In the end, consider how the different sizes and locations of the images impact the musicality of the song.

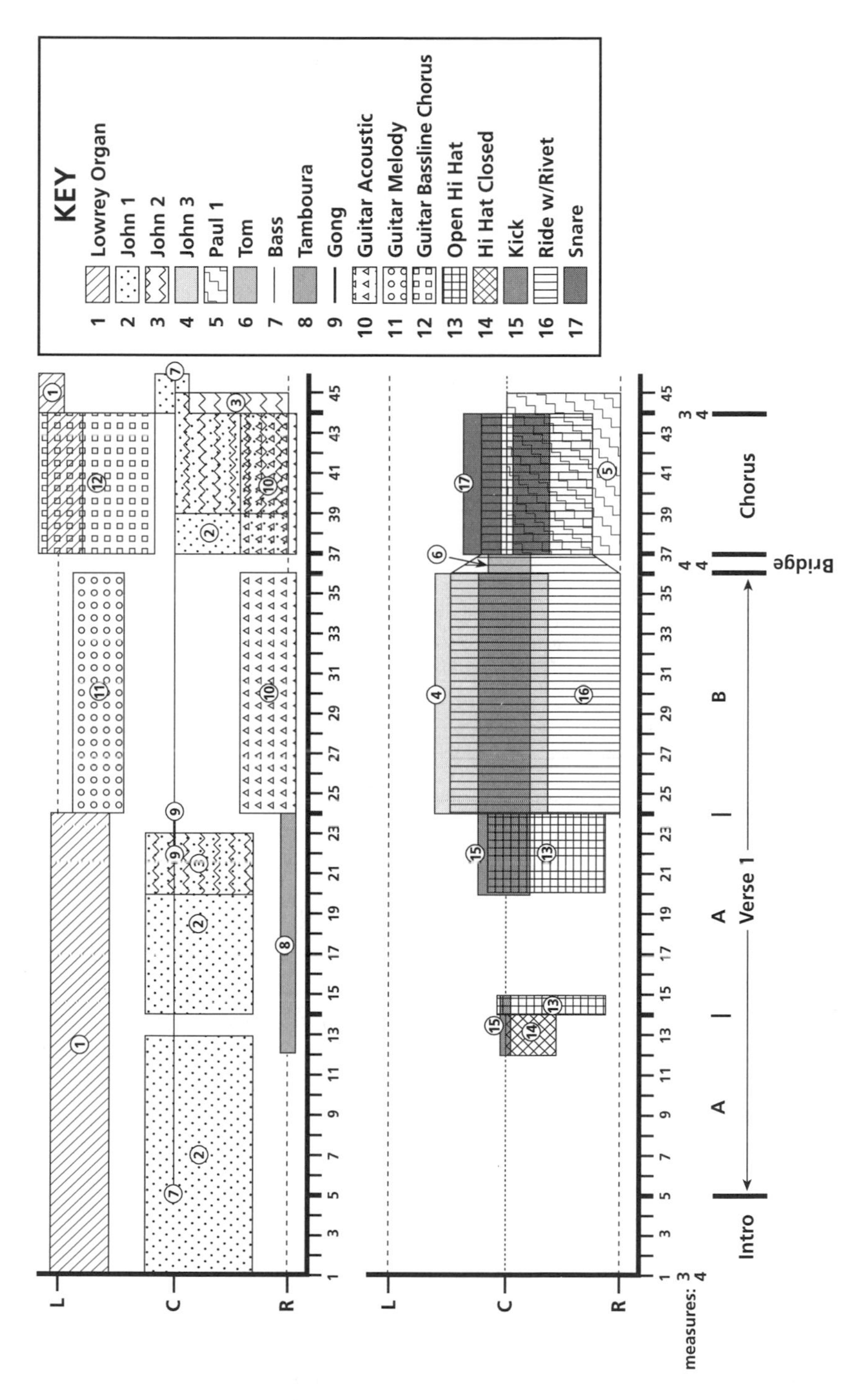

Figure 10-5 Stereo location graph—The Beatles' "Lucy in the Sky with Diamonds" (1999 *Yellow Submarine* version).

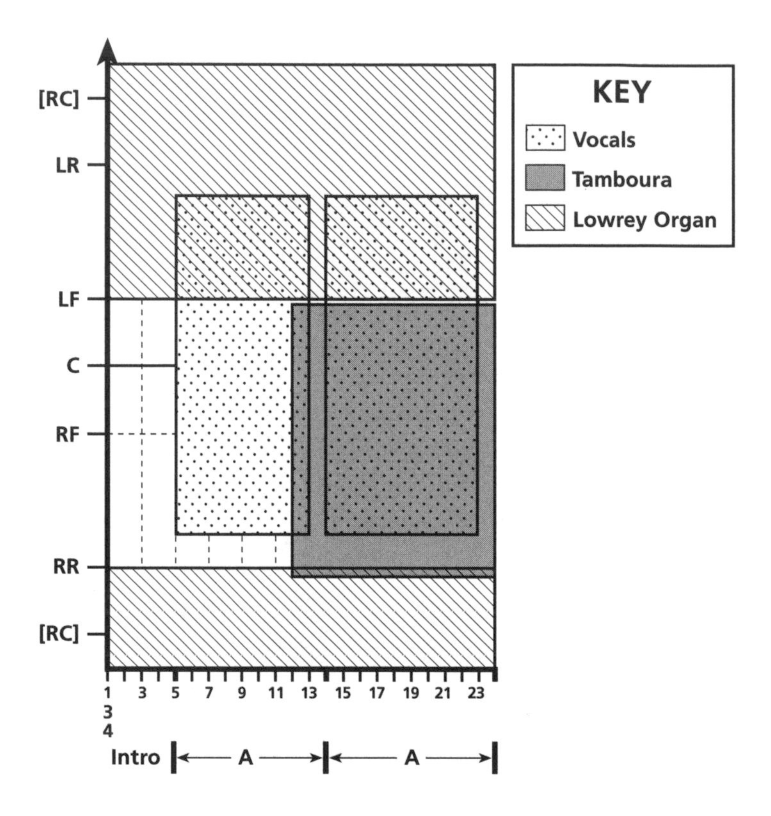

Figure 10-6 Surround sound location graph—The Beatles' "Lucy in the Sky with Diamonds," Lowrey organ, tamboura, and vocal images.

Figure 10-8 is a blank distance location for the beginning sections of "Lucy in the Sky with Diamonds." The graph can be used to plot the distance location of sources from any of the three versions of the song—or comparing one or more selected sources from all three versions. In listening to the three versions of the song, the reader will notice some striking differences in distance location. Between the two-channel versions sound source changes in distance locations are more pronounced in the *Yellow Submarine* version, and during the first verse the bass and lead vocal are closer than in the *Sgt. Pepper* version. Consider how distance cues enhance certain musical ideas and sound sources, and provide clarity or blending of images in various instances.

The sound quality of all of the sound sources shapes the pitch density of the song. It also contributes fundamentally to the performance intensity of the sound source, and in some cases may contribute to determining the song's reference dynamic level. The environmental characteristics of the source also contribute to its overall sound quality. They fuse with the source's sound quality to add new dimensions and

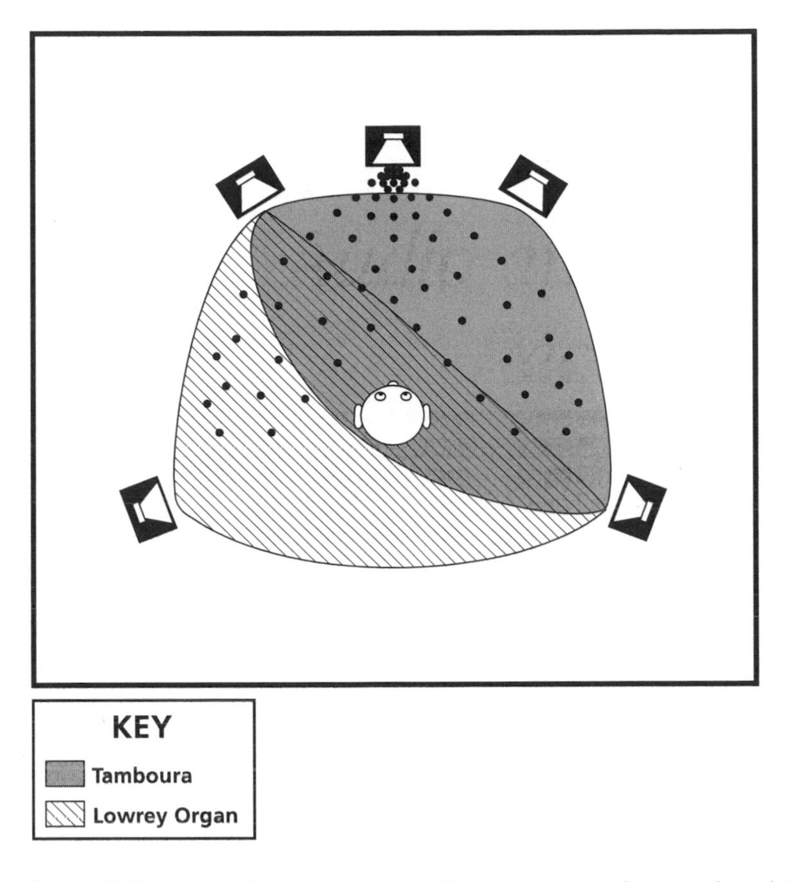

Figure 10-7 Surround image placements of Lowrey organ, tamboura, and vocal images—The Beatles' "Lucy in the Sky with Diamonds." Dots depict density and area of vocal image.

additional sound quality cues to the resulting composite sound. Lastly the sound quality of each source provides important distance information, as the amount of timbral detail primarily determines distance location.

This timbral detail carries over into clarity of sound source timbres in the mix. Sound sources can have well defined timbres with extreme detail and clarity. Conversely, their sound qualities can be well blended with details absorbed into an overall quality. Both extremes are desirable in different musical situations. Both would place the sound at different distances, and each would cause the sound source to be heard differently in the same mix—one would more likely be prominent in a musical texture than the other.

How sound sources "sound" is an important aspect of recordings. Models of instruments and specific performers have their own characteristic sound qualities. Sound qualities are matched to musical materials and the desired expressive qualities to create a close bonding of sound

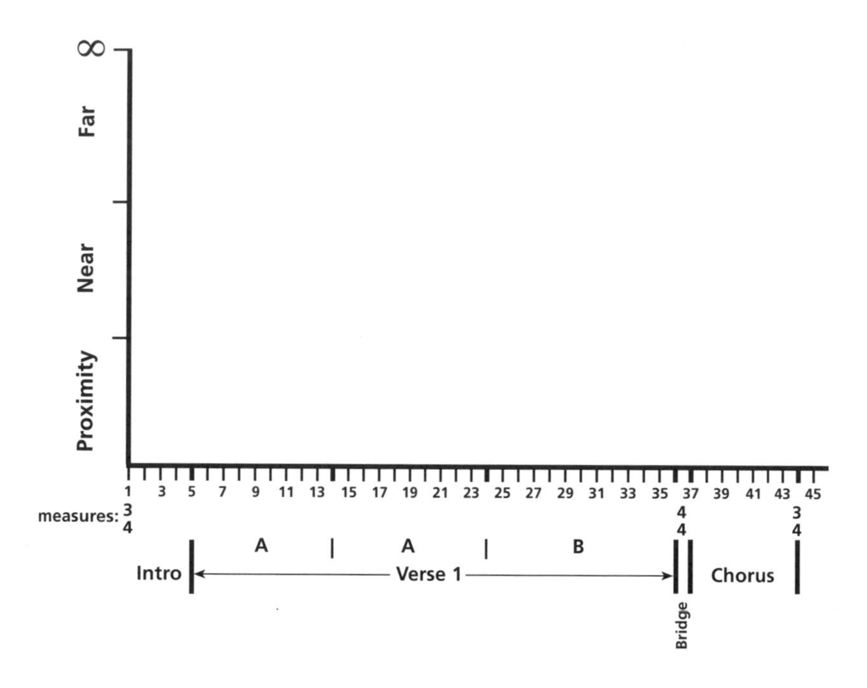

Figure 10-8 Distance location graph for The Beatles' "Lucy in the Sky with Diamonds."

quality and musical material. For instance, the opening of "Lucy in the Sky with Diamonds" would sound very different on a Hammond organ than on the Lowrey of the recording. A listener would recognize something incorrect about the source, even before they might identify the sound quality as being different. That musical idea is forever wedded to that original sound quality.

The Complete Evaluation

Great insight into productions can be found through an evaluation of all of the artistic elements in a particular recording/piece of music. This process will allow the listener to explore, in great depth, the inner workings of all sound relationships of a recorded piece of music. The listener would benefit from performing this exercise on a number of pieces, over the course of a long period of time. Not only will the listening and evaluation skills of the listener be refined, if recordings are carefully chosen, these evaluations will also provide many insights into the unique production styles of certain engineers and producers, as well as an understanding of the artists' work.

The project of performing a complete evaluation of a piece will be lengthy. It will take the beginner many hours of concentrated listening.

The demands of this project are, however, justified by the value of the information and experience gained. This project will develop and refine critical and analytical listening skills in all areas.

For greatest benefit, an entire work should be evaluated. The listener should be mostly concerned with evaluating all of the artistic elements. An evaluation of the traditional musical materials and the text might be helpful as well but is not necessary for most purposes. Upon an evaluation of the artistic elements individually, the listener should evaluate how these aspects relate to one another, and how they enhance one another.

This complete analysis of an entire recording is strongly encouraged. Many aspects of a recording will only become evident, when evaluations of several artistic elements are compared with one another, and compared to the traditional musical materials. The use of the artistic elements in communicating the musical message of the work will become much more apparent, when their interrelationships are recognized.

The listener will compile a large set of data, in performing the many evaluations of Chapter 6 through the pitch density of Chapter 10. These many evaluations will represent many different perspectives and areas of focus.

Some of this information will be pertinent to understanding the musical message of the work, and some of the information will pertain to its elements of sound (such as how stereo location is used in presenting the various sound sources). Also, some of the information will be pertinent to appreciating the technical qualities of the recording. All of this information will contribute to the audio professional's complete understanding of the piece of music, of how the piece made use of the recording medium, and of the sound qualities of the recording itself.

The sequence of evaluations that may prove most efficient in evaluating an entire work (depending upon the individual work, it may vary slightly) is:

- List all of the sound sources of the recording
- Evaluate the pitch areas of unpitched sounds
- Define unknown sound sources and synthesized sounds through sound quality evaluations
- Create a time line of the entire work
- Plot each sound source's presence against the time line
- Designate major divisions in the musical structure against the time line (verse, chorus, etc.)
- Mark recurring phrases or musical materials, similarly, against the time line; an in-depth study of traditional musical materials would be appropriate at this stage, if of interest

- Evaluate the text for its own characteristics and its relationships to the structure of the traditional musical materials, as appropriate
- Perform any necessary melodic contour analyses of those lines that fuse into contours
- Determine reference dynamic level
- Perform a program dynamic contour evaluation
- Perform a musical balance evaluation
- Perform a performance intensity evaluation
- Evaluate the work for stereo location or surround location
- Evaluate the work for distance location
- Perform environmental characteristics evaluations of all host environments of sound sources and the perceived performance environment
- Create a key of the environments to evaluate space within space information
- Perform a pitch density evaluation of the work
- Study these evaluations to make observations on their interrelationships and to identify the unique characteristics of the recording

The following materials may be coupled on the same graph (on separate tiers), or on similar graphs. They are all at the same perspective (at the level of the sound source):

- Performance intensity
- Musical balance
- Distance location
- Stereo location or surround location

The above four artistic elements will be interrelated in nearly all recording productions. Observing the interrelationships of these elements will allow the listener to extract significant information about the recording. In making these observations, the listener will continually formulate questions about the recording and seek to find solutions to those problems.

The questions of how artistic elements (and all musical materials) relate to one another will center on:

- Patterns of activity within any artistic element (patterns of activity are sequences of levels within the artistic element, and rhythmic patterns created by the relationships of those levels)
- Levels of any artistic element (how high-pitched, what loudness levels, etc.)

- Interrelationships of patterns between artistic elements (Do the same or similar patterns exist in more than one element?)

Music is constructed as similarities and differences of values and patterns of musical materials. This is also the way humans perceive music. People perceive patterns within music (its materials and the artistic elements). Listeners will perceive the qualities (levels and characteristics) of the elements of the music and will relate the various aspects of the music to one another.

At the same time, the listener should compare what they are hearing with what was previously heard, and to their previous experiences. Meaning and significance will be found in this information by looking for similarities and differences between the materials.

The listener should ask, "what is similar" between two musical ideas (or artistic elements); "what is different;" "how are they related?" These will be answered through observing the information that was collected during the many evaluations. The shapes of the lines on the various graphs may show patterns. The vertical axes of the graphs may show the extremes of the states of the materials and all of their other values.

The listener's ability to formulate meaningful questions for these evaluations will be developed over time and practice. They will be asking: "what makes this piece of music unique," "what makes this recording unique," "how is this recording constructed," "how is this piece of music constructed," and "what makes this recording effective?" Many other, much more detailed questions will be formulated during the course of the evaluation. The listener should finally ask, "which of these relationships are significant to the communication of the musical message; which are not?"

The use of the artistic elements in the recording can also be considered in their relationships to the traditional musical elements and materials. This brings an understanding of the importance of each musical idea, as related to the piece as a whole. Through these observations, the recordist will obtain an understanding of the significance of the artistic elements to communicating the message (or meaning) of the music. The recordist can then understand and work to control how the recording process enhances music, and how it contributes to musical ideas and the overall character of the piece.

This entire evaluation process will greatly assist the recordist in understanding how the artistic elements may be applied in the recording process to enhance, shape, or create musical materials and relationships.

The recording production styles of others can be studied and learned. By understanding the sound qualities of a recording and being able to recognize what comprises those sound qualities, the sound of another recording or type of music can be emulated by the recordist in their own work as desired.

Summary

Graphing the artistic elements may be time consuming, and at times tedious and perhaps frustrating. The graphing of the activity of the various artistic elements is important, however, for developing aural skills and evaluation skills, especially during beginning studies. It is also valuable for in-depth looks at recordings; providing insights into the artistic aspects of the recordist's own recordings and the recordings of others.

This process of graphing the activity of the various artistic elements is also a useful documentation tool. Graphs can be used to keep track of how a mix is being structured, or how the overall texture is being crafted. Many of the graphs or diagrams can even be used to plan a mix. For example, the imaging diagram can be used to plan the distribution of instruments on the sound stage and consider distance assignments before beginning the process of mixing sound—perhaps even before selecting a microphone to begin tracking. Working professionals through beginning students will find these useful in a variety of applications.

Graphing artistic elements is not proposed for regular use in professional production facilities and projects. The graphs are not intended for the production process itself. Audio professionals must be able to recognize and understand the concepts of the recording production, and hear many of the general relationships, quickly and without the aid of the graphs. The graphs are intended to develop these skills, and to provide a means for more detailed and in-depth evaluations that would take place outside of the production process.

Recordists who have developed a sophisticated auditory memory will also find these graphing systems of evaluation to be useful for notating their production ideas, and for documenting recording production practices. These acts will allow them to remember and evaluate their production practices more effectively, allowing them more control of their craft.

All of the exercises presented in the text are listed at the beginning of the book after the Table of Contents. The exercises are ordered to systematically develop the reader's sound evaluation and listening skills. Working through the exercises in this order will be most effective for most people. Readers with much experience and well-developed skills will still find at least a few exercises that are sufficiently advanced to test and improve their skills. Learning anything new requires effort and a willingness to reach into the unknown. Very sophisticated listening skills are required in audio and music. Developing such skills from the beginning will take practice, patience, and perseverance. At times the listener will be told to listen to things they have never before experienced. They will have no reason to believe such sound characteristics even exist, let alone can be heard, recognized, and understood. Faith will be required; a willingness to be open to possibilities and leap blindly into an activity, searching the sound materials for what they have been told exists.

Following the system will give the listener a refined ability in critical and analytical listening. The listener will learn to communicate effectively about sound, and be able to apply this new language to many situations.

Among other things, Part 3 will explore how these new listening skills and knowledge of sound can enhance the recordist's ability to craft recordings.

Exercises

Exercise 10-1
Pitch Density Exercise.

Select a recording with four to six sound sources, to graph the pitch density over the first two major sections of the work. The recording should contain pitch density changes within and between sections.

The pitch areas in pitch density are determined in a similar way to defining the pitch areas of nonpitched sounds. The material being plotted is more complex, however, and the process more involved.

The musical texture must (1) be scanned to determine the musical ideas present. The musical ideas might be a primary melodic (vocal) line, a secondary vocal, a bass accompaniment line, a block-chord keyboard accompaniment, and any number of different rhythmic patterns in the percussion parts. The number and nature of possible musical parts are limitless.

The musical ideas must (2) then be clearly identified with the sound source or sources performing it. This may be a simple process, or not. It is possible for a single instrument to be presenting more-than-one musical idea, such as a keyboard presenting arpeggiated figures in one hand and a bass line in another. It is possible for many instruments to be grouped to a single musical idea, then suddenly to have one instrument emerge from the ensemble to present its own material. The possibilities are much more complex than merely labeling each musical part with the name of an instrument, although this is often the case.

Each idea will (3) then have its pitch areas defined by a composite of the pitch area of the fused musical material, and the primary pitch area of the instrument(s) or voice(s) that produced the idea.

The process of determining pitch density will follow this sequence:

1. During the initial hearing(s), establish the length of the time line. Notice the entrances and exits of the sound sources/musical ideas as they relate to the time line.
2. Check the time line for accuracy and make any alterations. Clearly identify the musical ideas fused to sound sources and sketch their presence against the completed time line.
3. Make initial evaluations of the pitch density of the primary musical ideas by placing the material in its pitch area(s). Continue this for all parts.
4. Return to each idea, and consider and define the spectral information that should be fused with the material to finalize the pitch areas. Finalize these judgments. The process is complete when the all information perceived is incorporated into the graph.

Part Three

Shaping Music and Sound, and Controlling the Recording Process

11 Bringing Artistic Judgment to the Recording Process

Artistic judgment will be brought to the recording process by the recordist. How the artistic elements are used by the recordist shapes the recording in significant ways. The recordist will shape the sound in artistically sensitive ways, or not. Skilled recordists are able to make these decisions intuitively, if not consciously.

The recordist can bring artistry to the recording by learning the recording process well enough to use it creatively, by learning the sound qualities of the instruments of the recording studio (recording equipment and technologies), and learning all of the possible ways the recording processes and devices can transform sound. These will give the recordist the tool set needed to control how the recording process shapes sound. The recordist will then be able to make decisions on how to shape the recording. Those are the decisions that directly contribute to the music and create a characteristic sound to the recording.

Part 3 Overview

Part 3 will explore how the recording process can be used creatively. It will teach the reader to consider the sound qualities of recording devices and talk about how to evaluate sound during production—and before and after. The recording process will be considered, from beginning to end, in broad terms.

Part 3 will define basic concepts, aesthetic and artistic considerations, and some philosophies on recording. This creates a broad framework for the recordist, a way to orient any recording project toward fundamental principles. This big picture approach may at times seem to state the obvious, but it is intended to bring the reader (recordist) to keep perspective of the overall direction of a project and of basic concerns. All too easily we get overwhelmed by details and lose track of overall quality and direction. Our effectiveness gets diminished and our artistic vision blurred.

One of the goals of this section is to stimulate thought, to get the reader/recordist to consider how to use the recording process creatively, and how to create their own approach to working in the studio. Ultimately this should lead the reader to develop their own personal way of working, and perhaps even their own production sound.

While Part 3 explores certain aspects of production in detail, it does NOT present step-by-step sequences of instructions. It is designed to move the reader to think about what occurs and to consider the process as a creative endeavor.

Recording equipment is discussed in general terms. The concepts delivered are relevant to all devices and technologies, and should remain so with changing technologies. It is significant to note this discussion is not technology dependent, and should remain valid when applied to any technology or device.

Part 3 needs to be supplemented by readings on recording technologies, and on equipment use and production techniques. These areas must be learned for the reader to be able to control the recording process. This information is outside the scope of this writing and can be found in many excellent books (some of which are listed in the Bibliography).

The Signal Chain

The overall concept of the recording process involves the flow of signal through a chain of recording/reproduction devices. The recording-and-reproduction signal chain may take many forms, depending on the nature and complexity of the recording project. The stages of the signal flow are interrelated. While signal flow is sequential, it can be altered for the individual project or because of personal working preferences (developed over time and with experience). The flow of the signal through the chain, and the order of equipment and events, will normally be consistent with Figure 11-1.

The activities and associated devices of the recording-and-reproduction signal chain generally appear in the sequence outlined below. Some devices are used continually (or intermittently) throughout the recording process, such as the mixing console and the monitoring system. Other activities, such as editing, may occur at several different stages of the production process and within the signal chain, or may not occur at all.

- Microphones (or electronic instruments)
- Console (pre-processing, record levels and routing)
- Digital multitrack recorder or analog multitrack recorder (perhaps with noise reduction processors)

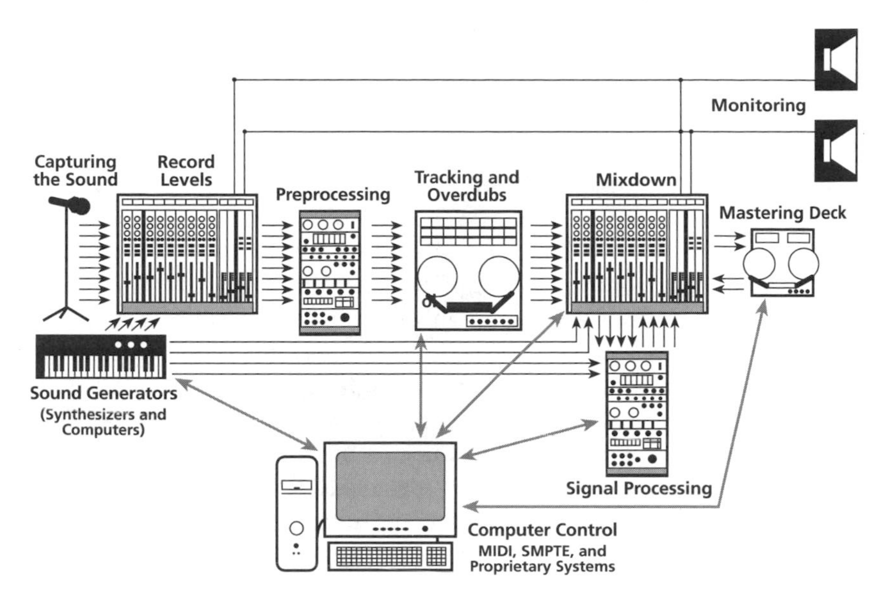

Figure 11-1 Recording-and-reproduction signal chain.

- Console (routing and mixing)
- Signal processors
- System control methods (automation systems, MIDI, SMPTE, and others)
- Master recorder (analog, digital, or computer-based)
- Editing (razor blade or computer-based)
- Monitoring

Many subcategories of these devices exist, and many of these devices can be used for multiple purposes. In keeping with the approach of remembering broad concepts, the reader is encouraged to think about equipment functions in basic terms. Specifically, what does the device do (for example, consoles only mix and route signals—anything else is an add-on). Specifically, how does it accomplish its task (it combines several signals into one with variable proportions and it sends signals from one place to another). Such general observations can keep the process simple, when it gets complicated.

The sonic imprint of recording devices is an important factor that will be covered in detail. The reader/recordist will learn to evaluate these imprints and learn to apply them appropriately. How the individual device shapes sound, both by how it functions and because of its own inherent sound quality, are central concerns. Choosing devices is an important artistic decision as well as a practical and technical one. It requires artistically sensitive judgment on the part of the recordist, as the recording devices and processes shape the aesthetic and artistic elements of sound.

Guiding the Creation of Music

The recordist can have many roles in making a recording. Among these roles is creating a working environment where the magic of making art can happen. How the recordist might participate in making art will be covered in the next chapter. The recordist usually sets the tone for recording projects, and can control many things including a project's pace and how people interact. The recordist is usually largely responsible for guiding recording projects, whether or not this is acknowledged by the others involved. They are often responsible for keeping the creative process moving effectively, efficiently, and invisibly; giving guidance to the artists or giving them enough space and support so they are free to be creative.

It has often been said that the recordist should first be a psychologist. While this statement may be somewhat extreme, the recordist needs to be sensitive of interpersonal relations. How they work with performers and others will shape the music as well as their actual recording duties.

The ways people normally treat one another in everyday living, and especially in standard business environments, are often nonproductive (at best) in the recording studio. Recordists should consider how they speak with the artists about the project, and how they interact with the artists socially and during the creative processes. The type of image the recordist presents to the artist will influence the artist's ability to function well professionally, and to keep them relaxed about their environment and focused on the project.

Recordists are trying to get creative people to do their best work, while attempting to perform their own tasks at the highest standard. The process of creating art (a music recording) is an emotional roller coaster of ecstasy of what has just been discovered and anguish over not having an equally brilliant answer to "what comes next?" Time and financial constraints further stress artists (clients).

Musicians/creative people are exposed and vulnerable in the recording process. The recordist must be certain to do nothing (or to allow anything to happen within the session environment, or by anyone else) to make the artists feel unprotected or, worse, threatened. Musicians must have the freedom to be creative around the recording studio, without feeling that their every move is watched, or evaluated. At times they will need to feel they are alone.

The expressive nature of performing music will often involve taking chances, stretching performance abilities to their limits or beyond, and making mistakes. These necessary activities can potentially embarrass confident, let alone less than confident, performers if they are critically judged at this vulnerable time. Performers need to be confident to perform well. The recordist is attempting to get the performers to exceed the height of their ability. Nothing should be allowed to happen that would diminish the confidence level of the performers and to take away from the trust that the performers must have in the recordist.

While evaluations have their place in the recording process, judgments—especially those of a negative nature—rarely can be used constructively.

Summary

The broad strokes of Part 3 will give the reader/recordist a method for applying artistic judgment to the recording process. It will lead them to envision the project as a whole before it begins, and to retain that vision while executing the many technical tasks and artistic decisions required to craft a music recording.

The reader will gain insight into how recording devices and techniques can be evaluated and applied to shape *The Art of Recording.*

12 The Aesthetics of Recording Production

The differences between the various approaches to the aesthetics of recording production can be reduced to two central issues: (1) the relationship of the recording to the live listening experience and (2) the relationship of the recordist to the creative and artistic decisions of the recording.

The recording process can capture reality, or it can create (through sound relationships) the illusion of a different world. Most recordists find themselves moving within the vast area that separates these two extremes, where they enhance the natural characteristics and relationships of sounds. The recordist will determine an appropriate recording aesthetic for their current project based on its planned material and function.

The recordist will use the recording process to support the production aesthetic of the project. The artistic elements of sound will be captured and shaped in relation to the live listening experience and to the musical message. Recording techniques and technologies may also be used to shape the performance, especially through editing, mixing, micing, processing, and over-dubbing techniques.

The recordist will play a more active role in shaping the materials of the music (or the presentation of those materials) in certain projects more than in others. This is often a result of the function of the recordist, in relation to the other people involved in the recording project, in the artistic decision-making process.

For each project individually, the recordist will define the proper recording aesthetic, and their own role in the creative process of *The Art of Recording*.

The Artistic Roles of the Recordist

The recordist must have a clear idea of their role in the creative process for each project. The project may include the composer of the music, one or many performers, a conductor of an ensemble, and/or a specific recording producer. The recordist must know their responsibility to the final artistic product, and the roles and responsibilities of the others involved.

Of the many possibilities, the recordist may be functioning to capture the music as closely dictated by the composer. They may be functioning to capture, as realistically as possible, the performance of an ensemble, as precisely directed by the conductor; they may be functioning to capture the interactions and individual nuances of a group of performers, without altering the performance through the recording process; or the recordist may be functioning to precisely execute a recording producer's instructions. In all of these cases, the recordist is allowing the artistic vision and decisions of others to be most accurately represented in the recording. The recordist's role then is to facilitate the artistic ideas of others, and not to directly impose their ideas onto the project.

The recordist's role sometimes might be to offer suggestions to the creative artists or even to take an active role in the artistic decision-making processes. The role of the recordist might be active in shaping a performance of an existing work, or in creating a new piece of music. The recordist might be active in determining the sound qualities of the instruments of the recording, or in determining the sound sources themselves. Vastly different levels of participation in the artistic process are often required from one project to the next.

The process of writing a piece of music for a recording is often a collaborative effort. It may be between many people (composer, performers, producer, recordist) or just a few (performer/composer and recordist/composer).

In many ways, the recordist functions as a creative artist and can serve the traditional functions of a composer, a conductor, and a performer. The recordist also shapes sounds in nontraditional ways. Recordists have unique controls over sound and live performances that allow for an additional musical voice. It is possible to compose with the equipment (instruments) of the recording studio, to shape sounds or performances through the use of recording and mixing techniques, or to create a new musical environment for someone else's musical ideas and performances.

As will be explored below, the recording studio can be thought of as a musical instrument or a collection of musical instruments. In this way, the recordist may conduct all of the available sound sources (for example, bringing sounds into and out of the musical texture through mixing);

may "perform" the musical ideas through the recording process; may alter or reshape the sounds of the sources, or "interpret" the musical ideas, in ways that are not possible acoustically; and may create (compose) new musical ideas or sounds.

The Recording and Reality: Shaping the Recording Aesthetic

The recordist has many potential roles in shaping the recording aesthetic. The role of the recordist might be to capture a live event as accurately as possible in relation to the dimensions of that real-life experience, or the recordist might seek to alter the artistic elements of sound to enhance the quality of that real-life experience. The recordist may even seek to create a new reality or set of conditions for the existence and relationships of sounds. Reality is simulated, enhanced, or created through the recording process.

The relationship of the recording to the live listening experience is central to the aesthetic quality of the recording. A recording may differ from the live listening experience by (1) the use of the artistic elements of sound in ways that cannot happen in nature, and (2) the presentation of impossible human performances and compilations of perfect performances.

The aesthetic and artistic elements that most influence the life-like qualities of the recording are environmental characteristics, the dimensions of the sound stage, and the relationships of musical balance to the timbres of sound sources.

Sound Stage and Environments

Sound exists in space. Humans conceptualize sound, especially in the context of music performances, in relation to the spaces in which the sound exists. The recording process must provide the illusion of space to convince listeners that the sound has been reproduced in a way that is associated with their reality. The recording will provide the illusion of a performance space or a physical environment for the performance.

This illusion of a performance space is a perceived performance environment wherein the recording can be imagined as existing during its re-performance (playback). The realistic nature of the performance of the recording will play a central role in establishing the relationship of the recording to the live listening experience. The listener will subconsciously scan the recording to establish environmental characteristics, an imaginary stage (sound stage), and a perceived performance environment. This information allows the listener to complete the process of establishing a reality for the listening experience of the recorded music performance.

It is necessary for the recordist to deliberately shape these important characteristics and to precisely determine this aspect of the recording's aesthetic. The imaginary environments will be either the captured reality of the original performance space, an altered or enhanced reality of the original performance space, or new realities that are created for the performance through signal processing. The listener will make fundamental judgments about the material of the recording based on the qualities of the environment(s) simulated in the recording and will match the musical material against the appropriateness of the environmental cues.

The listener will imagine the location of the sound sources relative to one another and to the overall environment. The listener will envision the sound stage of the recording. In so doing, the entire ensemble will be placed at a certain distance from the listener, and each individual sound source will be placed at a distance and at an angle from its perceived location. The relationship of these cues to the potentials of live performances will also define the aesthetic of the recording in relation to the possibilities of our physical existence. The recordist must give focused consideration to the makeup of the sound stage. The sound stage provides an opportunity to shape the aesthetic of the recording and to provide the musical materials and recording with many spatial dimensions.

The sound stage of the recording might place the sound sources in locations that purposefully resemble those of a live performance. In certain recording techniques, the integrity of this imaging is a primary concern. Certain stereo microphone techniques are designed to accurately capture the depth of the sound stage and the lateral location of the sound sources. Other techniques accurately capture the microphone-to-stage distance and stage width. Multitrack recordings can also deliberately create a sound stage that recreates the live performance relationships of the performers.

The recordist often alters the sound stage to enhance the musical material. One or several additional microphones may be used to accent certain members of an ensemble. The highlighted instruments are given a distance from the listener, width of image, or a specific location that provides them with more prominence in the musical texture. This can be performed subtly, so as not to dramatically alter the natural qualities of the recording, or it can be quite pronounced, depending on the aesthetic of the recording.

Sound stages are created for multitrack recordings and for recordings made with only (or mostly) synthesized sounds. These recordings were created outside a common environment and with minimal naturally occurring spatial cues captured with the sound sources. The recording will be given spatial cues by the recordist through the recording process. If not, the listener's imagination will generate these relationships. The recordist will control the sound stage for the recording by crafting the characteristics of environment(s), distance, and stereo/surround location.

The recordist can provide the sound sources with life-like environments and place them in natural physical relationships to one another, or they can purposefully create environmental, distance, and localization cues that would be impossible in nature.

These concepts have been reinterpreted in two common approaches to mixing for surround sound. The traditional use of the listener's front field can still be the focal point of the surround mix. Left/right/center channels can present the primary materials, and be enhanced by placing ambience and special effects in the rear field. This replicates the observed performance of traditional music listening experiences. Surround might also approach the sound field as a 360° environment, where instruments and mix elements can appear anywhere in the surround sound field. The listener is now surrounded by sound sources and may perceive themselves to be within the music, perhaps even right in the middle of the performance ensemble. This has the potential to be a strikingly new listening experience.

The sound stage diagrams of Chapter 9, and Figures 14-1 and 14-2 from Chapter 14 will help the recordist to craft sound stage relationships. They will prove useful in many ways, including evaluating the relationships of sound sources and keeping track of the locations of sources.

Musical Balance and Sound Quality

The interrelationships of musical balance and the differences of sound quality of sound sources played at different dynamic levels (performance intensity) are integral parts of live performances, and are easily altered by the recording process.

Recordings that attempt to capture the aesthetics of the live performance will seek to capture the musical balance of the performers as they (or the conductor) intended. The changes in the sound quality of the instruments will be precisely aligned with changes of dynamic levels in the musical balance of the ensemble and to changes in musical expression. It is important to maintain these relationships to keep the character of the live performance.

Recordings that sought to enhance the characteristics of the live performance may contain slight changes in musical balance that were not the result of the performers, but were rather the result of the recording or mixing process. These alterations will be heard as changes in dynamic levels that are not supported by changes in the sound qualities of the instrument(s). This enhancement might take place in only a few instruments, or it may be used extensively throughout the entire ensemble. This enhancement technique may be quite subtle and difficult to detect, or it may be prominent. A soloist with an orchestra is a common example of when this might occur.

Alterations in dynamic levels, and thus musical balance, that are not aligned with changes in performance intensities have become integral parts of music written for recordings. Multitrack mixes frequently exhibit changes in musical balance that were not caused by the performers. These changes in dynamic level are inconsistent with the sound qualities of the instruments in the final recording. This inconsistency of one element tracking another enhances the potential of each element to be used individually in shaping or enhancing the musical material. Further, the expressive qualities of sound quality/performance intensity can be incorporated into a mix without the impact of a louder or softer dynamic level than desired.

The relationship of musical balance to the timbre of sound sources in many multitrack recordings creates a wealth of contradictions between reality and what is heard. The aesthetics of this type of recording leans toward redefining reality with each new project and is a stark contrast to the aesthetic of trying to capture the reality of the live performance.

The recordist's approach to any project should include a conscious decision on a level of realism. How will the final sound relate to real life experiences, and how will the characteristics of sound be shaped? What is the listener intended to believe, and how can this be achieved?

The Recording Aesthetic in Relation to the Performance Event

The recording process will shape music performances in such a way that the sound qualities and relationships of live performances may be altered. How the process alters the live listening experience is central to the aesthetics of the recording.

Production Transparent Recordings

The recording medium is often called upon to be transparent. In these contexts, it is the function of the recording to capture the sound as accurately as possible, to capture the live performance without alteration. This type of aesthetic is common for archival recordings that function to document events. These *production transparent recordings* may or may not be sensitive to the performance environment. At times, these recordings attempt to capture the sound of the music performance without considering the artistic dimension of the relationship of the music (and musicians) and the performance space (and audience). In other instances, these recordings seek to negate any influence of the performance space on the sound of the recording.

Because these are recordings of live performances, the recordist is not involved with compiling the performance. The performance takes place in real time, and it will not be possible to back up and fix a certain section or idea. The recordist is primarily concerned with the technical aspects of the sound of the recording (critical listening) and the sound quality of the overall ensemble (sound quality at the highest level of perspective).

A limited number of microphones are often used in making this type of recording. Usually two microphones are used in some appropriate stereo microphone technique, placed fairly close to the ensemble. The microphones generally are sent directly to a two-track mastering recorder, with little or no signal processing. The recordist will exercise little immediate control over the quality of the sound and over the shaping of the performance.

The recording medium can also be transparent in documenting a performance, while placing the music in a complementary relationship with the host environment of the performance. Specific pieces of music are best suited to certain environments and are most accurately perceived from certain listening distances. The artistic message of a specific piece of music will be most effectively communicated in a certain environment and with the listener at an ideal distance from the ensemble.

Spatially Enhanced Production Transparent Recordings

Spatially enhanced production transparent recordings can ensure pieces of music will be perceived as having been performed in an ideal environment, with the listener located at an ideal distance from the ensemble, when listening to the recording. This approach locates the listener at the *ideal seat*, and can be accomplished without altering the performance itself and maintain transparency of the recording process.

The recordist will determine the type and amount of influence the acoustic performance environment will have on the final recording. Microphone selection, choice of stereo microphone array, and array placement within the performance environment are the primary determinants of the environment sound that is captured from the performance environment. Artificial reverberation units or other time processors may sensitively enhance the characteristics of the environment. The distance of the listener from the ensemble is determined primarily through microphone placement and through time processing.

This recording aesthetic attempts to present the music in the most suitable setting possible for that particular work and to simulate the listening experience in the concert hall. This recordist seeks to ensure that the sounds will be in the same spatial relationships as the live performance, that the recording process will not alter the balance of the musical parts, and that the quality of each sound source will be captured in a consistent manner. In these *live acoustic recordings*, the recording may seek to reproduce the sound of the performance space—surround sound systems can be used for great realism in this approach.

This aesthetic can have the recordist more involved with the decision-making process in some projects than in others and may be used for live concert recording as well as session recording. This aesthetic may be used for many types of music. While it is common in orchestral and other art music formats, it is equally appropriate for jazz, or any other music recordings where the performers are refined in their sensitivity to and control of their relationships to the whole ensemble. In session recordings, some (or much) editing may be a part of this aesthetic. A consistency of sound quality and spatial relationships between all portions of the work will nearly always be sought.

Enhanced Performances

The recording medium may *enhance the performance* in widely varying degrees. This aesthetic may be a slight extension of the concept of a transparent live recording, with the recording process slightly enhancing certain musical ideas, or this aesthetic may set another extreme of being a life-like session recording that was recorded out of real-time.

This aesthetic simulates a natural listening experience, by capturing or creating many of the inherent characteristics of a live, unaltered performance. The timbre and dynamic relationships, spatial cues, and editing techniques all serve to create the impression that the recording did indeed take place within reality—as an actual, live performance.

When this aesthetic is an extension of the concept of a transparent live recording, sounds are placed in the sound stage in the same relative positions as the instruments were in during the recording. The width and depth of the sound stage and image sizes are realistic, and the recording will usually have a single environmental characteristic applied to the overall program (a single soloist might be present with a slightly different environment). Dynamic changes are nearly always aligned with timbre changes, though some microphone highlighting might create a limited number of dynamic changes without timbre changes. The recording process is used to slightly enhance certain musical ideas from the live performance.

This aesthetic may be used for controlled live performances (those that have been rehearsed with the recordist) or in recording sessions, for a wide variety of musical styles. Minimal micing will usually be used, often an overall stereo array with a small number of accent microphones (or stereo pairs). Accent microphones allow this aesthetic to be adaptable to stage recordings of large classical ensembles or for musical theater and opera. The recording is usually mixed directly to a two-track mastering deck (or surround), with mixing decisions taking place during the rehearsals or during the recording session(s). Recording submixes to a multitrack recorder is also common, but many of the decisions related to the sound of the recording are still accomplished during the recording session or rehearsals.

Recording sessions will often be comprised of many takes of large and small sections of the work. As the ensemble balance is largely controlled by the performers, and the parts are not singled out (making re-recording of individual parts unavailable), ensemble problems of accuracy and sound quality often cause a lengthy recording session and a large set of session tapes.

The master tape usually ends up being a collection of the many takes. The editing of these many takes of the musical material becomes an integral part of the recording process. The best takes are selected based on musical and technical qualities. These are then edited together (cut and paste) to compile a *perfect performance* of the work. The master tape represents the final performance. The goal of this approach is usually to craft the best possible (perfect) performance, interpretation, and presentation of the music.

The aesthetic of slightly enhancing the reality of the performance may also be found in session recordings that simulate natural sound relationships. Although recorded out of real-time, the recordings will simulate the experience of live music. Some emphasis of certain musical materials (and/or artistic elements) over others will be unavoidable in the recording process and will diminish the naturalness of the relationships of the sounds. Some recordings may simulate reality only generally, but their inherent conception is still toward providing the illusion of a naturally occurring performance. Even with the complete control of the multitrack recording (micing, processing, and mixing) process, the goal is to provide the illusion of a live performance. In all other ways, music written for recording might follow the most unnatural recording processes and concepts.

Music written for the recording medium may be significantly different from live, acoustic music. It may be constructed in different ways, and it may contain additional artistic elements. Music written to be recorded, and especially music written during or through the recording process, is often composed and/or performed in layers.

The musical materials are often written and recorded one part at a time, or a small group of instruments at a time. The recordings use close micing techniques that ensure a separation of parts (and thus allow for precise control of the individual sound source) or will physically isolate the performers/performances from one another. The parts are continuously compiled on a multitrack recorder, with each new musical line added to the musical texture. Players often perform their parts many times; any number of versions may be recorded before the desired result is achieved. The recordist may be responsible for listening for performance mistakes, listening for the most interesting and successful performance, keeping track of which portion of which musical part was performed most accurately, on which take, etc.

The final piece may be a composite of any number of performances, and it may be a controlled integration of many different musical ideas and personalities. The performances may or may not have taken place at the same studio, or during the same day (or year), and the performers may or may not have met and discussed their musical intentions.

The recording medium can create the illusion of a performance that contains characteristics that cannot exist naturally. This aesthetic has become common since the early/mid-1960s. In this "new" aesthetic, the recording medium's unique sound qualities and creative potential are used. It becomes a musical ensemble with its own set of resources for shaping a performance or creating a musical composition.

Music written to be recorded may exploit environmental characteristics, musical balance and sound quality contradictions, sound stage depth and width, sound source imaging, or its other unique elements to create, define, or enhance its musical materials.

This aesthetic might purposefully create relationships that cannot exist in nature—a whisper of a vocalist might be significantly louder than a cymbal crash. This aesthetic will use the unique qualities of recorded sound in the communication of the work's musical message.

Recordings of this aesthetic might seek to create a new reality for each work or project. Unique relationships of sound are calculated and incorporated into the music. Recordists (engineers and producers) develop personal styles of the ways they shape aspects of balance, imaging, sound stage, and environment, while continuing to explore the expressive potentials of recording and the medium's relationships with reality.

Much of today's popular music falls within either the above aesthetic, or within the aesthetic of using the recording medium to enhance the illusion of a live performance. Many of the artistic considerations of the recording process are very apparent in these two aesthetics.

Altered Realities of Music Performance

The listener's perception of the reality of the music performance event itself is also altered by the aesthetics of the recording production. The recording provides an illusion of a live performance, and the content and qualities of the perceived music performance may vary from a slight improvement of our listening realities, to being a live performance that is existing in ways that are impossible in our known world.

Recording allows a music performance to be an object that can be precisely polished by the artists, physically held in one's hands, and owned by a member of the general public.

The reality of a live music performance as an experience witnessed in a fleeting instant in time, and as retained only in the memories of those who experienced the event, is significantly altered by the recording process. The recording is a permanent performance of the piece of music—one that can potentially live well beyond the artist's lifetime.

Additional pressures, ideals and aesthetics are placed on the artists responsible for any individual recording, as opposed to a live performance. A recording may transform the live listening experience: (1) through the presentation of humanly impossible performances, (2) by providing performance conditions that are inconsistent with reality, (3) through the presentations of error free and precisely crafted performances, and (4) by providing a permanent record of a music performance.

A recording is a *permanent performance* of a piece of music. It is a period of time that has been created or captured, and that may be preserved forever. The performance can be revisited (and observed at any level of detail) at any time, and any number of times, and by anybody.

Recordings can often become *definitive performances* of a piece of music. The definitive performance may be thought of as being either that of a certain artist or of the particular piece of music. An artist's performance/recording of a work might be what is widely accepted as the definitive performance (or reference) of how the work exists in its most suitable form, in relation to performance technique or to the communication of the musical message. A specific recording of a work can also serve as a definitive reference of how a work exists in its most suitable state, in relation to recording practice or to musical considerations.

Recordings not only are a means of creating an art form, they also preserve the artistic ideas of music performance and expression that do not rely on the recording process. Recordings may permanently preserve the music performances of an artist. They may provide historical documentation or archival functions by preserving the music performances of particular artists, ensembles, events, etc.

The great contradiction of producing a recording that is a permanent record is that the recording often becomes dated. Artists develop and grow. Their musical abilities, levels of understanding, artistic sensibilities, and their musical ideas change. The permanent performance that was previously created (perhaps only a few weeks before) may no longer be representative of the artists' abilities or aesthetic opinions.

The recording will often represent the artists' and recordist's idea of a perfect performance of the work. Theoretically, a perfect performance of any piece of music can be produced through the recording process. The definition of the perfect performance may vary considerably between performers, but the concept of the recording itself will be similar. It will be a presentation of what the performers and producer believe to be the most appropriate interpretation of the piece of music, under the most appropriate performance conditions (instruments used, performance space, etc.).

A perfect performance will combine the artists' desired interpretation of the music (and an absence of performance inaccuracies), with an illusion of the drama of a live concert, and as being experienced at the ideal listening location of an ideal performance environment for the ensemble and piece of music. Often practical considerations of the recording process will compromise the actual quality of the recording, but the goal of the recording remains constant.

The recording may present musical ideas, and sound qualities and relationships that are impossible to create in live performance. Musical materials may be presented in ways that are beyond the potentials of human execution. These might be rapid passages performed precisely and flawlessly, dynamics and sound quality expressions that change levels quickly and in contradictory ways, or the use of a single human voice to perform many different parts. These are only a few of the possibilities. Humanly impossible performance techniques and relationships are easily created in recording.

The reality of what is humanly possible in a music performance is often inconsistent with the music performance of the recording. The relationships of sound sources, the characteristics of sound sources, the interrelationships of musical ideas and artistic elements, and the perceived physical performance of the musical ideas may be such that they could not take place in nature. The music performance of the recording may be such that it could not be accomplished without recording techniques and technologies.

Recordings have greatly influenced our expectations of live music performance. The listening audience of a recording becomes accustomed to a recording as being the perfect performance of a piece of music. The audience may learn the subtleties of a recording quite intimately.

A particular performance of a piece of music is created or captured in a music recording. When an audience member owns a copy of the recording, or when a recording has received much exposure through media, the audience may have listened to a recording many, many times. The recording becomes the definitive performance of the music, for some audience members. The audience will carry this knowledge of the music recording into a live music concert, and may impose unrealistic expectations onto the performers and the event.

The artists' new and different interpretations of the music, the absence of recording production techniques, the inconsistencies of human performance, or other factors, might create differences between the live performance and the known recording of a piece of music. A potential exists for audiences to become less involved with the drama and excitement of the live performance of music. Audiences may attend concerts to publicly hear live performers and the music performances they have come to know well as recordings, heard privately many times. Audience members do not always accept the reality that the live performance was not the same as the studio-produced recording. A potential exists for the audience member to be dissatisfied that the definitive performance they know well, was not reproduced for them at the live event.

An audience may place unrealistic expectations on the performers. Performers may be expected to perform flawlessly, or with the same version and interpretation of the work as their released recording. An audience may expect the performing artist to provide the role of reproducing one particular performance of the work. By reacting to the audience, the performer may be restricted from allowing their interpretation of the music to evolve and change according to their growing experience, and may be restricted from creating a more exciting performance. The subtleties of artistic expression that are possible only through the artist and audience interaction, along with other unique qualities of live music performance, may be lost from an event and diminish the musical experience.

It is unrealistic to expect to hear a precise reenactment of a recording in a live concert environment. Many music recordings have been produced in such a way that a live performance of all musical parts, sounds, and relationships is impossible. These are potential negative outgrowths of the audience's familiarity with certain recordings and the new listening habits afforded by readily available music performances.

The general public hears much more recorded music than live music. They are often prone to judging live performances with the expectations of a perfect, recorded performance. Human inaccuracies are sometimes not easily tolerated, and new musical and expressive interpretations of a known piece of music may be heard as simply wrong. Further, people own their own personal copies of performances. A tendency to personalize or to become attached to those performances is common. When the performance is changed, something personal (their music) has been altered.

Summary

The recording aesthetic is determined by the relationship of the recording to the live listening experience. The recording aesthetic is arrived at through a careful consideration of the musical material, the function of the recording (type of music recording, sound track, and advertisement) and the desired final character of the recording. The recordist's role is defined by their contributions (or lack thereof) to the process of making the creative decisions of the recording.

The recordist's overall control of the many qualities of the final music recording is highly variable. The recordist is responsible for the overall characteristics of the recording and may be in control of (and responsible for) its most minute details, depending on the recording techniques being used. The recordist might have precise control over shaping or creating a performance, or they may be engaged in capturing the global aspects of a live performance.

The amount and the types of control used in the recording process will determine the degree of influence the recordist has on the final content of the music recording. The recording medium may be used to greatly influence the sounds being captured by the microphones, or the recording medium may shape sound much more subtly. The recording process will be used differently, depending on the particular project.

The aesthetics of recording production vary with the individual, with the musical material, and with the artistic message and objectives of a certain project.

An aesthetic position or approach may be appropriate for a certain context, or it may not. An approach might enhance the artist's conception of the music, or it may not. An approach may be consistent with other considerations of the project or the music, or it may not. Perhaps consistency is not desired.

The aesthetic approach to recording production creates a conceptual context of the artistic aspects of recording. The intangible aspects of the art can then be appreciated within this context. The recordist must seek to define the aesthetic of the recording, in order to successfully control and shape it.

13 Preliminary Stages: Defining the Materials of the Project

Ideally, the content and scope of recording projects are planned in detail before the recording process begins. Before any sounds are made in the studio is the best time for the recordist to formulate clear ideas of the dimensions of the project. The recordist will learn to anticipate the requirements of various types of projects, and the related logistical and technical decisions to be made. They will also understand the artistic vision of the piece. With this understanding, the recordist will then be able to correctly make the choices of the preliminary stages of a music production or audio project. Technical, logistical, artistic, and conceptual concerns will all be addressed.

These choices will affect the entire project. If they are made poorly, the entire project may be compromised. The recordist can anticipate the problems of each stage of the recording project by considering the state and sound qualities of the project at all of the various stages of the sequence of creating a music recording.

In the preliminary stages, the materials of the project are defined. These materials comprise the creative ideas and materials of the work. The recordist might be responsible for making many of these decisions, depending on their role in the decision-making process of the artistic aspects of the project.

The choices the recordist may be required to make during the preliminary stages of the project are choice of sound sources, choice of how to capture the sound sources (initial shaping of the sound sources), and choice of how the sound sources will be heard during the recording process (monitoring format).

The tracking process often must be planned during the preliminary stages of a recording project. This topic is covered in the next chapter. The decisions of tracking will, however, often occur during the preliminary stages before sounds are dedicated to tape. This encompasses defining the number of available tracks, placement of instruments/material on the multitrack tape, mixing of microphones during the tracking process, placing sync tones, performance order planning, planning reference tracks, and rehearsing the session.

When the entire recording can be envisioned at this planning stage, control of the recording and clear artistic direction can be established at the outset. Even the sound stage can be considered in planning the project, as microphone selection now may well limit sound source distance and image size options during the mixdown stage. These preliminary stages prepare for successful projects of quality that make efficient and effective use of studio time.

Sound Sources as Artistic Resources and the Choice of Timbres

The music to be recorded and a certain set of sound sources (voices and instruments) to perform the music are usually determined for the recordist as dimensions of the project. Clients usually dictate what is recorded and who performs. Some decisions on the selection of sound sources and on many of the supportive aspects of the music are, however, often made during the preliminary stages of the recording process (or during the production process itself). Sound sources are selected, created, or shaped. In all forms, their selection is an important decision in shaping the sound of a recording. It can also be important in determining the success of the project and the character of the recording.

Sound sources deliver the materials of the music production. The sound sources will be the vehicle that presents the musical ideas. They must be selected carefully and with attention to their anticipated roles in the production. Sound sources are often coupled with the musical ideas themselves, as the primary artistic resources of the music—the musical idea and the sound quality of the source are often melded into a single artistic impression.

In selecting sound sources, decisions are being made as to the most appropriate timbre to present the musical materials. In doing so, the sound source is considered in relation to:

- Suitability of its sound quality to the musical ideas
- Potential to deliver the required creative expression
- Pitch area information for anticipated placement in relation to pitch density

Performers as Sound Sources

Individual performers may be selected because of their unique sound qualities. Individual performers, themselves, are unique sound sources. This is especially true of vocalists, who are sought for their unique singing voice and styles, as well as for their speaking voices.

Accomplished instrumental performers that have developed their own style(s) of playing or that are skilled in performance techniques are also sought for their unique sound qualities. Individual performers often bring their own creative ideas and special performance talents to a project, and considerably aid the defining of the sound qualities of the sound sources.

The act of selecting particular performers for a recording is important for defining the sound quality of the sound source down to the minutest detail. Since the recording is a permanent performance of the piece of music, the selection of the performers for this performance is often an important consideration in determining the sound qualities of the sound sources.

Just as live, human performers function as unique sound sources in a music recording, nonhuman performers can function in the same ways. Computers and sequencers are nonhuman, mechanized performers.

Computers and sequencers have the potential to be programmed in great detail. They are capable of performing complex musical materials and ideas, by controlling (sending performance instructions to) sound sampling and synthesis devices. Computers and sequencers (both hardware and software) have the potential to give certain characteristic sound qualities to the music, but may also be very life-like. They are capable of very detailed and precise control over a performance (and of making related timbre changes), often providing very human-like results.

Creating Sound Sources and Sound Qualities

The sound manipulation and generation techniques of sound synthesis allow sound sources to be created with the design of new timbres. New sources are created with these new sound qualities. Many approaches to sound synthesis are available:

- Analog synthesis techniques
- Additive and FM digital synthesis techniques
- Many hybrid (analog + digital + sampling) synthesis techniques (such as waveshaping, phase distortion, wavetable, physical modeling, granular synthesis techniques, etc.)

- *Musique concrète* techniques
- Recording and performing techniques on sound samplers (sampled live or with commercially available sound libraries)
- Computer generated (often employing typical digital synthesis or sampling techniques)

The creation of sound sources allows the recordist great freedom in shaping sound qualities. The recordist will be functioning as a *sound designer,* whose goal is to create a sound (with a sound quality) that will most effectively present the musical materials and ideas of the music. Sound sources that precisely suit the contexts of the sound and the meaning of the music may be crafted or created by the recordist.

While an examination of the sound synthesis process is out of the scope of this writing, it is important for the recordist to be aware of the many creative options afforded by sound synthesis. The study of sound synthesis from the perspective of building timbres will greatly assist the recordist in understanding the components of sound, and how the components of sound may be used as artistic elements. Signal processing and sound synthesis share many common traits.

By creating sound sources, the recordist will be presenting the audience with unfamiliar "instruments." The sources (new instruments) may be performing significant musical material. The reality of the performance has been altered out of the direct experience of the listener. The recordist will create a new reality of sound relationships or might emphasize known sound relationships to reestablish known experiences. These relationships will be accomplished in such a way as to support the musical materials and ideas of the recording.

Environmental Characteristics

The human realities of sound relationships are most closely associated with acoustical environments. The listener will process the characteristics of the environment within which the sound source is sounding, and the location of the source within its acoustical environments, to imagine the reality of the performance. The acoustical environment itself will also function as a sound quality, shaping the sound source.

Fusion occurs combining the source's sound quality and environmental characteristics into a single impression. The selection of environment therefore may be as important for sound quality concerns as it is for shaping the spatial aspects of the recording.

On the extreme, a set of environmental characteristics may be so much a part of the sound quality of a sound source, that the actual components of timbre become secondary in the global impression of sound quality judgments. The acoustical environment, in essence, becomes the sound source that is projected by the instrument, which it contains. This is an unusual sound occurrence and is accomplished through a very high percentage of reverberant sound over direct sound. It often causes the sound to appear to be "other worldly."

Nonmusical Sources

Nonmusical concepts often find a place in a music project. As sound sources, speech and special effects require special consideration.

With speech as a sound source, a particular voice is selected to complement the meaning of the text and to complement the other sounds in the musical texture. The voice is carefully selected for the appropriateness of its sound quality, and thus its dramatic or theatrical impact, to the meaning of the text to be recited.

Special effects are sound sources that are used to elicit associated responses or thoughts from the listener. The associated thoughts generated by the special effects are not directly related to the context of the particular piece of music. Special effects pull the listener out of the context of the piece of music, to perceive external concepts or ideas. A horn sound occurring in a piece of music is an effect, used to elicit the mental image of an automobile. The same sound used as part of the musical material, used to complement the musical ideas of the work, would not be a special effect, but rather as a musical sound source.

Sounds for special effects and for sampling synthesis devices may be obtained from sound libraries. Sound libraries are collections of sounds that are commercially available. They are resources that provide a wide variety of sounds, from seemingly every conceivable source (natural and created sounds) and many different types of presentation (light through heavy footsteps of the same person, a child's footsteps, an elderly person's footsteps, high-heeled shoes, etc.). Individual sound libraries allow the recordist to obtain specific types of sounds for specific projects.

Microphones: The Aesthetic Decisions of Capturing Timbres

The interactions created by the selection of a particular microphone, placing the microphone at a particular location (within the particular environmental conditions), and matching those decisions to the characteristics of the particular sound source are major determinants of the sound quality of sound sources in music production.

A specific microphone will be used to make a certain recording of a sound source, because of the ways its performance characteristics complement the sound characteristics of the sound source. This interaction of the characteristics of the microphone and of the sound source allows the recordist to obtain the desired sound quality of the recorded sound.

A microphone will be selected for a particular recording because it is most appropriate for the desired, final sound quality. The recordist will determine which microphone is most appropriate by comparing the contributions of the sound quality of the sound source and the performance characteristics of the microphone, to the recordist's idea or conception of the final sound quality sought. Perhaps no other decision shapes sound quality more than microphone selection and placement. Even distance cues of the sound stage are determined by the timbral detail captured by the microphone.

The recordist will also evaluate their selection of microphone against the practical limitations of placing the microphone in the recording environment, and in relation to the sound source to be recorded and sound source the recordist does not wish to record.

No single microphone will be the best microphone for every sound source or for the same source for every piece of music. The microphones selected for recording the same sound source may vary widely depending on the above circumstances, the desired sound, and what microphones are available.

Performance Characteristics

All microphones can be evaluated by their *performance characteristics*. These characteristics are information on how the microphone will consistently respond to sound. Thus, through these characteristics, the recordist can anticipate how the microphone will transform the sound quality of the sound source, while it is being recorded.

As the microphone alters the sound source, it has the potential to contribute positively in shaping the artistic elements. If the recordist is in control of the process of selecting the appropriate microphone for the sound source and conditions of the recording, the selection and applications of microphones can be a resource for artistic expression. The artistic elements of the recording can be captured (recorded) in the desired form, and the microphone and its placement will become part of the artistic decision-making process.

A number of microphone performance characteristics are most prominent in shaping the sound quality. These characteristics are of central concern in determining the artistic results of selecting these microphones for certain sound sources. These microphone performance characteristics are:

- Frequency response
- Directional sensitivity
- Transient response
- Distance sensitivity

Frequency Response

Frequency response is a measure of how the microphone responds to frequency levels. Amplitude differences at various frequency ranges are determined throughout the audio range, to calculate the sensitivity of the microphone to frequency.

The frequency response of a microphone is often comprised of certain frequency bands that the microphone will accentuate or attenuate. The matching of a sound source with similar frequency characteristics may or may not provide the recordist with the desired sound. The microphone may cause accentuation of certain characteristics of the sound source, and perhaps a microphone with somewhat opposite frequency characteristics as the sound source will be a more appropriate choice. Again, this decision is dependent upon the final, desired sound.

Microphone frequency response adds new formant regions to the sound source. The frequencies emphasized and/or attenuated by the microphone act on all sounds equally, regardless of pitch-level. Microphone frequency response directly shapes, contributes to, and captures the spectrum of the sound source.

Directional Sensitivity

Microphones do not respond equally to sounds arriving at different angles to its diaphragm. The *directional sensitivity* of a microphone is its sensitivity to sounds arriving at various angles to the diaphragm. The *polar pattern* of a microphone depicts the sensitivity of a microphone to sounds at various frequencies in front, in back, and to the sides, and the actual pattern is spherical around the microphone. Directional response measures the microphone's sensitivity to sounds arriving from angles, but calculates this sensitivity at only a few frequencies.

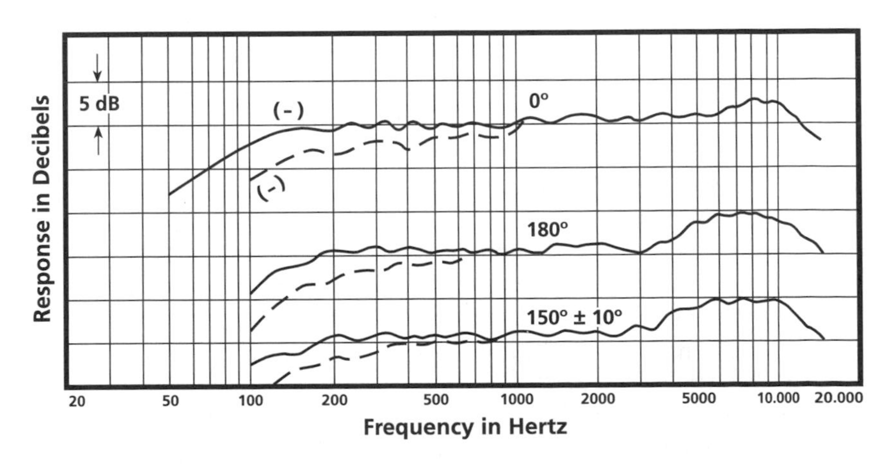

Figure 13-1 Microphone frequency response of a hypothetical microphone, *on-axis* (0°) and 180° and 150° from *off-axis*.

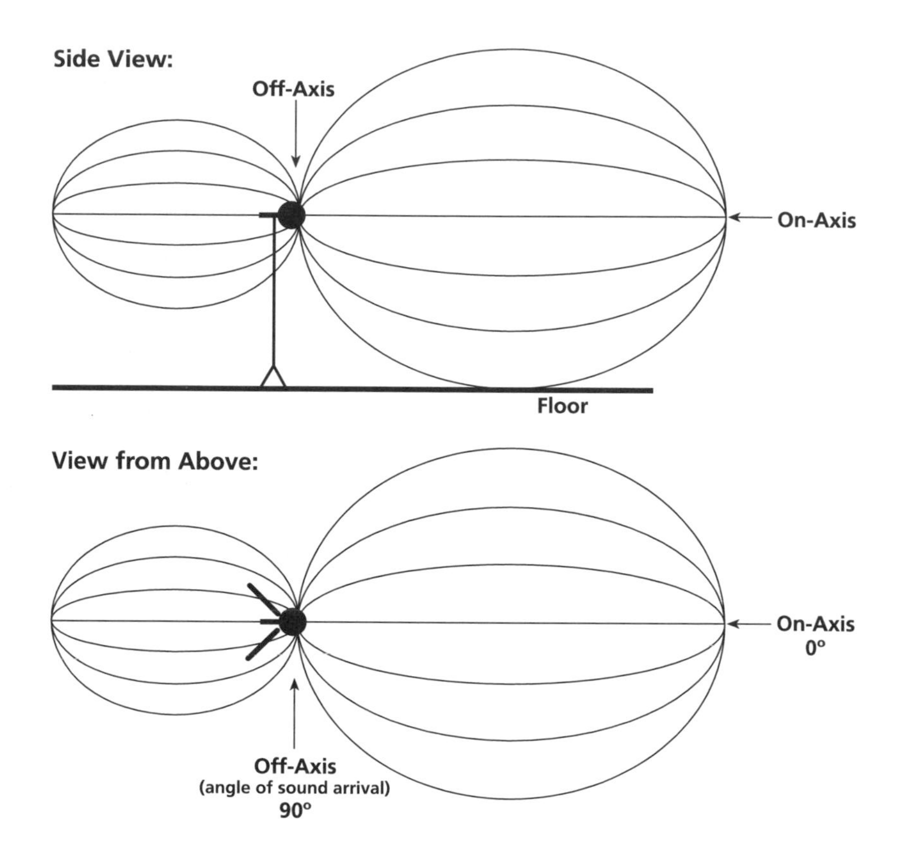

Figure 13-2 Polar pattern spheres and the microphone axis.

Sounds directly in front of the microphone diaphragm are considered to be *on-axis*. Sounds deviating from this 0° point on the polar curve are considered *off-axis* and are plotted in relation to the on-axis reference level. The frequency response of most microphones will vary markedly to sounds at different angles. Even microphones that show no pronounced frequency areas of accentuation or attenuation (flat frequency response) on-axis, will show an altered frequency response at the sides and the back of the polar pattern.

These variations in frequency response at different angles are commonly called *off-axis coloration*. This coloration alters sound source spectrum just as on-axis frequency response accentuates and attenuates specific frequency bands. These changes are more pronounced at or towards the attenuated angles of the microphone patterns. In the intermediate angles between directly on-axis and the dead areas of directional patterns, a slight change in the angle/direction of the microphone can make a substantial difference in the frequency response of the captured

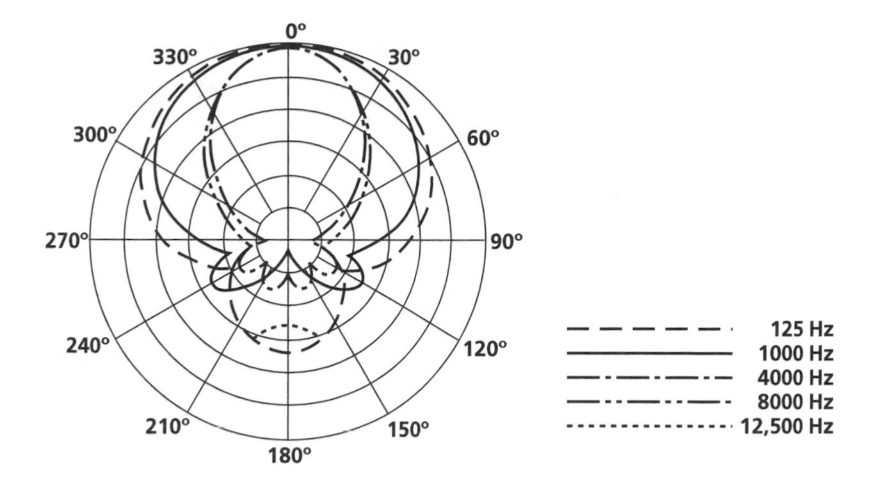

Figure 13-3 Polar pattern sensitivity at various frequencies.

sound source. Frequencies about 4 kHz are usually most dramatically altered by slight angle changes.

The amount of off-axis coloration is an important measure of the microphone's suitability to a variety of situations. This is especially pronounced in stereo microphone array recording techniques. Instruments at the edges of the array's pick-up pattern and the reverberant sound of the hall will arrive at the array mostly from angles that are off-axis. The sound qualities of those instruments and of the reverberant energy may be altered significantly by the off-axis coloration. Off-axis coloration has the potential to have a profound impact on the sound qualities of the recording.

Transient Response

Microphones do not immediately track the waveform of a sound source. A certain amount of time is required before the applied energy is transferred into movement of the microphone's diaphragm. Also, a certain amount of acoustic energy must be applied to initiate this movement, and some of this energy (which is the sound wave of the sound source) is dissipated in the action of getting the diaphragm moving.

Thus, microphones have different response times before they will begin to accurately track the waveform of the sound source. This *transient response* time distorts the initial, transient portions of the sound's timbre. Slow transient response is most noticeable when the microphone is applied to a sound source that has a fast initial attack time (in the dynamic envelope) and a large amount of spectral energy during the onset.

Transient response is a microphone performance characteristic that is not included in the manufacturer specification information that accompanies promotional literature and owner's manuals. Further, it is not calculated by a standard of measurement. It exists as an alteration of the original waveform's initial attack time in both the dynamic envelope and the spectral envelope. Furthermore, it often will affect different frequencies differently in the spectral envelope.

The developed hearing of the recordist will be used primarily to judge this microphone characteristic. The recordist must become aware of differences in timbre that are present in the early time field of the sound source, as compared with the live sound source. These differences are identified through the critical listening process and are vitally important to the recordist. Recognizing and understanding how the transient response is altering the source's sound quality will allow the recordist to be in control of capturing and shaping the sound as desired.

Two microphones with identical frequency response curves may have very different sound characteristics caused by different transient response times.

Distance Sensitivity

All microphones will respond differently to the same sound source, at the same distance and angle. The ability of each microphone to capture the detail of a source's timbre will be different. Microphones will have different sensitivities in relation to distance. The *distance sensitivity* of a microphone is often influenced by the polar response of the microphone and/or its transduction principle (condenser, moving coil, ribbon, etc).

Directional patterns often are able to capture timbral detail of a sound source at a greater distance than an omnidirectional microphone. This is primarily the result of the ratio of direct to indirect sound, and a masking of timbre detail, but it may also be attributed to transduction principle, depending on the particular circumstances. Similarly, condenser microphones will have a tendency towards greater distance sensitivity than dynamic microphones, due to their more sensitive transfer of energy. Many times a microphone with a small-sized diaphragm will have greater distance sensitivity than a microphone with a larger diaphragm, all other factors being equal.

The concept of the distance sensitivity (sometimes called "reach") of a microphone is an important one. The recordist must also judge this microphone characteristic through acute listening and experience. They must become aware of differences in timbre detail that are present between the miked sound source and the live sound source. Distance sensitivity is a microphone performance characteristic that is not included in the manufacturer specification information, although it might be measured scientifically if an appropriate scale were devised. Distance sensitivity is a characteristic that must be learned from experience and must be anticipated for the individual environment and recording conditions.

Microphone Placement

Many variables must be considered during the process of selecting and placing microphones. The primary variables for microphone selection were presented above. The variables of *microphone placement* will directly influence the selection of a microphone, even after an initial selection has been made. The placement of the microphone in relation to the sound source and the performance environment will greatly influence the sound quality of the recording. At times this influence may be as great as the selection of the microphone itself.

The recordist must consider the following when deciding on placing the microphone in relation to the sound source:

- Distance relationships between the microphone, the sound source, and the reflective surfaces of the recording environment (performance space)
- Distance of the microphone from the sound source to be recorded and other sound sources in the performance space
- Height and lateral position of the microphone in relation to the sound source
- Angle of the microphone's axis to the sound source
- Performance characteristics of the microphone selected, as altered by the above four considerations

Microphone placement interacts with microphone performance characteristics to create the recorded sound quality of the source. The distance of the microphone will be largely determined by the distance sensitivity of the microphone. The angle of the microphone will be largely determined by the frequency response of the microphone in relation to its polar pattern, the characteristics of how the sound source projects its sound, and the characteristics of the environment. The height of the microphone is also a result of the directional characteristics of the sound source (as instruments radiate different spectral information in different directions), the environment, and the microphone's frequency response and distance sensitivity.

Controlling the Sound of the Performance Space

The sound characteristics of the environment in which the sound source is performing may or may not be captured in recording the sound source. The recordist may or may not wish to include them in the sound quality of the sound source. In either instance, the recordist must be in control of the indirect sound of the environment arriving at the microphone from reflective surfaces.

The recordist may seek to control the balance of direct and indirect sound. Through the selection and placement of a microphone with a suitable polar pattern and distance sensitivity, the desired characteristics of the environment may be captured, in the desired amount, along with the sound of the sound source. This will allow the recordist to record the sound of the sound source within its performance environment. The distance cues of the initial reflections in the early time field, and the amount of timbre detail, are evaluated by the recordist when deciding on the ratio of direct to indirect sound, and distance placement is adjusted accordingly.

The recordist's objective may be to capture the sound source without the cues of the environment. The sound source may be physically isolated (with gobos or in isolation booths) from other, unwanted sounds from the environment and other sources, or the leakage of unwanted sounds to the recording microphone might be minimized with microphone pattern selection and microphone placement. This will allow the recordist the flexibility of being in complete control of the sound source, with environmental characteristics later applied to the sound through signal processing.

The sound of the environment will be controlled by the relationships of the microphone to the sound source and any other sound sources that may be occurring simultaneously, including the sound of the environment itself.

Reflective Surfaces

The *reflective surfaces* of the environment (or of any object in the environment) can cause the sound at the microphone to be unusable. Interference problems may be created when the sound from a reflective surface and the direct sound reach the microphone at comparable amplitudes. The slight time delay between the two sounds will cause certain frequencies to be out-of-phase (with cancellation of those frequencies) and certain frequencies to be in-phase (with reinforcement of those frequencies).

The frequencies that will be accentuated and attenuated can be determined by the difference of the distance between the reflective surface and the microphone (D_1), and the distance between the sound source and the microphone (D_2), in relationship to air velocity. The amount of reinforcement and cancellation of certain frequencies that will occur when the two signals are combined will be determined by their amplitudes. Constructive and destructive interference are most pronounced when the two signals are of equal amplitudes. As the difference in amplitude values between the two signals becomes larger, the effect becomes less noticeable.

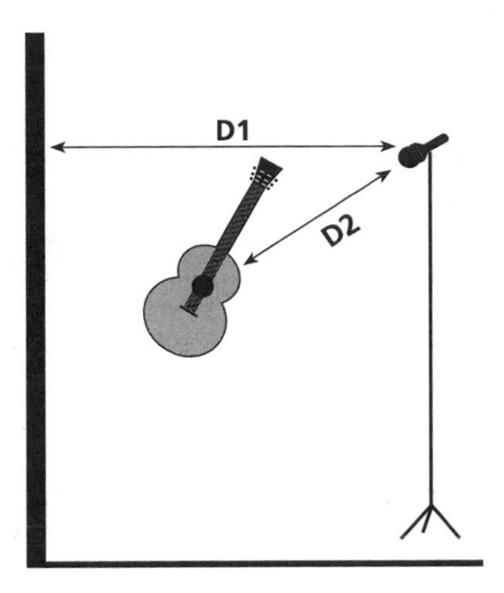

Figure 13-4 Distances between microphone, sound source, and reflective surfaces.

The constructive and destructive interference of the combined signals result in a frequency response with emphasized and attenuated frequencies (peaks and dips). These peaks and dips of the frequency response curve of the sound's spectrum have been compared, in analogy, to the tines of a comb. Thus, the term *comb filter* has been applied when a signal is combined with itself, with a slight time delay between the two signals.

Microphone Positioning

The distance of the microphone to the sound source, and other sound sources in the environment, along with the height, lateral position, and angle of the microphone, alter the sound quality of the sound source. As such, they are variables that may be used as creative elements in shaping sound quality (and the musical material), in the same way as microphone selection. These positioning variables will most significantly influence the following aspects of sound:

- Environmental characteristics (and secondary distance) cue—ratio of direct to reflected sound
- Environmental characteristics cue—early time field information created by initial reflections
- Distance cue (definition or amount of timbral detail)
- Capturing of the blend and shaping the overall quality of the source's timbre

The distance from the microphone to the sound source is the primary factor that controls the ratio of direct to reflected (indirect) sound. As the microphone is placed closer to the sound source, the proportion of direct sound increases in relation to reflected sound.

Reflective surfaces that are located near the sound source will cause inconsistencies with this general rule. If at all possible, sound sources should be placed at least three times the distance of any reflective surface, as the distance from the microphone to the sound source. The height and angle of the microphone in relation to the sound source, coupled with polar patterns, may allow the microphone to keep from picking up the reflected sound off surfaces close to the sound source (such as the floor or a gobo).

Distance cues are established by microphone placement when an audible amount of sound from the recording environment is present in the source's sound quality. While this is not usually the case with close miked sound sources, many sound sources in multitrack projects are recorded from a distance that will capture certain information from the recording environment. Of course, all distant micing techniques will capture a significant amount of the sound of the environment.

The reflection information in the early time field is especially prominent in recordings made with a moderate distance between the microphone and the sound source. The floor or objects immediately around the sound source will often create reflections in recordings of this type. While only a few reflections may be present in the final sound, and the reflections may be of significantly lower amplitude than the direct sound, they will impart important environment information. These reflections often provide environmental cues that conflict with the perceived stage-to-listener position distance cues.

Microphone placement location and performance characteristics will play vital roles in establishing distance cues, through defining the amount of timbre detail present in the recorded sound source.

Generally, the recorded sound source will have greater timbral detail the closer the microphone to the sound source and/or the more sensitive the microphone in terms of distance sensitivity and transient response. This provides a distance cue that may place the listener of the recorded sound near the reproduced image of the recording, sometimes unnaturally near the source. This level of timbral detail directly establishes the level of intimacy of the recording. The depth of sound stage will also be greatly determined through these cues.

Distance cues are often contradictory when a high degree of timbre detail and a large amount of reverberant energy in relation to direct sound are present simultaneously. This unrealistic sound may be desired for artistic purposes.

The sound quality of the sound source may potentially be altered by close microphone placements. As above, these alterations can be used to creative advantage, or they can interfere with obtaining the desired sound quality. The sound quality may exhibit the following alterations when the source is close miked:

- An increase of definition over known, naturally occurring degrees of timbre detail
- Changed spectral content of the sound source
- An unnatural blend of the source's timbre, caused when the source has not had enough physical space to develop into its characteristic sound quality

A microphone placed within two feet of a sound source may alter the source's spectral content. The frequency response of certain microphones is altered by the *proximity effect*. Response in the low frequency range rises relative to response in higher frequencies, for cardioid and bi-directional microphones.

Blend

The sound source itself, and the way it radiates sound, must also be considered in the context of the sound's environment.

Sound sources require physical space for their sound quality to develop and coalesce. The sound quality of instruments and voices is a combination of all of the sounds the instrument produces. When a microphone is placed in a physical location near the sound source, it will often be within this critical distance necessary for the sound to develop into a single sound wave. The sound source will not have had the opportunity to *blend* into its unique, overall sound quality. The microphone will capture only a portion of the sound. This resulting sound quality can be very different from how the sound source exists in acoustic environments.

It is not unusual for recordists to capture a sound before its characteristic acoustic sound has fully developed. This can be accomplished in such a way as to contribute positively to the music, or it may work against it. It is important that the recordist be aware of this space and in control of recording the desired blend of the source's sound.

Many instruments and voices are commonly recorded with close microphone techniques, capturing only a portion of the source's characteristic timbre. Many times the resulting sound quality is desired. At other times additional microphones can be positioned to capture the missing components of the sound, and are then mixed to blend the sound quality of the instrument or voice. Care must be taken to obtain the desired sound quality with close micing. Performers and their instruments can create many undesirable noises, causing difficulties in obtaining a suitable sound.

In close micing sound sources, all or most natural environmental cues are absent from the sound. Any environmental characteristics will be added to these sources through signal processing. Electronic or amplified instruments that are plugged directly into the mixing console (through a direct box) are similar in their lack of environmental characteristics. If the listener is to be provided with a realistic listening experience (which may or may not be desirable), environmental characteristics need to be added to these sounds.

Recordists often capture the sound of a sound source from a number of close microphone locations and blend the sound themselves. This can be performed to great creative advantage. They may control the blend of the various portions of the source's sound, instead of relying on the sound source's interaction with the performance environment to blend the sound as desired. Pianos are commonly recorded in this way, as are acoustic guitars. This technique can be applied to a wide variety of sound sources.

Stereo Microphone Techniques

Stereo microphone arrays are comprised of two or more microphones, in a systematic arrangement. They are designed to record sound in such a way that upon playback (through two channels or surround) a certain sense of the spatial relationships of the sound sources present during the recorded performance is reproduced. These techniques can capture the sound qualities of the live performance and of the hall, the performed balance of the instruments of the ensemble, and the spatial relationships of the ensemble (stereo location and distance cues), with minimal alteration.

Many stereo microphone techniques have been developed. Among the most commonly used are:

- X-Y coincident techniques
- Middle-Side technique (M-S)
- Blumlein X-Y (crossed figure-eights)
- Near-coincident techniques (NOS and ORTF)
- Spaced omnidirectional microphones
- Spaced bi-directional microphones
- Binaural system (artificial head)
- Sound field and other specialized microphones and systems

Stereo microphone techniques are often used in recording large ensembles, in large acoustic spaces, and from a rather distant placement. The techniques are very powerful in their accuracy and flexibility, and may be applied to either a single sound source or to any sized ensemble. They may be used from a rather distant placement (perhaps 15 or more meters, depending on the pertinent variables), to within about a meter of the sound source. Close placements of stereo arrays are commonly applied to drum sets, for example, sometimes supplemented with accent microphones, sometimes not.

Stereo microphone techniques can be thought of as a pre-processing of the recorded sound. Pre-processing a signal is the altering of components of the sound source's sound quality before it reaches the routing and mixing stages of the recording chain. The stereo microphone technique will add and capture spatial information to the recording.

The stereo microphone array can significantly alter the sound source in a number of ways. All stereo microphone techniques have their own unique characteristics, and their own inherent strengths and weaknesses. The inherent sound qualities of the stereo microphone arrays can be used to great advantage if the recordist understands and is in control of their sound qualities. The sound qualities of each of the various arrays can be learned by evaluating their unique sound qualities, as they function on the following list of artistic elements.

- Perceived listener to sound stage distance
- Amount and sound quality of the environmental characteristics of the performance space
- Perceived depth of the ensemble, sound source, or sound stage
- Perceived width of the sound source or the sound stage
- Definition and stability of the lateral (stereo) imaging
- Musical balance of the sound sources in the ensemble
- Sound qualities of the entire ensemble or of specific sources within the ensemble

The microphone placement variables are also factors in stereo microphone techniques, as they determine where the array is to be placed. Placement of the array will considerably impact the quality of sound. The microphones used in the stereo microphone techniques impart their own unique sound characteristics, as described. Matching specific microphones (type of transducer or model/manufacturer) to the stereo microphone array is important in ensuring the effectiveness of the array's capturing the desired sound qualities of the performance.

The recordist will often envision an *ideal seat* for the performance when determining the placement of a stereo microphone array. This type of placement of the array will seek the sound qualities that are most desirable for the particular ensemble, performing a specific piece of music in the performance space. The recordist will seek to balance the hall sound with that of the ensemble, capture an appropriate amount of timbre definition from the ensemble, retain all performed dynamic relationships, and establish desirable and stable spatial relationships in the sound stage. The balance of the total energy of the direct sound and that of the reverberant sound will be carefully considered in determining the placement of the microphone array. The point where the two are equal has been identified as the *critical distance*. The recordist will focus on this ratio of the reflected to direct sound to identify a desired balance.

Accent microphones are often used to supplement stereo microphone techniques. These are microphones that are dedicated to capturing a single sound source, or a small group of sound sources, within the total ensemble being recorded by the array. The accent microphones are placed much closer to the sound sources than the array and may cause the recordist to consider some of the close micing variables discussed above.

At times, secondary stereo arrays can suitably function as a set of accent microphones. This is especially usable with large ensembles (such as an orchestra plus chorus). The recordist must compensate for the delay times between the arrivals of the accent array(s) and the overall array signals when they are combined. These time differences may be minimized through applying delay units to the secondary array(s), thus avoiding comb-filtering distortions.

When using accent microphones in conjunction with stereo microphone arrays, the accent microphones are most often used to complement the array. They assist the overall array by bringing more dynamic presence and timbre definition to certain sound sources in the ensemble, and they allow the recordist some control over the musical balance of the ensemble. Accent microphones also create noticeable time differences between the arrival of the sound source(s) at the stereo array, and the arrival of the sound source(s) at the accent microphone.

Adding accent microphones will diminish the realistic sound qualities of the stereo microphone array. The sound relationships of the performance will be altered, by adding the dynamics, sound quality, and distance cues from the accent microphone(s). The stereo microphone array's ability to accurately capture the performance will be diminished in using accent microphones to improve the sound present in the hall. Accent microphones will alter time cues and dynamic relationships noticeably, making the sound source(s) captured by the accent microphone significantly more prominent in the musical texture.

Equipment Selection: Application of Inherent Sound Quality

Recording/reproduction equipment all impart a unique sonic imprint on the sound. As was discussed with microphones above, individual recorders, mixing consoles, signal processors and any other devices in the recording chain all have unique sound qualities, which are the result of their performance characteristics. They all modify the original signal; some do this almost imperceptibly, and others quite profoundly. Any individual device will be evaluated for how it transforms the frequency, amplitude, and time components of the original signal. The modifications of the original signal caused by the basic performance characteristics of a device create the device's *inherent sound quality*. Further, different technologies have distinct sound characteristics, or the potential for sound characteristics.

The recordist will need to determine which pieces of recording equipment to use to realize the desired sound qualities of the individual sound sources and the project's overall sound. The recordist may approach this problem directly by evaluating and understanding the inherent sound characteristics of the available individual devices, and the inherent sound characteristics of the technologies of those devices, against the unique needs of the individual project. A compatible match will be sought for the device, technology, and sound source to arrive at a desired sound quality.

Technology Selection: Analog versus Digital

Digital recording, processing, and editing equipment is not necessarily better than analog equipment, nor is the opposite true. Both have great potential for artistic expression, and both can generate recordings of impeccable quality. No technology is inherently better-suited, than another, for generating, capturing, shaping, mixing, processing, combining, or recording sound. Some devices may have functional features that are more attractive to an individual recordist, and some people develop personal sound preferences. These are, however, matters of taste. Technologies are simply different in how they sound—how they retain the original signal characteristics and how they alter the sound source.

Analog technology has certain inherent sound characteristics. Digital technology has certain inherent sound characteristics. The characteristics of one technology may or may not be appropriate for a particular project. Inherent sound qualities are inherent sound deficiencies if they work against the sound quality the recordist is trying to obtain. Inherent sound qualities are desirable if they produce the sound quality the recordist is looking for.

It is difficult to make generalizations as to the characteristics of analog versus digital technology. The sound qualities of both technologies vary widely depending on the particular unit and the integrity of the audio signal within the particular devices. An eight-bit digital system is significantly less accurate and flexible than a 32-bit system. A consumer-model analog system is significantly noisier and less accurate than a professional unit.

Differences often exist between the two technologies in:

- Accurately tracking the shape of the waveform (especially the initial transients of the sound wave, in both technologies)
- Processing all frequencies equally well (especially frequency response linearity in analog or quantization issues in digital)

- Storing the waveform without distortion from the medium (especially tape noise floor or A/D and D/A conversion accuracy)
- Altering the waveform in precise increments and precisely repeating these functions (a measure of signal processors)
- Performing repeated playing, successive generations of copying, and long-term storage with minimal signal degradation (a measure of recording formats)

Many other, more subtle, differences exist, especially between specific devices of each technology.

Selecting Devices and Models

No recording device is inherently better than another similar device. While certain devices might be more flexible than others, and certain devices are certainly of higher technical quality than others, the true artistic concern for a piece of equipment is its suitability to the particular needs of the project, at a given point in time.

The advantages of any particular device in one application, may be a disadvantage in another. The sound quality of one device may be appropriate to one musical context, and not to another. The measure of the device will be in how its inherent sound qualities can be used to obtain the desired sound qualities of the recording. Pieces of recording equipment should be evaluated, by the recordist, for their sound qualities, and their potential usefulness in communicating the artistic message in the piece of music. This evaluation is performed through a critical listening process similar to that used to evaluate microphone performance characteristics.

Pieces of recording equipment are tools. The tools may be applied to any task, with consistent results. The recordist needs to decide if the particular tool (piece of equipment) is the appropriate piece of equipment for the sound quality that is required of the particular project.

Musicians often carry with them a number of musical instruments. They will use a different model of the same instrument (perhaps made by a different manufacturer) to obtain a different sound quality of their performance, depending on what is required by the musical material. The recordist should recognize this is similar to their situation.

In selecting recording equipment, the recordist is, in essence, selecting a musical instrument. The sound quality of sound sources, or of the entire recording may be markedly transformed by the piece of equipment, while the sound is under the control of the recordist. This is the way a traditional musical instrument is applied by a traditional performer.

Recordists will develop sound quality preferences and working preferences for particular pieces of equipment, and perhaps for a particular technology. Developing such preferences may or may not be artistically healthy. The recordist may become inclined to consider a certain technology to be better than another simply because it is the one they are most familiar with, not because it is the one that is most appropriate for the project. Personal preferences (or personal experiences) might become confused with the actual quality or usefulness of a device or a technology.

Personal preferences are not a measure of quality, in and of themselves. Audio devices have inherent sound qualities that are determined by technology and the device's unique performance characteristics. The recordist will be using these devices to shape the sound of the music recording. Learning the inherent sound qualities of as many devices as possible will increase the creative tools of the recordist, and provide them with more options in obtaining the sound they want. These devices are the musical instruments that are used in the art of recording.

Monitoring: The Sound Quality of Playback

During the preliminary stages, the recordist will decide on how they will monitor the recording process. Monitoring will be the means through which the recording itself will be evaluated. This most often takes place in a recording control room—whether a commercial facility, a home studio, a closet used for a remote recording, or something else. All of the sounds and relationships of the recording project are presented to the listener through the monitor system.

The monitor system is more than a pair of loudspeakers. In terms of hardware, the monitor system also includes power amplifiers, crossover networks, connector cables, and perhaps the monitor mixer of the console and additional processors. All hardware must function efficiently and effectively in relation to the other components, in a complementary manner in terms of sound qualities, to reproduce a waveform that has not been unfavorably altered.

The monitor system also encompasses the listening room itself. The placement of the loudspeakers in the room, and the interactions of the room and the loudspeakers become part of the sound quality of the monitor system. Room acoustics and sound reproduction equipment must also be matched to create an accurate playback system.

The monitor system has the potential to transform all sound qualities and relationships in the recording. The monitor system should not alter the original signal, or at the very least the recordist must know how the sound is being altered. The recordist must evaluate the project from many different perspectives, focusing on many elements of sound. Each approach to listening will require the monitoring system to accurately reproduce sound quality on all hierarchical levels, the spatial relationships of sound sources, and the frequency, amplitude, and time information of the recording.

If the recordist is to control the recording process, they need to be able to evaluate the recording itself, not the recording modified by the monitor system. The recording will only have the same sound relationships in another (neutral) listening environment if the sound quality is not originally altered through the recordist's monitors.

The most desirable monitoring system is transparent. The loudspeakers reproduce sound in the control room without altering the qualities of sound. In practice, this is nearly impossible, but the sound alterations caused by monitor systems can be minimized. Many complexities are present in the selection of an approach to monitoring. The immediate considerations for the monitoring process itself are:

- Loudspeaker and control room interaction
- Effective listening zone
- Sound field: near field versus room monitoring
- Monitoring levels
- Stereo and mono compatibility

Loudspeaker and Control Room Interaction

The control room itself will alter the sound emanating from the loudspeakers. The acoustics of the control room can cause radical changes in the frequency response and time information of the sound. Ideally, any listening room would have a constant acoustical absorption over the operating range of the loudspeakers, and would appropriately diffuse the sound from the loudspeakers to create a desirable blend of direct and reflected sound at the mixing/listening position.

Nonparallel walls; nonparallel floor and ceiling; acoustical treatments to absorb, reflect and diffuse sound where needed; careful selection, placement and installation of loudspeakers; and a sufficient volume (dimensions of the room) will minimize the influence of the control room on the sound coming from the loudspeakers.

The room should absorb and reflect all frequencies equally well, should produce very short decay times that are at substantially lower levels than the direct sound from the loudspeakers. The room should not produce resonance frequencies and should not produce reflections that arrive at the listening position at a similar amplitude as, or within a small time-window (2–5 ms) of, the direct sound.

As needed, rooms may be tuned for uniform amplitudes of all frequencies by room equalizers, and tuned for the control of reflections (time) with diffusers, traps, and sound absorption materials. The ideal control room has been designed to include very specific acoustical treatments, and in such a way that room equalization is not needed.

Designers of recording studios have widely divergent opinions on the most desirable acoustical properties of control rooms. Conflicting information and opinions are common. The objective of all designers is very similar; however, to produce a listening environment that is most suitable for the listening of reproduced sound.

Loudspeaker placement is part of the design of control rooms. One common design approach dictates that the loudspeaker should be a part of the wall, mounted within the wall itself so the front of the loudspeaker is flush with the face of the wall. This negates the usual boost of low frequencies that results when a loudspeaker is placed near walls, ceilings, and floors (and especially in corners). Freestanding loudspeaker placement away from side walls by 4 ft. and from the front wall by about 3 ft. will minimize the boost of low frequencies (that occurs when the omnidirectional low frequencies reflect off the wall surfaces and combine with the direct sound from the speaker).

The loudspeakers should be aligned on the same vertical plane, usually at ear-height for a seated person or slightly higher. It is important

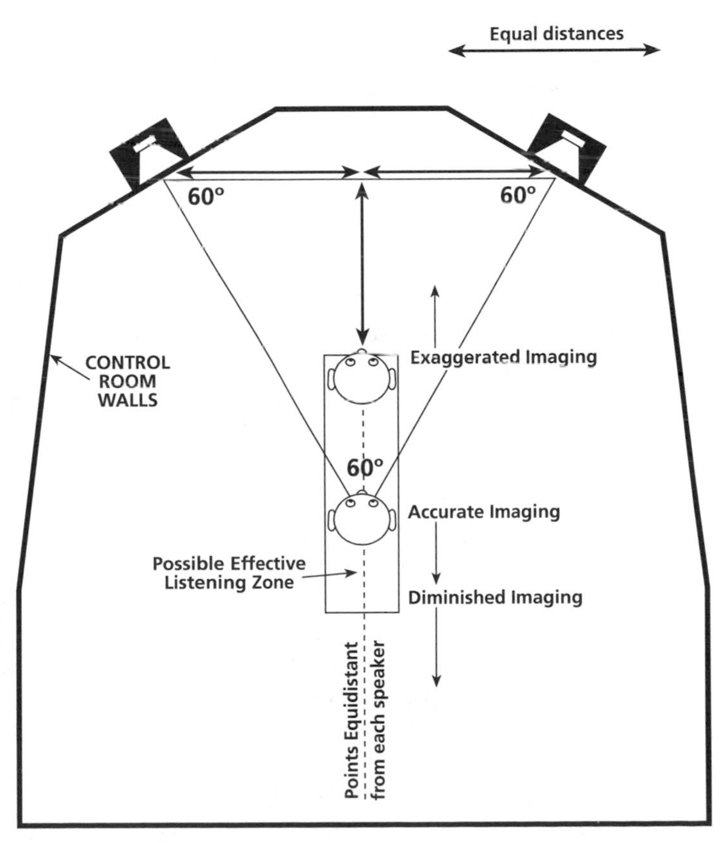

Figure 13-5 Loudspeakers as part of the control room and the effective listening zone.

that the meter bridge of the console not be in the path between the loudspeakers and the mix position, and that strong reflections off the console are minimized. The loudspeakers should be aligned symmetrically with the sidewalls.

The *effective listening zone* is an area in the control room where the reproduced sound can be accurately perceived. In a control room, the mix position and/or the producer's seat are located in the effective listening zone. In most control rooms the size of the effective listening zone is usually quite small. It is an area that is equidistant from the two loudspeakers in stereo and all loudspeakers in surround. The area is located at roughly the same distance from each speaker as the speakers are from each other. Angling loudspeakers correctly provides optimal imaging in the effective listening zone, given complementary room acoustics. The reader should review stereo and surround loudspeaker positioning described earlier.

For accurate spatial perception, it is necessary for the effective listening zone be carefully evaluated. The listener must be seated in the proper location, and the volume level must be the same at each speaker when identical signals are applied to each channel. Moving the listening location closer to the speaker array exaggerates imaging and moving away from the loudspeakers diminishes imaging relationships. Rooms are built and/or speakers are placed so few strong reflections arrive at this area. The control room should be virtually transparent (add or subtract no characteristics to the sound) in the listening zone.

Near Field and Room Monitoring

The control room monitoring system must provide accurate reproduction of the sound qualities of the recording being made. The loudspeakers of the system are of high quality and are carefully engineered. They are designed to provide unaltered, detailed sound, while working within the acoustic characteristics of the control room. They are inherently different from loudspeakers intended for home use (even high-end speakers intended for the audiophile market).

Most people who purchase the final recording will have playback (monitoring) systems of somewhat (to significantly) less quality than the recording studio. The sound of the recording in the audience's listening environment (the individual's home, automobile, etc.) will be different from that of the recording control room. Most consumer monitor systems seek to blend sound in pleasing ways—pro audio monitors seek to provide the recordist with extreme clarity of sound, minimizing any added fusion of sound elements. It is common for recordists to use several sets of monitor speakers (sometimes in and often outside the studio) to obtain an idea of how their project will sound over home-quality playback systems.

Bookshelf-type speakers are often used in the studio to provide a reference to this type of playback environment. They can represent the sound of typical, moderately priced speaker systems. They often tend to have a narrow frequency response, de-emphasizing high and low frequencies areas. These speakers are placed on the meter bridge of the mixing console, about 3 to 5 ft. apart. This is called *near field* monitoring. The sound is heard near the speakers, before the acoustics of the control room can act upon the sound. The reflected sound in the control room has little or no impact on what is heard in near field monitoring because of the listener's close position to the loudspeakers.

Since the control room has very little influence on the sound quality of near field monitoring, this approach is often the exclusive monitoring system of control rooms that have poor room acoustics. In this case, high-quality monitors specifically designed for professional near field monitoring replace the bookshelf speakers described above. These speakers are preferred over the large studio monitors by some recordists, especially during long sessions when the more intense energy of (some) large speakers can fatigue the ear.

The size of the effective listening zone for near field monitoring is very small. The listener must be precisely centered between the two loudspeakers, and the listener cannot be beyond the distance from the two loudspeakers as they are from one another, or the effects of the room will come into play. The listener should be located approximately the same distance from each loudspeaker, as the two loudspeakers are from each other.

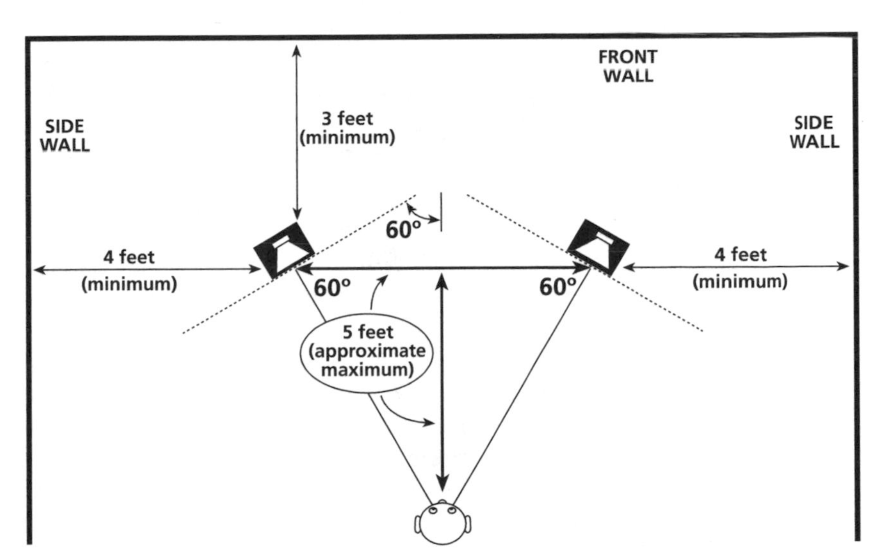

Figure 13-6 Loudspeaker relationships to the listening environment and the listener in near field monitoring.

It is not unusual for recordists to periodically switch back and forth between room and near field monitors. This allows the recordist to evaluate the sound qualities and relationships of the recording from different listening locations and with loudspeakers that have different sound qualities. The recordist will be attempting to create a consistency between the sounds of the two monitor systems in as much as this is possible. The speakers are usually switchable at the mixing position, with each pair having a dedicated amplifier(s), set to match monitoring level while switching speakers.

Control room monitors also tend to provide more detail of sound quality than near field monitors. This added detail allows for greater accuracy in critical listening applications.

Monitor Levels

The sound pressure level (SPL) at which the recordist listens will influence the sound of the project. Humans do not hear frequency equally well at all amplitudes, as discussed in Chapter 1. A recording created at a *monitor level* of 100 dB (SPL) will sound considerably different when it is played back at the common, home listening, monitoring level of 75 dB or lower. The bass line that was present during the mixdown session will not be as prominent in playback at the lower level. Similarly, the tracks that were recorded at 105 dB during a tracking session (because someone wanted to "feel" the sound) will have quite different spectral content when they are heard the next day at 85 dB, during the mixdown session.

The recordist will need to develop consistency of listening levels. Ideally, monitoring levels should be reasonably consistent throughout a project—tracking, mixing, and mastering.

The most desirable range for monitoring is 85 to 90 dB. Recordings made while monitored in this range will exhibit minimal changes in frequency responses (≤5 dB at the extremes of the hearing range) when played back as low as 60 dB SPL. Monitoring at this level will also do much to minimize listening fatigue during prolonged listening periods (recording and mixing sessions).

The possibility of hearing damage is a very real job hazard for recordists. If the recordist primarily listens in the 85-90 dB range, they will have accurate hearing much longer than if they consistently monitor at a level 10-15 dB higher—they may very likely also have a longer career.

Stereo and Mono Compatibility

The recordist will need to be sensitive to how a recording will transfer to other environments and other playback formats. Aside from the varying quality of home playback systems described above, the recordist may be concerned about the playback of a recording over radio or the Internet.

The targeted audience for some recordings may require that the recordist be concerned with *mono compatibility* of a stereo recording. The recordist will monitor for mono by using a single-driver mini-monitor speaker in the near field. This type of speaker is intended to simulate the listening conditions of a radio or an automobile playback system. The recordist may determine the mono compatibility of the stereo program (being mixed on the studio monitors) by routing the monitor signal to the single loudspeaker, or by summing the two channels through the console's "mono" switch and using both of the studio's usual monitor speakers.

When the two channels are summed, conflicting phase information causes comb filtering and other signal-integrity problems. In addition, the balance of the sound sources can be radically altered. Center-image sound sources may be raised as much as 3 dB when the signals of the two channels are combined.

For these reasons, it is not uncommon for separate mixes of the same work to be made for different purposes. Mixes made for radio broadcast, and some Internet uses, will often have altered frequency information to somewhat compensate for the inadequacies of the media, and will seek to maintain many of the dynamic level relationships achieved in the stereo mix.

Separate surround mixes and stereo mixes are also made for music intended for both formats. A surround recording can suffer greatly when reduced to stereo—especially by the natural tendency of summing the surround channels with adjacent front channels, and/or by dividing the center channel signal and combining it with the left and the right channels. The *stereo compatibility* of any surround recording will be determined greatly by the content of the surround channels and center channel, and how they are combined with the left and right channel material. Many compatibility issues will need to be clearly defined—and resolved by recordists—as surround recordings begin to establish a larger share of music recording sales.

Headphones

Headphones are required for creating and listening to binaural recordings and may be the only feasible monitoring option for a remote recording. Otherwise, monitoring is most accurately accomplished over loudspeakers, except in the rarest of circumstances.

Headphones will distort spatial information, especially stereo imaging. The listener will perceive sound within their head, instead of as occurring in front of their location. This distorts depth of imaging. The interaural information of the recording is not accurately perceived during headphone monitoring, causing potentially pronounced image size and stereo location distortions.

Further, the frequency response of headphones will not match that of loudspeakers and is inconsistent. Frequency response of headphones will vary with the pressure of the headphones against the head and the resulting nearness of the transducer to the ear.

Sound evaluations can only be accurately performed over quality loudspeakers, in a transparent (or complementary) listening environment.

Summary

Before sounds are recorded, the recordist makes many decisions that shape the recording project. Many of these choices will limit how the music might be shaped later in the recording process and must be approached with knowledge and sensitivity. Selecting and defining sound source timbres, capturing those sought qualities with effective microphone selection and placement, crafting new timbres with synthesis, selecting recording devices throughout the signal chain, and selecting monitoring systems and conditions are all decisions that can profoundly impact any project.

When the entire project is considered at this planning stage, decisions in many areas can be made that will not limit the recordist as the project progresses. A clear sense of artistic direction will be established that will allow the music's potential to be realized and a high-quality recording to be made.

These preliminary stages prepare for successful projects of quality that make efficient and effective use of studio time. They also allow recordists to be in control of their craft and shape the artistic qualities of music.

14 Capturing, Shaping, and Creating the Performance

With objectives clearly defined and a vision of the sound and content of the project, the music recording can now be most effectively accomplished.

The recordist will capture the performance through recording or tracking the musical parts and instrumentation, and will shape sound qualities and relationships of the performance through equipment selection and signal processing. The mixing process will continue the shaping of the performance and will result in the creation of the performance (that is the recording).

The actual act of recording sound begins here. The preliminary stages that were discussed in the previous chapter define the dimensions of the creative project and of the final recording. In addition to these, certain other preparations must take place before a successful tracking or recording session can begin.

Whenever possible, the musicians should be prepared for their performance at recording and tracking sessions. Unfortunately, it is possible some musicians will get their music or performance instructions when they arrive for a recording session. This limits their potential to contribute to the musicality of the project and can minimize the quality of their performance.

Ideally, music should be given to performers well in advance of a session. Solo performers should have the opportunity to learn their parts before they join the other musicians, and all musicians expected to perform simultaneously as a group should be well rehearsed together (as an ensemble). As much as possible, musicians should be given the opportunity to arrive at the session knowing what will be expected of them, know their parts, and be ready to perform. Often the best case is that the music to be recorded will have been performed several times in concert—in front of an audience. In the reality of music recording, this ideal is not always feasible.

Some final instrumentation decisions and many of the supportive aspects of the music are, however, often decided upon during the preliminary stages of the recording process (or during the production process itself). The recordist will learn to anticipate that some musical decisions will be made during tracking. They will become aware that dimensions of creative projects can tend to shift (sometimes markedly) as works evolve—especially between types of clients and types of music. The recordist must keep as many options open as possible, to keep creative artists from being limited in exploring their musical ideas. They must not find themselves in the position of not being able to execute a brilliant musical/production idea (without hours of undoing or redoing) because of a recording decision made a few minutes earlier. Allowing for flexibility may be as simple as leaving open the option of adding more instruments or musical ideas to the piece, by leaving open tracks on the multitrack tape, or in console layout, cue mix changes and signal routing, or in mixdown planning.

The recording studio is the musical instrument of the recordist. The recording process is the musical performance of the recordist. In order to use recording for artistic expression, the recordist must be in complete control of the devices in the studio, and must understand their potentials in capturing and altering the artistic elements of sound.

Recording and Tracking Sessions: Shifting of Focus and Perspectives

Recordings are made with performers recorded individually or in groups, or with all parts of the music performed at once. "Recording sessions" as used here are recordings of all the parts being played at once; the entire musical texture is recorded as a single sound.

Tracking is the recording of the individual instruments or voices (sound sources) or small groups of instruments or voices, onto a multitrack tape. This is done in such a way that the sounds can be mixed, processed, edited, or otherwise altered at some future time, and without altering the other sound sources on the tape. For this isolated control of the sound source to be possible, it is imperative that the sound sources are recorded with minimal information from other sound sources and at the highest, safe loudness level the system will allow.

Recording and tracking sessions require the recordist to continually shift focus and perspective, while listening to live musicians. Further, they will move between analytical and critical listening processes. The musical qualities of the performance will be constantly evaluated at the same time as the perceived qualities of the captured sound. The recordist is responsible for making certain both aspects are of the highest quality and exist in a suitable state on any storage medium.

Focus will have the listener's attention moving freely but deliberately between all of the artistic elements of sound and all of the perceived parameters of sound. The recordist will need to shift focus between artistic elements (perhaps shifting attention between dynamic levels, pitch information, or spatial cues), then immediately shift to a perspective that evaluates program dynamic contour (of the overall sound). The recordist is required to continually scan the sound materials to determine the appropriateness of the sound (and its artistic elements) to the creative objectives of the project, and to determine the technical quality of the perceived characteristics of the sound. The listening skills developed in Part 2 will be used in these activities.

Within the recording and tracking sessions, the recordist is concerned about the aesthetics of the sound quality that is going to "tape," and they are concerned about the technical quality of the signal. Depending on their function in the particular project, they may not be in the position to make decisions related to *performance quality*. They should none-the-less be aware of the performance and be ready to provide their evaluations when asked or to anticipate the next activity (repeat a take, or move on to the next section, as examples). It is imperative that the recordist always be aware of the *technical quality* of the recording. It is their responsibility for high-quality sound to be recorded.

Analytical Listening Concerns

There are many performance quality aspects of music that will engage the recordist in analytical listening at this stage. Most of these aspects are related to:

- Intonation
- Control of dynamics
- Accuracy of rhythm
- Tempo
- Expression and intensity
- Performance technique

The performance technique of musicians is critical in recording. The performer's sound cannot be covered up by the other players or assisted by the acoustics of a performance environment, and will be apparent in the final sound. The ways performers produce sound or the instruments themselves can create the desired musical impact, or it may not. The person responsible for the project will need to make evaluations and to offer necessary alternatives. The recordist should know any performance technique problems and the natural acoustic sound properties of all of the instruments (while well outside the scope of this writing).

The analytical listening skills gained in Part 2 will be used continually at this stage. Issues of musical balance, performance intensity, distance location, spatial imaging, environmental characteristics, sound quality, pitch density, etc., will be incorporated into the recordist's listening activities.

Critical Listening Concerns

The technical quality of the recording will be reflected in the integrity of the recording's sound qualities. The recording process must not be allowed to alter the perceived parameters of sound, unless there is a particular artistic purpose or technical function for the alterations. The perceived pitch, amplitude, time elements, timbre, and spatial qualities of the recorded sound sources should be accurately captured in the recording, and should be the only sounds present in the reproduced recording. Any extra sounds are noise, and any unwanted alterations to the sound are distortions. Accurate recording and reproduction of the audio signal must be consistently present for professional quality music recordings.

Other critical listening applications during recording and tracking sessions include:

- Isolation of sound sources during tracking
- Undesirable sound quality of the sound source
- Unwanted sounds created by performers or instruments

Sounds will have a certain degree of isolation from one another. If the sounds are to be altered individually in the mixing stage, they must be isolated. When a group of sounds function as a single unit, it may not be appropriate to isolate the sounds from one another—they are often blended by the performance space. Problems arise when unwanted sounds leak onto tracks that contain sound sources that were intended to be isolated from all other sound sources. This leakage can be the cause of many problems later on in mixing directly to two-track or surround, or in the multitrack mixdown process.

Sound quality should be carefully evaluated during this stage. The amount of timbral detail, as well as many aspects of the precise sound quality, of the sound sources will be determined now and greatly shape the stage. These are major decisions that will have a decisive impact on the sound of the final recording and are largely determined by microphone selection and placement discussed in Chapter 13.

Many beginning recordists rely on equalization and other processing to obtain a suitable sound quality during the initial recording. This use of processing alters the natural sound quality of the sound source and does not alter the timbre equally well throughout the instrument's range. It can be used effectively when the processed sound is desired over the source's natural sound, or when practical considerations limit microphone selection and placement options. Often it can compromise the recording. Processing—especially equalization—is often used at this stage to compensate for poor microphone selection or placement.

Related to performance technique, the ways performers produce sound on their instrument, or the instruments themselves, can create unwanted sound qualities that must be negated during the tracking process. Often these aspects of sound quality are very subtle and go unnoticed until the mixing stage (when it is too late to correct them).

Instruments are capable of making unwanted sounds, as well as musical ones. The sound of a guitarist's left hand moving on the fingerboard, the breath sounds of a vocalist or wind player, and the release of a keyboard pedal are but a few of the possible nonmusical and (normally) unwanted noises that may be produced by instruments during the initial recording and tracking process.

These sounds are easily eliminated during the initial recording and tracking process through altering microphone placement, through slight modifications in performance technique, or through minor repairs to the instrument. The multitrack tape should be as free of all unwanted live-performer sounds and sound alterations as possible. These sounds will be much more difficult to remove later in the recording process. They may be comprised of certain performance peculiarities that (depending on the situation) can only be alleviated by signal processing (such as the use of a de-esser on a vocalist).

Anticipating Mixdown in Tracking

The recordist will be anticipating the mixdown process while compiling basic tracks. They will seek to have complete control of combining sounds during the mixing process.

Any mixing of microphones during the tracking process will greatly diminish the amount of independent control the recordist will have over the individual sound sources during the final mixdown process. Some mixing will often occur during the tracking stage, as submixes, to consolidate instruments and open tracks, or to blend performers that must interact with one another for musical reasons.

Submixes will be carefully planned at the beginning of the session, with a clear idea of how the sounds will be present in the final mix. Drums are often condensed into submixes (either mixed live, or through overdubbing and bouncing). Other mixes that will occur during the tracking process include the combining of several microphones (and/or a direct box) on the same instrument(s). The recordist must be looking ahead to how these sounds will appear in the anticipated final mix—especially for musical balance, sound qualities, distance cues (definition of timbral detail), and pitch density.

Pre-processing (such as adding compression while recording) alters the timbre of the sound source before it reaches the mixing stages of the recording chain. Pre-processing also diminishes the amount of control the recordist will have over the sound, during the mixing process. At times, it is desirable to pre-process signals; often it is not.

Desirable pre-processing might include stereo microphone techniques used on sound sources, effects that are integral parts of the sound quality of an instrument (distorted guitar), or processors that are used to provide a specific sound quality (compressed bass). At times pre-processing is used to eliminate unwanted sounds during tracking (such as noise gated drums). Once a source has been pre-processed, the alterations to the sound source cannot be undone. The recordist should be confident that they want the processed sound before recording it.

Some initial planning of the mixdown sessions will begin during the tracking process. Certain events that will need to take place during the mixdown session will become apparent as the tracking process unfolds. Keeping a tally of these observations will save considerable time later on and may help other tracking decisions.

Examples of items that should be noted for the mixing process are:

- Sudden changes in the mix that may be required because of the content of the tracks
- Certain processing techniques that are planned
- Any spatial relationships or environmental characteristics that may be desired for certain tracks
- Track noises or poor performances of certain sections that will need to be eliminated (muted) during the mix

These are just a few examples of the many factors that may become apparent during the tracking process.

Signal Processing: Shifting of Perspective to Reshape Sounds and Music

Among the most commonly applied audio devices are signal processors. They are important tools (instruments) for the recordist. These devices may play a large role in shaping the individual project. Specific devices are chosen because their individual, inherent sound qualities lend themselves to the particular project. They each control one of the three basic properties of the waveform—frequency, amplitude, or time.

Three types of signal processors exist, each functioning on a particular dimension of the waveform. As discussed in Part 1, an alteration in one of the physical dimensions of sound will cause a change in the other dimensions. Furthermore, alterations of the physical dimensions will cause changes in timbre (sound quality). The three types of processors do not only cause audible changes in the three characteristics of the waveform, they may also alter the timbre of the sound source. If considered according to how processors alter sound, signal processing can be simplified and approached with clarity.

- Frequency processors
- Amplitude processors
- Time processors

Frequency processors include equalizers and filters. Compressors, limiters, expanders, noise gates, and de-essers are the primary amplitude processors. Time processors are primarily delay and reverberation units. Effects devices are hybrids of one of these three primary categories. Some examples of these specialized signal processors include flanges, chorusing devices, distortion, fuzz, pitch shifters, and many more.

Uses of Processing

Signal processing can be used to shape sound qualities. It is applied to the sound source to complete the process of carefully crafting sounds. This is done for the character of the source's sound quality, and to shape sounds to complement the functions and meanings of the musical materials and creative ideas.

In the recording chain, signal processing can occur in a number of locations. It may be incorporated in the tracking as pre-processing, and is most often used to bridge the tracking and mixdown processes. It can also be added subtly in mastering.

Signal processing often occurs separately between the tracking sessions and in preparation for the mixdown session(s)—usually without performing actual re-recordings of the basic tracks. Tracks are evaluated and signal processors are applied to the tracks to determine final sound qualities. Processor settings are noted in session documentation for incorporation into the final mix. In effect, the recordist performs signal processing in real time during the mixdown. The possibility exists to change processor settings in real time during mixdown. Although this does not often happen in practice, it is common to alter the ratio of processed signal to unprocessed signal during the course of the mix.

Listening and Processing

While performing signal processing, the recordist will focus on the component parts of the sound qualities of the sound sources. Small, precise changes in sound quality are possible with signal processing, requiring the recordist to listen at the lowest levels of perspective, and to continually shift focus between the various artistic elements (or perceived parameters) being altered. These changes are often subtle, and can be barely noticeable to untrained listeners. Often beginning recordists are not able to detect low levels of processing. This is a skill that must be developed (and that will be realized through the exercises of Part 2).

Most signal processing involves critical listening. The sound source is considered for its timbral qualities out of context and as a separate entity. In this way, the sound can be shaped to the precise sound qualities desired by the recordist, without the distractions of context.

Knowledge of the physical dimensions of the sound and perception is vital for successful signal processing.

Signal processing alters the electronic (analog or digital) representation of the sound source. In this state, the sound source exists in its physical dimensions. The various signal processors are designed to perform specific alterations to the physical waveform, which will cause changes in the perceived timbre of the sound source. Signal processors are only useful as creative tools if the recordist is in control of these changes in the physical dimensions. Immediately after altering the sound source's physical dimensions, the recordist will shift focus to place the sound in the context of the music, as an artistic element.

After the sound has been reshaped, the listener will use analytical listening to evaluate the sound. The altered characteristics of the sound source and the overall sound quality of the source will be evaluated as they relate to the other sound sources and to their function in the musical context. They will ask "do these changes 'work' for the music, or musical instrument?" The sound quality shifts of processing will be evaluated according to their appropriateness to the musical idea.

The Mix: Performing the Recording

The mix is the piece of music—nearly in its final form. In effect, the mix delivers and supports the essence of the piece of music and is comprised of all its details. The form, reference dynamic level, perceived performance environment, imaging, and pitch density that create the spirit of the piece of music are found in the mix's overall sound. The many subtleties from distance to melody, drum sounds and performance intensity, and much more, are found in its details.

The mixdown of the multitrack tape can be considered an actual live performance of the recording. The sounds that were stored on the multitrack tape are fed to the mixing console; perhaps signal processed for sound quality, assigned spatial properties, and combined with the other sound sources. The mix of all of the sound sources is "composed" by the recordist by planning how sounds will be combined. The mix will then be performed in real-time, after considerable preparation and perhaps some rehearsal.

The mixdown session will result in a two-channel, mono, or surround version of the work, which will become the master of the recording. This version will have all the dimensions of the overall sound. The recordist will be aware of how their actions on individual sound sources affect these characteristics. This is accomplished by attention to shifting focus on various elements and levels of perspective.

Shaping the artistic elements creates the desired overall qualities of the final recording (mix):

- Combining the sound sources with attention to sound quality
- Performing the individual dynamic levels of the sound sources
- Providing each sound source with suitable spatial properties

Mixing is a process that encompasses:

1. Creating an artistic blend of timbres and dynamic levels, and assigned spatial location, distance, and environment qualities (using the skills and concepts of a traditional composer or orchestrator),
2. Rehearsing and coordinating the precise changes that will occur to the sound source tracks (functioning similarly to a traditional conductor), and
3. Actually performing the changes that were determined and rehearsed above in real time (similar to a performance on a traditional musical instrument).

Pitch Concerns in Combining Sound Sources

The individual sound qualities of all of the sound sources are combined in the mix. The recordist will be focused on the dimensions of the individual sound qualities, making any alterations to assist in the dramatic qualities of the sources and to assist in making more pleasing combinations of timbres. As the sound qualities of the sources are combined to create the sound qualities of ensembles (groups of instruments) and the overall sound quality of the program (the piece of music as a whole), the focus of the recordist will shift perspective between these various levels while continuing to scan between the components of timbre.

Throughout the process of compiling (composing) the mix, the recordist's attention will return to timbre and sound quality, listening to the timbres as separate entities (out of time) and listening to sound quality in the musical contexts of all hierarchical levels. The functions of timbre and sound quality that will be of concern are final shaping of the sound qualities of each source to define their unique character and register placement of the sound source in terms of pitch density.

The sound quality of the sound source plays a significant role in the successful presentation of the musical idea. It will receive final shaping during mixdown by signal processing and placement in the sound stage. A source's sound qualities may remain constant throughout the piece, or the qualities may make sudden changes or be gradually altered in real time during the mix. Many possibilities exist for shaping and controlling sound quality.

The recordist will consider pitch density when mixing sound qualities. Pitch density is the placement of pitch/frequency information throughout the hearing range, and the amount of pitch information in specific pitch areas. It emphasizes certain pitch areas over others, shaping the overall quality of the recording. This aspect of pitch relationships will directly contribute to the character and momentum of the piece of music.

A piece of music may have a high concentration of musical material (and/or sound qualities) in specific pitch areas. In one common approach, a recording is characterized by a high-concentration pitch material in the low pitch area. Certain works will alternate between a number of emphasized pitch areas (areas with a high concentration of musical material). This can add great consistency or a level of interest to the music.

Pitch density may be used in innumerable ways to assist in shaping the overall shape of the music and in defining the relationships of the individual sound sources. The pitch density graph of Chapter 10 can be used (with or without a time line) to plan the mix and to keep track of sound source registers. It can be a useful tool in composing the mix or in evaluating the recordings of others. The graph can be created by very carefully identifying pitch levels and areas, or it can include quick, general impressions.

Crafting Musical Balance

Sound sources in the mix are related to one another by dynamic level. The entire mixdown process is often envisioned as the process of determining the dynamic level relationships of the sound sources. As we have noted, the mixdown process is actually much more. It combines many complex sound relationships, of which dynamics is only one.

A *musical balance* of the individual musical ideas and performers is determined by the mix. This is the relationship of the dynamic levels of each instrument to one another and to the overall musical texture. The individual sound sources are combined into a single musical texture, each source at its own dynamic level. The mixing console allows the recordist to perform the individual dynamic levels of the sound sources and to make any changes in level in real time. This significantly shapes the mix and the music.

As noted before, a difference between the actual perceived loudness of the sound source, and the perceived intensity at which the musical material was performed, may be created at this stage of the production process. This difference between musical balance and *performance intensity* may be used to great creative advantage. The mixing process itself can take on dramatic and creative dimensions in altering the realities of sound quality and dynamic level relationships.

Creating a balance of dynamic relationships is one of the primary activities of mixing. Small changes in level can be difficult to detect, especially in the beginning, and this is an important skill to be developed. Perception of dynamics is easily distorted by activities in other aspects of sound. Prominence or importance of materials or elements is often confused with loudness. Dynamic relationships are most accurately perceived at a higher level of perspective—one that compares sources to one another in the overall texture.

Creating a Performance Space for the Recording

The spatial properties of the recording are created during mixdown. These properties will provide each sound source with unique characteristics and will give the entire recording its performance space.

The spatial properties created in the mix provide an illusion of a space within which the performance takes place. Sounds are placed at locations within this perceived space in stereo or surround location, and in distance relationships. Further, sounds are perceived as existing within their own environments (spaces).

The recordist creates space relationships for the reality of the recording. An illusion of a space is created and the performance (recording) is perceived as taking place within this space. The dimensions of this space are defined by the recordist in the mix.

Spatial properties contribute to the recording in a number of ways:

- Applied to each source to further define its character and sound quality (as environmental characteristics fuse with the direct sound to form the source's timbre)

- Applied to each source to suit the requirements of the mix (blend of instruments into ensembles or make sources more prominent in the musical texture)
- Provide illusions of space for the sound source (or groups of sources)
- Provide a set of spatial relationships for the entire program

The mixdown process will place each sound in space. Each sound source will be at a specific stereo or surround location, will be at a perceived distance from the listener, and will be perceived as emanating from a particular environment. The recordist must determine these locations and characteristics as part of the mixdown process, and/or in the signal processing sessions that precede it.

These spatial properties may be used to delineate the sound sources into having their own unique characteristics, or they may be used to cause a group of sound sources to blend into a sense of an ensemble. It is possible for a group of instruments to be grouped in one of the three dimensions, but to have very different characteristics in others. For example, all sources may be located in a specific area on the left side of the stereo array, but at very different distances and having distinctly different environmental characteristics. These similarities provide a connection between the sources and a unifying element in the sound stage. The differences distinguish the sound sources and can add clarity to their musical material.

The recordist will also determine the spatial properties of the overall program. The individual sources and any groups of sources will be perceived as located within a single area (the sound stage), within a single performance space (*perceived performance environment*). The dimensions of the sound stage will be created by lateral location and distance placements of the sound sources.

The dimensions of the perceived performance environment may be applied during the mastering process (with planning occurring during the mixdown). Often it will be the result of the listener's perception of a composite environment created by elements of the predominant environmental characteristics of the primary sound sources of the work. In this way, the shaping of the environmental characteristics of the sound sources that present the most important musical materials will have a direct and marked impact on the listener's impression of the performance environment within which the recording itself appears to take place.

Sound stages can be planned and evaluated using the diagrams of Figures 14-1 and 14-2. These will prove helpful in balancing the sound stage and in creating variety and interest—as desired—of image locations and distances. These diagrams are snapshots of time and may represent any time unit from a moment to a complete song. They allow imaging to be recognized and understood. The dimensions of the sound stage can be

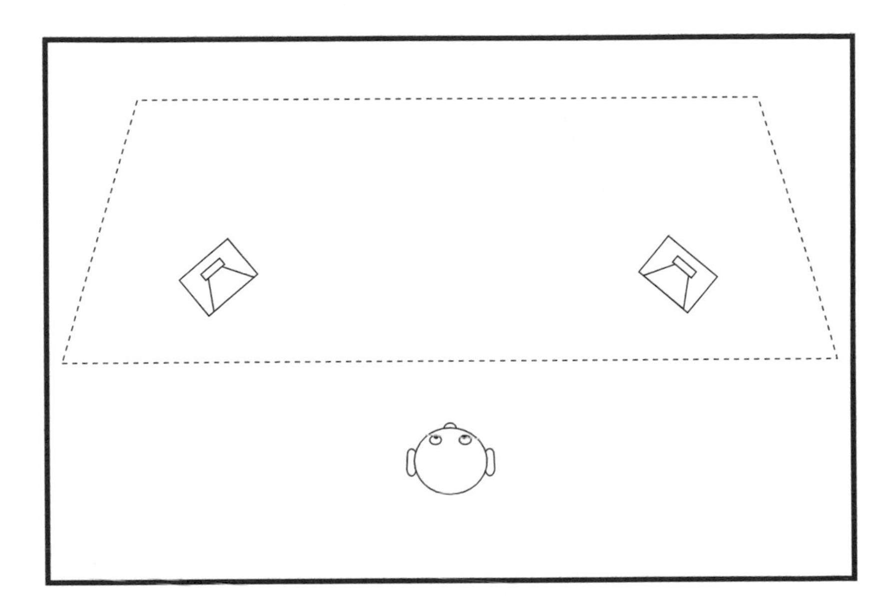

Figure 14-1 Sound stage diagram for two channel recordings.

drawn around the loudspeakers in each graph, and the listener location will be determined in two-channel recordings. The diagrams will show the front edge of the sound stage, giving the recordist a reminder of the sense of intimacy of the recording, and the depth of the sound stage providing important environment size and distance information. A significant set of sound relationships can be planned, crafted, and evaluated with these diagrams.

Composing and Performing the Mix

The mix is a performance of the recording. Performances must be adequately rehearsed to be accurate, and to be artistically expressive and interesting. During the mix, many things can happen.

The three artistic elements that are determined in the mix will be the "musical material" that is "performed" during the mixing process. These three artistic elements of sound qualities, dynamic levels, and spatial properties may be altered in real time, or may be at pre-determined levels. The recordist will thoroughly plan the relationships of all sounds throughout the recording.

The recordist will prepare for mixdown by composing the mix. In doing so, the following are carefully crafted:

- Specific dynamic levels determined for each sound source/track
- Specific sound qualities determined for each sound source

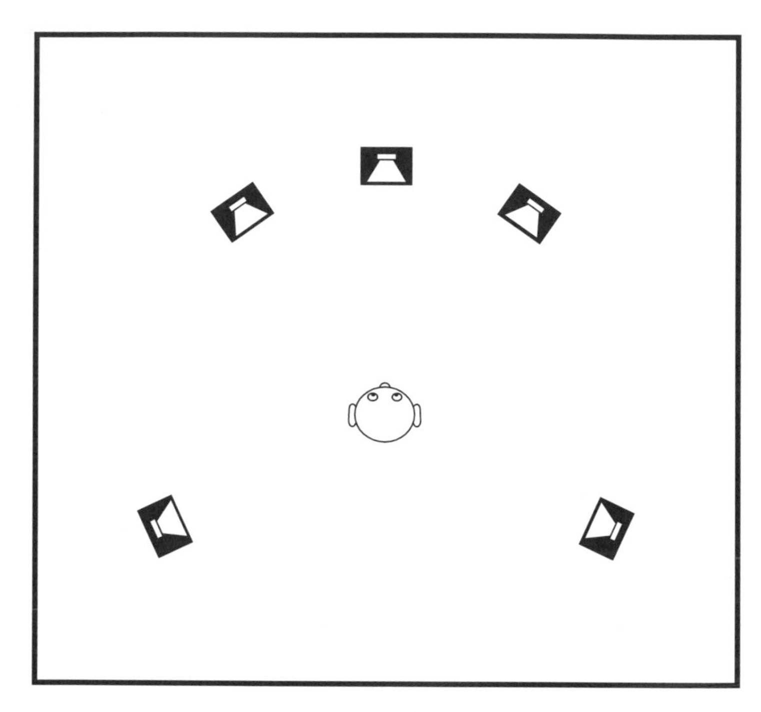

Figure 14-2 Sound stage diagram for surround sound recordings.

- Specific environmental characteristics determined for each sound source
- Specific surround or stereo locations determined for each sound source
- Specific distance locations determined for each sound source

These levels or placements may remain constant throughout the piece. It is quite common for certain artistic elements of certain sound sources to remain constant throughout a mix. It is also common for changes to occur in specific elements, for certain sound sources. These changes may take place in the sound sources presenting the secondary musical materials, as well as in those sources presenting the primary musical ideas. The types of changes, amounts of change, timing of changes, etc., will directly shape the piece of music.

The musical materials will be given a degree of prominence, separation, or blend in the mix. Prominence may be caused by any element (location, pitch register, performance intensity/sound quality, etc.), not only dynamic level. Equivalence dictates that any element can be significant in shaping the musical message—or in providing significant character to the music and recording.

The recordist will plan out all changes to the sound sources and the music, and rehearse them in dry runs of the mixdown process. The changes in the mix will be related to two basic ideas: (1) changes between two or more specific, pre-determined levels of any of the three artistic elements, occurring at specific points in the piece, and (2) changes between two or more specific, pre-determined levels of any of the three artistic elements, occurring at a specific rate of speed and between specific points in the piece.

The recordist will identify and control when the changes that need to start and end (in relation to the music), the beginning and ending levels of the changes in the mix, and the speed of the changes between levels.

The recordist will often need assistance in performing a complicated mix. With a little practice, several people can be involved in the process of changing fader levels, muting channels, altering processor settings, rerouting signal paths, etc. Successfully executing complex mixes often requires help from qualified assistants. With planning the mix and rehearsing all changes of levels, etc., a group of people can easily function as one—as long as the person leading the mix has a clear idea of what is to happen when.

Mixes are often created for individual sections of a song. This may be required if the recordist is short on assistants, in the case of complex mixes, or if large-scale changes in signal processor routing and assignments, or many changes in fader-levels, occur in the mixdown.

The beginning and ending points of the mixes will be planned for ease of editing and for musicality. The different mixes will move one-to-another smoothly and unnoticeably. Large-scale changes in the mix are striking, and quite dramatic; they occur most frequently at major structural divisions of the piece, such as between verses and choruses.

All submixes must be at similar loudness levels. The mixes of the various sections will be edited together to create a master tape of the song. Sudden changes in level will be very noticeable and can detract for the musicality of the piece.

An automation system can be of great assistance to the recordist in the mixdown process. Most importantly, an automation system can allow the recordist to refine dynamic level relationships of the tracks (sound sources), and can free up the recordist to perform other tasks during the mixdown process. An automation system may perform some of the changes to the artistic elements, that were described above. Segments of the mix can be rehearsed, refined, and be consistently repeated with automation. Weaknesses in the mix (perhaps masking of parts, poor stereo location distribution, or countless others) can be identified and corrected more easily when the recordist is not occupied with moving levels and processor settings while listening to the complete mix.

The applications of automation are largely dependent upon the sophistication of the individual system. Automation systems typically are able to accurately change dynamic levels of channels and will mute channels, with consistency, in real time. Some systems are able to change equalization through the automation or to alter the panning of the channel. Some are also able to route signals and much more.

Summary

The mixdown is the performance of the recording. Preparation for the mix begins with selecting sound sources suitable for the musical materials and selecting the appropriate microphones to capture the sound sources. The sound sources are then recorded onto multitrack tape, with their sound qualities carefully shaped, in a planned sequence of events. The sound sources are shaped for sound quality, dynamic level, and spatial properties, in making the detailed plans for performing the mixdown.

The mix results in the final version of the piece of music. This version may be transformed a final time in making the master of the recording.

15 The Final Artistic Processes and an Overview of Music Production Sequences

Multitrack and direct-to-master recording processes can represent the extremes of the different recording aesthetics, and are very different in their production sequences and their utilization of recording techniques. The artistic use of the recording medium, the importance of the performers and their interactions, and the amount of input the recordist has on the artistic aspects of the recording may be very different between multitrack recordings and direct-to-master recordings.

The differences in approach to production (often linked to aesthetic approaches discussed in Chapter 12) are reflected in the sequence of events that occur in creating the music recordings. They are directly related to the amount of control and influence the recordist will have over the musical relationships in the recording, and the extent to which the musicians determine the musical relationships (by their performances as individuals and by their interactions as an ensemble).

The final, creative processes of editing and mastering the music recording will be different between individual approaches to multitrack recordings. Those differences will be even greater between multitrack recordings and the various approaches to direct-to-master recordings. Finally, the relationships of editing and mastering to the final recording will be considered, and lead to the final presentation of the recording and music.

An Overview of Two Sequences for Creating a Music Recording

Every music recording project is unique. Some generalities can be made, but details will create differences between projects. The sequence of events in the individual recording production and the use of the recording chain will be adapted to suit the needs of the individual music recording.

Some projects will require more session preplanning than others will. One project may require more mixdown preparation and rehearsal than another may. Other projects might have very different requirements from any conventional project. The order of events will be mostly consistent with the outline below; some overlapping of activity between the events will be common, as well as some alterations to the orders of the events—or portions thereof.

Multitrack Recording Sequence

A complete sequence of events for a multitrack recording might be:

1. Session preplanning—conceptualize project and pieces of music to be recorded (writing the music if necessary), rehearse musicians, define sound sources and their timbres, select microphones, plan track assignments, determine recording order of the tracks, plan soundstage;
2. Tracking session—record reference tracks (vocals and accompaniment, etc.), followed by recording the basic tracks (primarily the rhythm tracks);
3. Editing of basic tracks for out-takes, and to create the basic structure and length of the piece; reorganize tracks;
4. Overdub sessions—adding solo parts and secondary ideas to the basic tracks, refining the musical material; composing and recording any additional parts to fill newly discovered requirements of the piece;
5. Processing and mixdown preparation sessions—finalize the sound qualities of sound sources; edit the source tape for mixdown (reorganize tracks, remove unwanted sounds);
6. Mixdown rehearsal sessions—"compose" the mix by defining the artistic elements of dynamic levels, spatial properties and sound quality for each sound source, and by considering the interrelationships of the mix and the musical materials of the piece; rehearse the mixdown sequence with people who will assist in the session;
7. Mixdown session—perform the mix(es), mixing the multitrack tape down to two tracks (often occurring during the same session as Step 6.); and
8. Mastering session—assemble and process a master tape by combining the section mixes and by applying any global signal processing.

Direct-to-Master Recordings

Direct-to two-track recordings (or direct-to surround or direct-to mono) are common in music recordings for film, television, and advertising. These recordings are mixed to final relationships (master) during the recording session—and the mix/recording processes impose minimal alterations to the ensemble's sound. The applications of this approach to recording are not limited to recording art music (such as orchestral, choral, or chamber music) though common in those areas. This approach may be suitable to jazz, folk, ethnic musics, popular, rock, or any other music when the musicians (or the conductor) want to be in control of the musical relationships within their performance, or when the function of the recording is best served by having all of the musical parts performed at once (often the case for film scoring or archival recordings, as examples).

The process of making direct-to-master recordings is strikingly different from the multitrack recording process discussed above. Nearly all of the recording process considerations of Chapter 14 are not directly relevant to this approach.

Further, the act of defining the sound quality, as presented in Chapter 13, is shifted from the sound source to the perspectives of the overall ensemble, of groups of instruments within the ensemble, or to the perspective of a limited number of individual soloists.

A complete sequence of events for a direct-to-master recording might follow the outline below. These events will be discussed in detail and in the form of a commonly occurring sequence in the following paragraphs. This sequence, and the details that follow, are guidelines that are altered for the individual project— sometimes markedly.

1. Session preplanning;
2. Creating the sound quality of the recording;
3. Consultations with the conductor (musicians);
4. Recording session;
5. Selection of takes; and
6. Editing to compile a master tape.

1. Session preplanning always begins the production sequence. Once the music to be recorded is known, the recordist will need to know the performance level of the musicians (performers) and the location of the recording session (if it will not take place in the studio). This will allow suitable microphones to be selected, an appropriate stereo microphone technique to be identified (if desired), and microphone placements to be planned. The acoustics of the recording environment will also need to be evaluated when the recording is to take place in a space unknown to the recordist.

The recordist may then meet with the primary performers (or the conductor of the ensemble) to determine how the music can be effectively divided into sections, or do this alone. Problems of editing sections together into the master tape, and issues in stopping and starting the ensemble will be considered when making these divisions. The order in which the sections will be recorded may then be determined. The recordist's and conductor's scores, and the musicians' parts, should be marked to identify these sections, to make starting and stopping the ensemble during the recording session clear and efficient. A discussion between the recordist and the conductor (or performers) should also clarify the recordist's artistic role in the project.

Further discussion, and perhaps some recorded rehearsals in the recording space, should clearly define the sound qualities that will be sought for the recording project.

2. Crafting the recording's actual sound quality can be obtained by monitoring a final rehearsal of the ensemble, in the performance space in which the recording will take place. Alterations to microphone selection and placement will be made to achieve the desired sound quality of the recording, previously discussed. The microphone selection and placement will largely determine the spatial properties, dynamic level relationships, and the sound quality of the recording. Balancing of microphones and signal processing will do the final shaping of the sound quality of the recording. Any necessary signal processing (environmental characteristics, time delay, and dynamic processing being most common) for accent microphones and the stereo array (or arrays) will be added and tuned at this stage of the production sequence.

After the final sound quality has been established, portions of the rehearsal are recorded for later reference and discussion. Any changes in the mix that may be required for the recording session are determined. These changes will be thoroughly rehearsed during this rehearsal of the ensemble.

3. The recordist and conductor (or musicians) will listen to the reference tape that was made during the dress rehearsal. Often their discussion will be solely on the subject of sound quality; all musical considerations may be determined between the conductor and the musicians, or amongst the musicians themselves. Any alterations that must be made to the recording's sound quality are determined during this discussion; the recordist needs to obtain a clear idea of the sound qualities required for the project.

All requested changes to the sound quality of the recording are worked into the recording process. The microphone and recording equipment set-up for the recording session will reflect these changes. Sound quality is rechecked during the musicians' warm-up period, before the beginning of the recording session. The recordist and conductor (musicians) might now make final evaluation of the changes that were made to the sound quality and confirm that the sound quality is correct. The recordist, conductor, and musicians will briefly clarify the logistics of the recording session (how stopping, starting, slating, etc., will be handled).

4. The recording session follows. During the recording session, the sections of the piece are performed in the pre-arranged order. Many takes may be performed of each section of the work until two suitable takes (of each section) are recorded. Each take of each section is monitored by the

recordist, with a focus on consistency of loudness levels, tempo, intonation, performance quality, and the expressive qualities of the performance. An assistant engineer or second engineer may be used to assist in sound evaluation. This person would focus their attention on the technical and critical listening aspects of the sound of the recording. If possible, another assistant will be used to maintain a record of the content of the session tapes, making notes on the recordist's observations of each take, and of the observations of the musicians' spokesperson (usually the conductor, if one is present).

Any changes in the mix that are required in the recording were choreographed during the rehearsal session(s). These changes in the mix are performed, in real time, during the musician's performance in the recording session. The recordist may coordinate the activities of one or more assistant engineers, who would physically perform the actual changes in the mix. The recordist would remain focused on the accuracy level of all of these changes, as they are made, as well as their relationships to the performance.

A multitrack recording of the session may be made simultaneously with the reduction mix to make a safety recording of the session. This will allow the recordist to perform a remix of the session at some future time, should this be necessary. All microphones will usually be sent from the console directly to the multitrack recorder, often routed from the console's direct or patch outputs. No balancing of dynamic levels or extra signal processing would normally be performed on these tracks. This would defeat the purpose of recording the multitrack tape.

5. The conductor (musicians) will often listen to the session tapes with the recordist. Often they will listen to only a few takes of each section of the piece. These takes will have been preselected by the recordist, using the conductor's observations (that were written down by the production assistant) during the recording session as a guide for the selection of takes. The takes that will be used in the final recording are determined during this conference between the recordist and the conductor. Both parties will discuss their perception of the sound qualities of the takes, and the recordist may or may not be asked to evaluate the musicality or accuracy of the performances of each take. Any specific aspects of the sound quality that are undesirable will be identified. The recordist might determine signal processing alterations to attempt to solve (or minimize) any sound quality problems and may play some of the possible alterations for the conductor (performers) during this session. A remix from the safety multitrack tape might be considered at this point as a last resort in the event of very poor session results.

6. Any signal processing or mixing alterations that were determined in the review of takes are performed by the recordist in the mastering session. These changes may be performed before or after the master tape has been compiled, depending on the type of alterations that need to be made. A master tape is created by splicing together the selected takes—at the

correct locations and in the correct order. Any global signal processing will be applied to the overall program after the edited tape has been assembled. In this case, a master tape will be made by playing the spliced session tape through any signal processing device(s), to another mastering deck, which will record the actual master of the work. The recordist will arrange for the conductor (musicians) to hear the master tape, for final observations and approval. Any final alterations to sound quality (etc.) requested by the conductor will be performed by the recordist and will complete the project.

Editing: Rearranging and Suspending Time

The recordist can physically hold time in their hands and move it around. Audio recording transfers sound, which can only occur over time, into a storage medium where the sound is physically located, suspended out of time. This may seem obvious, but it is very significant. The sound can then be changed and reordered by physically altering the storage medium itself (as in cutting analog tape), or by altering the way the storage medium reproduces the sound (i.e., replaying a portion of a digital recording/sound file). The sound may be altered at any time, present or future, and may be replayed forwards, backwards, at any speed (even at uneven speeds).

In *editing* sound, the recordist is able to precisely shape material out of real time. Editing usually combines or joins several different time segments; each time segment being comprised of a group of any number of sounds. The time segments may exist as pieces of analog tape or as computer data.

In joining the sound segments, the recordist can significantly alter the piece of music and its artistic message. These alterations to the music that are made possible by editing serve many functions, see next page. The edit must be accomplished in an artistically sensitive manner and must be inaudible in all areas of technical quality.

It may be impossible to perform technically inaudible or artistically sensitive edits under some circumstances and in some locations in the piece of music. The recordist will identify potential edit points to carefully calculate an edit before it is made. In analog, this may even involve rehearsing the edit on a copy of the master tape.

Editing is often used to compile a master tape of a recording session. In this process, a few or a good many separate sections of the piece of music are joined into a single performance. The most appropriate material, or the most accurate and/or pleasing performances, will be selected for the master tape. A single performance is compiled from the many takes, of segments of a piece, of a direct-to-master recording, or a single performance is compiled from joining the few, individual mixes of multitrack recording.

It is possible to reorder sounds through editing techniques. The major sections of a piece of music may be rearranged. Entire measures may be exchanged, or sounds within a measure may be reordered.

The editing process will alter all sounds present. A reordering of sounds cannot occur unless they are isolated. It will not be possible to reorder the sounds of instruments in a drum fill without also moving the sounds that occur simultaneously with the drum sounds. Likewise, it is impossible to cut a sound source into numerous time segments, and reorder the sound, unless that sound is isolated from other sounds.

Identifying Edit Points

Edit points (also called splice locations) are calculated by anticipating the sound that will be created when the two segments are joined. A critical listening process of evaluating sound quality is used. Each segment to be joined will be evaluated for its sound qualities to determine the most appropriate location of the splice. Beginners often find the edit points in the music through trial and error. With developed skill, the listener will readily identify these locations by listening carefully to the sound qualities of the two segments, remembering what was heard, and comparing the two sound events. How the edit impacts the musical materials will also be considered by the recordist.

Audible edits are nearly always unacceptable and may be created by many factors. Both the critical listening concerns of audio quality and the analytical listening concerns of the musical materials must be considered in determining suitable edit points. The sound must be evaluated for any changes that might be caused by the edit process itself, and for any noises that may have been added. In calculating the edit, the recordist will scan all artistic elements, or perceived parameters of sound, at all perspectives, to determine a usable edit point.

It is not possible to perform an inaudible edit when large differences exist in any of the elements of sound, between the two time segments. Such a splice would result in a sudden alteration of a component of sound at the point where the two segments meet; the sudden change would be audible and unacceptable. As soon as an edit has been made, it will be checked for accuracy and to be certain it is inaudible, and that no noises where added in making the edit.

Under unique circumstances sudden changes between segments may be desired, as in creating a master tape where the splice actually joins very different musical ideas. In these instances, the recordist must make certain that the editing process does not create noise at the edit point, and that the sudden changes are presented as a part of the musical materials (have significance and are handled artistically).

Edits are most easily made at points where loud attacks are performed by prominent instruments or the entire ensemble, or immediately before or after (not during) areas of silence.

Sound sources that are sustaining over the edit point or that are present in each time segment make the edits more difficult. Changes in the sound source will make the edit point audible.

Among the most common of inconsistencies that are present between two time segments are differences in loudness levels. Even subtle changes can be quite audible. Calculating the loudness levels between various takes of an entire ensemble can be quite difficult, but is developed through learning to focus on program dynamic contour. Beginning recordists will often only notice problems in this area after the edit has been made.

Tape noise is part of analog recording. The amount of noise on the tape may or may not be consistent throughout a recording. Changes in noise floor at edit points are very noticeable.

Differences in sound quality of individual instruments and of the overall ensemble are easily overlooked. The potential exists for sound sources and an ensemble to undergo significant changes in sound quality from the beginning of a recording session to the end. Performer fatigue, performance intensity, artistic expression, or a change in temperature or humidity in the performance space may cause these changes in sound quality. Even subtle changes of sound quality can have a marked impact on the technical quality and musicality of the recording.

Changes in pitch between the two segments can be the most noticeable of all changes. The recordist must be well aware of any inconsistencies in this element. Inconsistencies may occur within a particular sound source, or it may be a change of the reference pitch-level (tuning) of the ensemble. Care must be taken to monitor the tuning of the ensemble and the intonation of the performers.

No changes in spatial properties should occur at the edit point, unless they are planned. It is common for spatial properties to be considerably different between time segments, when they represent different mixes of a multitrack master. Sudden shifts of distance locations are common, and have the potential to create few technical problems. Although sudden shifts of stereo location are equally common between time segments, musical and technical problems can be created. Among these are unstable images and phase differences between similar sounds at the edit point.

Sounds sources or environments that have a lengthy decay may need to be carried over across the edit point. This may or may not be possible, depending on the musical context and the nature of the sounds themselves. Edits at these points are sometimes possible when other sources mask portions of the sound, but must be carefully handled to avoid audible changes of sound quality. These edits may need to be planned before the recording session, with suitable alterations made to the performances at the session—for instance starting the performers a bar before a planned edit point to have reverberation present across the edit.

The musical material must remain in rhythm. It is possible to add or subtract time in making an edit. Rhythm changes are very noticeable in their affect on the performance, and measures will appear to be extended or shortened by fractions of a beat.

Tempo changes or inconsistencies can occur between takes. The tempo of the performances will be carefully monitored during the recording process, but like all of the above it must be reevaluated during editing. Any tempo differences that are present between segments will make the edit point very noticeable, and will also make significant changes to the music. An entire take may be unusable, solely because of tempo inconsistencies.

Editing Techniques and Technologies

Analog and digital recording systems have some different characteristics specifically related to their technology. The inherent qualities of each format create advantages or disadvantages depending on the application of the recording, and the specific nature of the recording session. Either analog or digital recording may be the most appropriate choice, depending on the individual recording project. Sound is edited very differently in the two technologies.

Analog Tape Editing

In an *analog recording*, a physical image of the sound is present as oriented magnetic particles on tape, and the physical characteristics of the image are directly proportional to the soundwave. In editing an analog tape, the tape itself is physically cut with a razor blade. Two cut ends of magnetic tape are joined (usually at 45°) with an adhesive tape.

Splice locations are found by slowly moving the tape across the playback head of the recorder. By rocking the tape across the head, the recordist is able to identify the edit point. The edit point is physically located on the tape at the playback head. The tape is marked, removed from the recorder's tape path, placed in an editing block, and is cut.

Once an analog tape has been spliced, it is difficult to redo an edit. Splices are difficult to separate without causing damage to the magnetic tape (which contains the sound—music). If the recordist is successful in undoing the splice without damaging the tape, it is difficult to cut thin time segments (pieces of tape) off the end of a magnetic tape (should the original splice be just a bit too far to the left of the desired edit point). It is almost impossible to add a small piece of magnetic tape onto the beginning of a tape segment (should the original splice be a bit too far to the right of the desired edit point). Identifying analog edit points, and the actual cutting and taping activities required of analog editing all require significant skill.

Difficult edits are sometimes rehearsed. Copies of portions of the session tapes are made, and the copies are edited. The recordist gains confidence, or finds the precise edit points that are usable, on the copies of the tape. Thereby allowing most errors to be made on tape that will not be used in the final version of the project. Obviously, this is a time consuming process.

Digital Sound Editing

Digital recording formats are quite different from analog. Sound exists as digital information, stored as data files. Specialized computers or specialized software for personal computers are used to edit the waveform. The digitized waveform can be altered by modifying and/or rearranging its digital information, but this need not be so. In many systems edits are simply "play lists" of select portions of select files at precisely defined starting and stopping points.

The primary disadvantage of digital editing is that the sound cannot be held in the recordist's hand. The recordist does not know the physical location of the recording and its component sounds. All editing is accomplished on a computer and must be conceptualized more abstractly than analog editing practices.

Conversely, the primary advantage of digital editing is that the sound is not physically present in the recordist's hands. The sound exists as computer information and may be acted upon in ways that are not limited by physical limitations. The following items are the most commonly used among the many functions of most digital editing systems:

- Precise edit points may be identified and saved for future use, with great time resolution
- An edit may be heard, changed, reheard, and evaluated by the recordist; in many systems an edit might never need be permanent
- Edits can be undone, quickly and easily
- The edit does not alter the original material; the original recording is not edited; a copy of the original recording is edited, as a computer file (with no generation loss)
- Overall dynamic levels of the time segments on either side of the edit may be controlled to match at the edit point
- Edits may be made by cross fading from one segment to the other, or by suddenly switching from one take to another (called a butt edit)
- Some systems allow the signal to be heard as the recordist moves the cursor point (simulating the rocking of an analog tape across the playback head)
- Time, dynamics, and frequency processing are usually available to address specific types of inconsistent sound quality and relationships between the two time segments
- Special effects, such as looping and reversing sounds are common

It should be evident that digital formats allow more flexibility in and control over the editing process than is available in analog editing.

Mastering: The Final Artistic Decisions

The final, master tape of the piece of music is the result of a mastering process. The sequence of the process is:

1. Creating a final version of the piece of music, by bringing the musical materials and their relationships to a final format (usually two-channel, but also surround);
2. Creating a final sound for the piece of music by shaping the overall characteristics of the recording (this is the mastered stage of the piece of music); and
3. Finally, all master tapes of all of the songs (pieces of music) of an album project (or film sound track, etc.) are compiled and shaped for consistency between all selections.

The piece of music in the final format may be a live mix, or it may be a source tape. The source tape, which leads to the master tape, may be created in a number of ways. It may be a mix from a multitrack recorder, or it may be a direct-to-master recording from a live performance. It may be an assembled tape from a number of independently mixed sections of a multitrack production, or it might be a compilation of any number of session takes that were mixed direct-to-master. At this stage, all of the musical materials and individual sound sources within the recording are in their final form. No additional alterations will be made to the musical materials and sound sources.

The source tape of the piece will now be shaped in terms of its overall quality. In doing this, a *master tape* will be created. The overall sound quality of the recording will be the global impressions created by the recording and any sound characteristics that are consistent throughout the recording.

The overall sound quality of the recording is a global impression of the emphasized sound characteristics within the recording. These sound qualities may be the result of (1) emphasized sound characteristics of the sound sources that present the primary musical materials, (2) signal processing that is applied to the overall program, or (3) a combination of the above. All three of these factors are used in the mastering process, depending on what is appropriate for the context of the project.

Signal processing applied to the overall program, *global signal processing,* is common. It may take the forms of altering the recording's dynamic contour, frequency response, or environmental characteristics. Nearly always, this is subtle and very carefully applied.

A consistent loudness level reference is required in certain contexts. It may be appropriate to expand or compress the dynamic range, to alter the dynamic contour or to limit certain dynamic-level peaks. Subtle applications of equalization, to alter the frequency response of the program, are sometimes used in the mastering process. Environmental characteristics may be applied to the overall program, to apply a perceived performance environment onto the entire program.

Signal processing may be added to the piece to create the impression of a sense of ensemble. In multitrack recordings, it is common to get a performance that sounds like the musicians are not reacting to one another; the group lacks unity. Adding a characteristic to the sound that affects all sounds equally, such as compression, may provide a consistency of sound that the performance lacks. Adding an environment to the overall recording can also provide this unity and consistency, but it can also dramatically alter the time cues of all distance and environmental relationships of the recording, and can alter the overall sound quality of the recording in terms of frequency response.

The mastering process of the piece of music may be accomplished during mixdown. Mastering is commonly interrelated with the mixdown process, in both multitrack and direct-to-master recordings.

In direct-to-master recordings, the tape that is created by editing together all of the selected takes will often be the master.

The master recording may be created in the same step as the mixdown of the multitrack tape. All global processing may be accomplished between the mix output of the console and the input of the mastering deck, making mastering during mixdown possible under certain circumstances.

With digital technology (and the minimal amount of sound degradation with succeeding generations of signal), it is possible to spend more time on the mastering process itself without diminishing the technical quality of the recording. The changes made to the overall program, in the mastering process, may be crafted much more carefully if the mixdown and mastering are handled in separate steps. It is useful to separate these two stages of the music production sequence, when circumstances allow.

Individual works and their master tapes are compiled into a master tape of the recording (album project). A consistency of sound quality and dynamic levels should exist throughout a project, and the individual works must be related properly to one another. An entire album project will often be bonded by an overall artistic concept. It will also often have an overall sound, comprised of characteristic sound qualities found throughout the album.

This overall sound, in an artistic sense, is the product of the musical ideas, of the production styles and techniques, and of the common sound qualities (those present or those implied) throughout the album. It is largely influenced by the order of the songs (or musical works) on the recording, the timing of silences between the pieces, and the perceived performance intensity of the pieces. For this to be realized it is imperative that the album move artistically from one song to another. Little or no space between pieces (suddenly moving to a new work), or lengthy pauses between pieces may be artistically correct—depending on the material, the context of the recording, and the intended overall impression. Some pieces of music end in silence. This time is used for listener reflection, for a sense of drama, or to allow the music to reach its own sense of conclusion. This silence might even represent the song's reference dynamic level. The lengths of silences between pieces will be carefully calculated to effectively serve the individual pieces of music and the overall project.

Some slight adjustments of the overall loudness between individual works (songs) of an album project are common. The loudness level of a piece will be adjusted to match an established reference level for the entire album. These adjustments are made during the silences between songs. The master tape of the entire project must, however, be reasonably consistent in terms of loudness. This will allow for the dynamic relationships of the recordings to be correctly transferred to any format (CD, LP, cassette, DVD-A, MP3, television, radio, etc.). In certain applications, dynamic peaks are limited to allow the project to transfer into other formats (such as media broadcast).

Frequency response of the master may also be altered. This imparts an overall timbre to the recording that will change the character of each track. In effect, formants or formant regions are added to each piece of music—and all their sound sources. Much potential exists to cause detrimental changes to those mixes and pieces of music. Good mastering engineers are very skilled at making these subtle changes to enhance these relationships, rather than blur them.

The Listener's Alterations to the Recording

The listener may shape the final sound of the recording. This may be through a conscious altering of the original characteristics of the recording or by accident. The listener may alter the sound qualities of a recording to align with their own personal preferences, changing the sound qualities that were crafted by the recordist.

Further, the sound reproduction systems of the final listener to the recording may be significantly different than the system that was used as a reference during the production of the recording. The listening environment and equipment used for home playback (sound reproduction) are almost never similar to (let alone the same as) those used in determining the final sound of the recording. The differences between studio and home listening environments, and studio and consumer sound reproduction systems cause great changes to be made in the sound qualities of the recording during playback in home listening environments. This situation has worsened with home surround systems that need to be set up carefully and more or less carefully calibrated in order for the intended sound to be heard.

Further, the listener alters the original sound characteristics of the recording through a number of activities: adjusting playback equalization and loudness level, their selection of playback equipment, the location of the playback system (especially loudspeakers) in their homes (where often visual aesthetics and the logistics of everyday living win out over sound quality), and through the playback of the recording in small listening rooms. Listening in automobiles, radio and Internet delivery formats, and headphone listening all have equal potential to transform the original recording into something very different.

The music recording will not have the same characteristics to the consumer as in the recording studio. The recordist will hope the music recording will not be radically altered as it is delivered to each individual listener. At the same time the recordist must acknowledge reality—such alterations will take place, to varying degrees, much more often than not. Ultimately, the recordist is not in control of how their art is heard.

Concluding Remarks

This book should not end on a fatalistic and negative note of recordists ultimately not controlling how their artwork is heard. Indeed, usually when the consumers care enough about a recording to alter it to make it their own, it means they care deeply about the recording, its music, and its message. In a not so odd way, it is a complement that the listener wishes to make the music even more to their personal taste—although most recordists would wish this alteration of the sound qualities (which they poured their soul into) would not happen.

Recording music is a wonderful endeavor. Being part of the making of a piece of music is often a very unique privilege. One can witness magic, and be part of something that surpasses the sum of all individuals of the project. One can certainly feel as though they have contributed to the making of great music—whatever the type of music or the level of accomplishment of the musicians. One can also be blessed with opportunities to use their skills to help others realize their dreams, or to use their skills to create music recordings of their own.

Recordists do shape and create art. They are artists in the truest sense of the term—composers, performers, and recording engineers/producers. *The Art of Recording* can be realized with:

- An understanding of what makes recording an art
- The listening skills to recognize those things and to make a professional quality recording
- The craft to use the recording process and its devices to shape sound and music creatively and with artistic sensitivity

Bibliography

Alten, Stanley R. *Audio in Media*, 5th edition. Belmont, CA: Wadsworth Publishing Company, 1999.

Backus, John. *The Acoustical Foundations of Music*, 2nd edition. New York: W.W. Norton & Co., Inc, 1977.

Ballou, Glen. *Handbook for Sound Engineers: The New Audio Cyclopedia*. Indianapolis, IN: Howard W. Sams & Company, 1987.

Bartlett, Bruce. *Recording Demo Tapes at Home*. Indianapolis, IN: Howard W. Sams & Co., Inc, 1989.

Bartlett, Bruce, and Jenny Bartlett. *Practical Recording Techniques*, 2nd Edition. Boston: Focal Press, 1998.

Bartlett, Bruce, and Michael Billingsley. 1990. An Improved Stereo Microphone Array Using Boundary Technology: Theoretical Aspects. *Journal of the Audio Engineering Society* 38 (7/8): 543–552.

Beatles, The. *The Beatles Anthology*. San Francisco: Chronicle Books, 2000.

Beatles, The. *The Beatles Complete Scores*. Milwaukee, WI: Hal Leonard Corporation, 1993.

Bech, Søren, and O. Juhl Pedersen, eds. Proceedings of a Symposium on *Perception of Reproduced Sound*; Gammel Avernæs, Denmark, 1987. Peterborough, NH: Old Colony Sound Lab Books, 1987.

Benson, K. Blair, ed. *Audio Engineering Handbook*. New York: McGraw-Hill, 1988.

Beranek, Leo L. *Acoustics*. New York: American Institute of Physics, Inc, 1986.

Bergson, Henri. *Matter and Memory*. New York: Humanities Press, 1962.

Berry, Wallace. *Form in Music*. Englewood Cliffs, NJ: Prentice-Hall, 1966.

Blauert, Jens. *Spatial Hearing*. Cambridge, MA: The MIT Press, 1997.

Blauert, Jens. Sound Localization of the Median Plane. *Acustica* 22 (1969/70): pp. 205–213.

Blaukopf, Kurt. Space in Electronic Music. In *Music and Technology, Stockholm Meeting June 8–12, 1970*, pp. 157–172. New York: Unipub, 1971.

Borwick, John, ed. *Loudspeaker and Headphone Handbook*, 3rd edition. Oxford: Focal Press, 2001.

Borwick, John. *Sound Recording Practice*, 4th edition. Oxford: Oxford University Press, 1994.

Butler, David. *The Musician's Guide to Perception and Cognition*. New York: Schirmer Books, 1992.

Camras, Marvin. *Magnetic Recording Handbook*. New York: Van Nostrand Reinhold Company, 1988.

Chowning, John. The Simulation of Moving Sound Sources. *Computer Music Journal* 1 (3), 1977: pp. 48–52.

Clifton, Thomas. *Music as Heard: A Study in Applied Phenomenology*. New Haven, CT: Yale University Press, 1983.

Cooper, Grosvenor W., and Leonard B. Meyer. *The Rhythmic Structure of Music*. Chicago: The University of Chicago Press, 1960.

Cooper, Paul. *Perspectives in Music Theory*. New York: Dodd, Mead & Company, 1973.

Davis, Don, and Carolyn Davis. *Sound System Engineering*, 2nd edition. Boston: Focal Press, 1986.

Davis, Don, and Chips Davis. The LEDE™ Concept for the Control of Acoustic and Psychoacoustic Parameters in Recording Control Rooms. *Journal of the Audio Engineering Society* 28 (9), 1980: pp. 585–595.

Davis, Gary, and Ralph Jones. *Sound Reinforcement Handbook*, 2nd edition. Milwaukee, WI: Hal Leonard Publishing Corporation, 1989.

Dell, Edward T., Jr. *Of Mockingbirds and Other Irrelevancies*. Francestown, NH: Marshall Jones Company, 1993.

Deutsch, Diana. *The Psychology of Music*. Orlando, FL: Academic Press, Inc, 1982.

Deutsch, Diana, and J. Anthony Deutsch. *Short-Term Memory*. New York: Academic Press, 1975.

Dodge, Charles, and Thomas Jerse. *Computer Music: Synthesis, Composition and Performance*. New York: Schirmer Books, 1985.

Dowling, William J. *Beatlesongs*. New York: Fireside, 1989.

Eargle, John. *The Microphone Book*. Boston: Focal Press, 2001.

Eargle, John. *Handbook of Recording Engineering*, 3rd edition. New York: Chapman & Hall, 1996.

Eargle, John. *Music, Sound and Technology*. New York: Van Nostrand Reinhold, 1995.

Eargle, John, ed. *An Anthology of Reprinted Articles on Stereophonic Techniques*. New York: Audio Engineering Society, Inc., 1986.

Erickson, Robert. *Sound Structure in Music*. Berkeley, CA: University of California Press, 1975.

Everett, Walter. *The Beatles as Musicians*. Oxford: Oxford University Press, 1999.

Fay, Thomas. Perceived Hierarchic Structure in Language and Music. *Journal of Music Theory* 15 (1–2), 1971: pp. 112–137.

Federkow, G., W. Buxton, and K. Smith. A Computer-Controlled Sound Distribution System for the Performance of Electronic Music. *Computer Music Journal* 2 (3), 1978: pp. 33–42.

Hall, Donald E. *Musical Acoustics: An Introduction*. Belmont, CA: Wadsworth Publishing Company, 1980.

Handel, Stephen. *Listening: An Introduction to the Perception of Auditory Events*. Cambridge, MA: MIT Press, 1993.

Harley, Robert. *The Complete Guide to High-End Audio*. Albuquerque, NM: Acapella Publishing, 1994.

Harris, John. *Psychoacoustics*. New York: The Bobbs-Merrill Company, 1974.

Hawking, Stephen W. *A Brief History of Time: From the Big Bang to Black Holes*. New York: Bantam Books, 1988.

Helmholtz, Hermann. *On the Sensations of Tone*. New York: Dover Publications, Inc, 1967.

Hertsgaard, Mark. *A Day in the Life: The Music and Artistry of the Beatles*. New York: Delacorte Press, 1995.

Holman, Tomlinson. *5.1 Surround Sound Up and Running*. Boston: Focal Press, 2000.

Holman, Tomlinson. *Sound for Film and Television*. Boston: Focal Press, 1997.

Howard, David M., and James Angus. *Acoustics and Psychoacoustics*, 2nd edition. Oxford: Focal Press, 2001.

Huber, David Miles. *Microphone Manual: Design and Applications*. Indianapolis, IN: Howard W. Sams & Company, 1988.

Huber, David Miles, and Robert E. Runstein. *Modern Recording Techniques*, 5th edition. Boston: Focal Press, 2001.

James, William. *Principles of Psychology*. New York: Dover Publications, Inc, 1950.

Karkoschka, Erhard. *Neue Musik / Analyses*. Herrenberg: Doring, 1976.

Karkoschka, Erhard. Eine Hörpartitur elektronischer Musik. *Melos* 38 (11), 1971: pp. 468–475.

Koffka, Kurt. *Principles of Gestalt Psychology*. New York: Harcourt, Brace, and World, 1963.

Kuttruff, Heinrich. *Room Acoustics*, 2nd edition. London: Applied Science Publishers Ltd., 1979.

LaRue, Jan. *Guidelines for Style Analysis*. New York: W.W. Norton & Company, Inc., 1970.

Leeper, Robert. Cognitive Processes. In *Handbook of Experimental Psychology*, S. S. Stevens, ed., pp. 730–757. New York: John Wiley & Sons, Inc., 1951.

Letowski, Tomasz. Development of Technical Listening Skills: Timbre Solfeggio. *Journal of the Audio Engineering Society* 33 (4), 1985: pp. 240–244.

Lewisohn, Mark. *The Beatles Recording Sessions*. New York: Harmony Books, 1988.

Martin, George. *All You Need Is Ears*. New York: St. Martin's Press, 1979.

Martin, George, with William Pearson. *With a Little Help from My Friends: The Making of Sgt. Pepper*. Boston: Little, Brown and Company, 1994.

Massey, Howard. *Behind the Glass: Top Record Producers Tell How They Craft the Hits*. San Francisco: Miller Freeman Books, 2000.

McAdams, Stephen, and Albert Bregman. Hearing Musical Streams. *Computer Music Journal* 3 (4), 1979: pp. 26–43.

Meyer, Leonard B. *Explaining Music: Essays and Explorations*. Berkeley, CA: University of California Press, 1973.

Meyer, Leonard B. *Music, the Arts and Ideas*. Chicago: The University of Chicago Press, 1967.

Meyer, Leonard B. *Emotion and Meaning in Music*. Chicago: The University of Chicago Press, 1956.

Miller, George. The Magical Number Seven, Plus or Minus Two. *Language and Thought*, Donald C. Hildum, ed., pp. 3–31. Princeton, NJ: Van Nostrand Company, Inc., 1967.

Mills, A. W. On the Minimum Audible Angle. *Journal of the Acoustical Society of America* 30, 1958: pp. 237–246.

Moulton, David. *Total Recording: The Complete Guide to Audio Production and Engineering*. Sherman Oaks, CA: KIQ Production, Inc., 2000.

Moulton, David. *Golden Ears: Know What You Hear*. Sherman Oaks, CA: KIQ Production, Inc., 1993.

Moylan, William. *The Art of Recording: the Creative Resources of Music Production and Audio*. New York: Van Nostrand Reinhold, 1992.

Moylan, William. *A Systematic Method for the Aural Analysis of Sound Sources in Audio Reproduction/Reinforcement, Communications, and Musical Contexts.* Paper presented at 83rd Convention of the Audio Engineering Society, New York, NY, 1987.

Moylan, William. *Aural Analysis of the Spatial Relationships of Sound Sources as Found in Two-Channel Common Practice.* Paper presented at 81st Convention of the Audio Engineering Society, Los Angeles, CA, 1986.

Moylan, William. *Aural Analysis of the Characteristics of Timbre.* Paper presented at 79th Convention of the Audio Engineering Society, New York, NY, 1985.

Moylan, William. *An Analytical System for Electronic Music*. Ann Arbor, MI: University Microfilms, 1983.

Neve, Rupert. *Design and the Designer: A Point of Reference.* Paper presented at the 99th Convention of the Audio Engineering Society, New York, NY, 1995.

Newell. Philip. *Recording Spaces*. Oxford: Focal Press, 2000.

Nisbett, Alec. *The Use of Microphones*, 2nd edition. Boston: Focal Press, 1983.

Nisbett, Alec. *The Technique of the Sound Studio*, 4th edition. Boston: Focal Press, 1979.

Olson, Harry F. *Music, Physics and Engineering*, 2nd edition. New York: Dover Publications, Inc., 1967.

Pellegrino, Ronald. *The Electronic Arts of Sound and Light*. New York: Van Nostrand Reinhold Company, 1983.

Plomp, Reinier. *Aspects of Tone Sensation: A Psychophysical Study*. New York: Academic Press Inc, 1976.

Pohlmann, Ken. *Principles of Digital Audio*, 3rd Edition. New York: McGraw-Hill, Inc., 1995.

Polanyi, Michael. *Personal Knowledge: Towards a Post-Critical Philosophy*. Chicago: University of Chicago Press, 1962.

Pousseur, Henri. *Outline of a Method.* In *die Reihe, Nr. 3,* Herbert Eimert and Karlheinz Stockhausen, ed., pp. 44–88. Bryn Mawr, PA: Theodore Presser, Co., 1959.

Randall, J. K. Three Lectures to Scientists. *Perspectives of New Music* 3 (2), 1967: pp. 124–140.

Reynolds, Roger. Thoughts of Sound Movement and Meaning. *Perspectives of New Music* 16 (2), 1978: pp. 181–190.

Reynolds, Roger. *Mind Models: New Forms of Musical Experience*. New York: Praeger Publishers, 1975.

Reynolds, Roger. It(')s Time. *Electronic Music Review* 7, 1968: pp. 12–17.

Risset, Jean-Claude. *Musical Acoustics*. Paris: Centre George Pompidou Rapports IRCAM No. 8, 1978.

Roads, Curtis, ed. *The Music Machine*. Cambridge, MA: The MIT Press, 1989.

Roederer, Juan G. *Introduction to the Physics and Psychophysics of Music*, 2nd edition. New York: Springer-Verlag, 1979.

Rossing, Thomas D. *The Science of Sound*, 2nd Edition. Reading, MA: Addison Wesley Publishing Company, 1990.

Russ, Martin. *Sound Synthesis and Sampling*. Oxford: Focal Press, 1996.

Schaeffer, Pierre, and Guy Reibel. *Solfège de l'objet sonore*. Paris: Editions du Seuil, 1966.

Schaeffer, Pierre. *Traité des objets musicaux*. Paris: Editions du Seuil, 1966.

Schaeffer, Pierre. *A la recherche d'une musique concrète*. Paris: Editions du Seuil, 1952.

Schouten, J.F. The Perception of Timbre. *Report of the 6th International Congress on Acoustics*, 90, 1968: pp. 35–44.

Smith, F. Joseph. *The Experiencing of Musical Sound: Prelude to a Phenomenology of Music*. New York: Gordon and Breach Science Publishers, Inc., 1979.

Stravinsky, Igor. *Poetics of Music: In the Form of Six Lessons*. Cambridge, MA: Harvard University Press, 1970.

Streicher, Ron, and F. Alton Everest. *The New Stereo Soundbook*. Pasadena, CA: Audio Engineering Associates, 1998.

Stevens, Stanley Smith, and Hallowell David. *Hearing: Its Psychology and Physiology*. New York: Acoustical Society of America, 1938, 1983.

Stevens, Stanley Smith, and E. B. Newman. The Localization of Actual Sources of Sound. *American Journal of Psychology* 48, 1936: pp. 297–306.

Stockhausen, Karlheinz. The Concept of Unity in Electronic Music. *Perspectives of New Music* 1 (1), 1962: pp. 39–48.

Talbot-Smith, Michael, ed. *Audio Engineer's Reference Book*, 2nd edition. Oxford: Focal Press, 1999.

Tenney, James. *META≠HODOS and META Meta≠HODOS*. Oakland, CA: Frog Peak Music, 1986.

Varèse, Edgard. The Liberation of Sound. *Perspectives of New Music* 5 (1), 1966: pp. 11–19.

Warren, Richard M. *Auditory Perception: A New Synthesis*. New York: Pergamon Press Inc., 1982.

Watkinson, John. *The Art of Digital Audio*, 3rd Edition. Oxford: Focal Press, 2001.

Watkinson, John. *The Art of Sound Reproduction*. Oxford: Focal Press, 1998.

Wertheimer, Max. Laws of Organization in Perceptual Forms. In *A Source Book of Gestalt Psychology*, Willis Ellis, ed., pp. 71–88. London: Routledge & Kegan Paul, 1938.

Wilson, David. Do You Hear What I Hear? *Mix Magazine* 8 (6), 1984: pp. 132–134.

Williams, David, and Peter Webster. *Experiencing Music Technology*, 2nd edition. New York: Schirmer Books, 1999.

Winckel, Fritz. *Music, Sound and Sensation: A Modern Exposition*. New York: Dover Publications, Inc., 1967.

Winckel, Fritz. The Psycho-Acoustical Analysis of Structure as Applied to Electronic Music. *Journal of Music Theory* 7 (2), 1963: pp. 194–246.

Woram, John M. *Sound Recording Handbook*. Indianapolis, IN: Howard W. Sams & Company, 1989.

Discography

Beatles, The.

A Day in the Life. *Sgt. Pepper's Lonely Hearts Club Band*. EMI Records Ltd., 1967, 1987. CDP 7 46442 2.

Carry That Weight. *Abbey Road*. EMI Records Ltd., 1969, 1987. CDP 7 46446 2.

Come Together. *1*. EMI Records Ltd., 2000. CDP 7243 5 29325 2 8.

The Continuing Story of Bungalow Bill. *The Beatles* (White Album). EMI Records Ltd., 1968. CDP 7 46443 2.

The End. *Abbey Road*. EMI Records Ltd., 1969, 1987. CDP 7 46446 2.

Every Little Thing. *Beatles for Sale*. EMI Records Ltd., 1964. CDP 7 46438 2.

Golden Slumbers. *Abbey Road*. EMI Records Ltd., 1969, 1987. CDP 7 46446 2.

Here Comes the Sun. *Abbey Road*. EMI Records Ltd., 1969, 1987. CDP 7 46446 2.

Hey Jude. *1*. EMI Records Ltd., 2000. CDP 7243 5 29325 2 8.

It's All Too Much. *Yellow Submarine* Songtrack. EMI Records Ltd., 1999. CDP 7243 5 21481 2 7.

Let It Be. *1*. EMI Records Ltd., 2000. CDP 7243 5 29325 2 8.

Let It Be. *Let It Be*. EMI Records Ltd., 1970, 1987. CDP 7 46447 2.

Let It Be. *Past Masters, Volume Two*. EMI Records Ltd., 1988. CDP 7 90044 2.

Lucy in the Sky with Diamonds. *Sgt. Pepper's Lonely Hearts Club Band*. EMI Records Ltd., 1967, 1987. CDP 7 46442 2.

Lucy in the Sky with Diamonds. *Yellow Submarine* Songtrack. EMI Records Ltd., 1999. CDP 7243 5 21481 2 7.

Lucy in the Sky with Diamonds. *Yellow Submarine* (Dolby Digital 5.1 Surround). Subafilms Ltd., 1968, 1999.

Maxwell's Silver Hammer. *Abbey Road*. EMI Records Ltd., 1969, 1987. CDP 7 46446 2.

Penny Lane. *Magical Mystery Tour*. EMI Records Ltd., 1967, 1987. CDP 7 48062 2.

She Came in Through the Bathroom Window. *Abbey Road*. EMI Records Ltd., 1969, 1987. CDP 7 46446 2.

She Said She Said. *Revolver*. EMI Records Ltd., 1966, CDP 7 46441 2.

Something. *1*. EMI Records Ltd., 2000. CDP 7243 5 29325 2 8.

Strawberry Fields Forever. *Magical Mystery Tour*. EMI Records Ltd., 1967, 1987. CDP 7 48062 2.

Tomorrow Never Knows. *Revolver*. EMI Records Ltd., 1966, CDP 7 46441 2.

While My Guitar Gently Weeps. *The Beatles* (White Album). EMI Records Ltd., 1968. CDP 7 46443 2.

Wild Honey Pie. *The Beatles* (White Album). EMI Records Ltd., 1968. CDP 7 46443 2.

You Never Give Me Your Money. *Abbey Road*. EMI Records Ltd., 1969, 1987. CDP 7 46446 2.

Parsons, Alan. "Blue Blue Sky." *On Air*. Parsonics Ltd., 1996. HDS 4414.

Yes. "Every Little Thing." *Yes*. Atlantic Recording Corporation, 1969. 8243-2.

Index